中国科学院科学出版基金资助出版

国外数学名著系列(影印版)　34

Algebraic Geometry II

Cohomology of Algebraic Varieties. Algebraic Surfaces

代数几何 II

代数簇的上同调,代数曲面

I. R. Shafarevich(Ed.)

科学出版社
北京

图字:01-2006-7386

I. R. Shafarevich(Ed.): Algebraic Geometry II: Cohomology of Algebraic Varieties. Algebraic Surfaces

图书在版编目(CIP)数据

代数几何Ⅱ:代数簇的上同调,代数曲面=Algebraic Geometry Ⅱ: Cohomology of Algebraic Varieties. Algebraic Surfaces/(俄罗斯)沙法列维奇(Shafarevich, I. R.)编著.—影印版.—北京:科学出版社,2007
(国外数学名著系列)

ISBN 978-7-03-018300-2

Ⅰ.代… Ⅱ.沙… Ⅲ.①代数几何-英文②代数簇-上同调论-英文③代数曲面-英文 Ⅳ.O187

中国版本图书馆CIP数据核字(2006)第154745号

责任编辑:范庆奎/责任印刷:安春生/封面设计:黄华斌

科学出版社 出版
北京东黄城根北街16号
邮政编码:100717
http://www.sciencep.com

中国科学院印刷厂 印刷
科学出版社发行 各地新华书店经销

*

2007年1月第 一 版 开本:B5(720×1000)
2007年1月第一次印刷 印张:17 1/4
印数:1—3 500 字数:323 000

定价:50.00元

(如有印装质量问题,我社负责调换〈科印〉)

《国外数学名著系列》(影印版)序

要使我国的数学事业更好地发展起来，需要数学家淡泊名利并付出更艰苦地努力。另一方面，我们也要从客观上为数学家创造更有利的发展数学事业的外部环境，这主要是加强对数学事业的支持与投资力度，使数学家有较好的工作与生活条件，其中也包括改善与加强数学的出版工作。

从出版方面来讲，除了较好较快地出版我们自己的成果外，引进国外的先进出版物无疑也是十分重要与必不可少的。从数学来说，施普林格（Springer）出版社至今仍然是世界上最具权威的出版社。科学出版社影印一批他们出版的好的新书，使我国广大数学家能以较低的价格购买，特别是在边远地区工作的数学家能普遍见到这些书，无疑是对推动我国数学的科研与教学十分有益的事。

这次科学出版社购买了版权，一次影印了 23 本施普林格出版社出版的数学书，就是一件好事，也是值得继续做下去的事情。大体上分一下，这 23 本书中，包括基础数学书 5 本，应用数学书 6 本与计算数学书 12 本，其中有些书也具有交叉性质。这些书都是很新的，2000 年以后出版的占绝大部分，共计 16 本，其余的也是 1990 年以后出版的。这些书可以使读者较快地了解数学某方面的前沿，例如基础数学中的数论、代数与拓扑三本，都是由该领域大数学家编著的“数学百科全书”的分册。对从事这方面研究的数学家了解该领域的前沿与全貌很有帮助。按照学科的特点，基础数学类的书以“经典”为主，应用和计算数学类的书以“前沿”为主。这些书的作者多数是国际知名的大数学家，例如《拓扑学》一书的作者诺维科夫是俄罗斯科学院的院士，曾获“菲尔兹奖”和“沃尔夫数学奖”。这些大数学家的著作无疑将会对我国的科研人员起到非常好的指导作用。

当然，23 本书只能涵盖数学的一部分，所以，这项工作还应该继续做下去。更进一步，有些读者面较广的好书还应该翻译成中文出版，使之有更大的读者群。

总之，我对科学出版社影印施普林格出版社的部分数学著作这一举措表示热烈的支持，并盼望这一工作取得更大的成绩。

王　元

2005 年 12 月 3 日

List of Editors, Authors and Translator

Editor-in-Chief

R. V. Gamkrelidze, Russian Academy of Sciences, Steklov Mathematical Institute, ul. Vavilova 42, 117966 Moscow, Institute for Scientific Information (VINITI), ul. Usievicha 20a, 125219 Moscow, Russia
e-mail: gam@ipsun.ras.ru

Consulting Editor

I. R. Shafarevich, Steklov Mathematical Institute, ul. Vavilova 42 117966 Moscow, Russia

Authors

V. I. Danilov, Central Mathematical Economy Institute, ul. Krasnikova 32, 117418 Moscow, Russia

V. A. Iskovskikh, Department of Mathematics, Moscow State University, Leninskie Gory, 117333 Moscow, Russia

I. R. Shafarevich, Steklov Mathematical Institute, ul. Vavilova 42, 117966 Moscow, Russia

Translator

R. Treger, 1004 White Pine Circle, Lawrenceville, NJ 08648, USA
e-mail: tregrob@aol.com

Contents

I. Cohomology of Algebraic Varieties

V. I. Danilov

Translated from the Russian
by R. Treger

Contents

Introduction

The subject of this survey is cohomology. This survey concludes the exposition of the foundations of algebraic geometry begun in (Danilov (1988)).

Many years ago it became apparent that the underlying topological structure of a complex algebraic variety influences, in an essential way, its geometric and arithmetic properties, and until recently the topological concepts made sense only for such varieties. Moreover, the development of the topological concepts such as manifold, Betti numbers, homology, cohomology, and fundamental group was motivated by the needs of algebraic geometry.

Historically, the first and most simple objects were algebraic curves or compact Riemann surfaces. A compact Riemann surface has only one topological invariant, which we may take to be its genus (the number of handles or half the 1st Betti number). From the algebraic point of view, the genus is the number of linearly independent regular differentials (i.e., $\dim H^0(X, \Omega^1_X)$), and from the geometric point of view, the genus is the dimension of the system of linearly independent cycles. These facts were already known to Riemann. At the beginning of this century, Castelnuovo, Enriques, and Poincaré obtained similar results for algebraic surfaces. By the middle of this century, Lefschetz and Hodge had developed the higher-dimensional cases. In particular, the cohomology groups of a smooth projective variety carry a Hodge structure, which is not available for an arbitrary topological space. Thus, from the topological point of view, algebraic varieties are more accessible than arbitrary topological spaces. In the early seventies, Deligne extended these results to arbitrary complex algebraic varieties.

The influence of topology of a variety on its arithmetic can be illustrated by the example of Mordell's conjecture that was recently proven by G. Faltings (Faltings (1983); see also Parshin-Zarkhin (1986)): Let X be an algebraic curve defined over the field $\mathbb{Q}$ of rational numbers; if its genus is greater than 1, than the set $X(\mathbb{Q})$ of its $\mathbb{Q}$-rational points (as well as $X(K)$ for an arbitrary finite extension K of $\mathbb{Q}$) is finite. This result is similar to a more elementary assertion due to de Franchis: Let X and Y be Riemann surfaces, with X of genus greater than 1; then there are only finitely many nonconstant holomorphic maps from Y to X. This topic is discussed in a recent remarkable survey on arithmetic of curves (Mazur (1985)).

As we mentioned above, the topological concepts such as cohomology were first developed for complex algebraic varieties. Later, it turned out to be convenient to employ not only constant coefficients but also variable coefficients as well as cohomology with sheaves as coefficients. In fact, many interesting geometric objects can be described as global sections or cohomology of coherent sheaves. From the algebraic point of view, the turning point was a theorem of H. Cartan. He proved the vanishing of the cohomology of coherent analytic sheaves on affine varieties (Theorem B). It allows us to compute cohomology using arbitrary affine coverings. This, in turn, led Serre to a definition of cohomology of coherent sheaves on an arbitrary abstract algebraic variety. Grothendieck put this theory into its complete form. We discuss it in Chap. 2 of the present survey.

Yet another tremendous impact on the development of topological concepts for abstract algebraic varieties originated from the work of A. Weil. Based on various facts concerning the number of solutions of algebraic equations over finite fields, he suggested that all these facts might be explained by the existence of a cohomology theory for varieties over finite fields similar to the classical cohomology theory. Inspired by this idea, Grothendieck invented the etale topology and etale cohomology. Drawing upon this theory, Deligne established the Weil conjectures; see Chap. 4.

In this survey, we have not touched upon the most recent theory of universal cohomology. This theory remains, for the most part, at a conjectural level, and we refer the reader to (Beilinson (1984, 1986), Beilinson-MacPherson-Schechtman (1987)). The three closely related great cohomology theories (complex, coherent, and etale) are based on some general facts and concepts from homological algebra which we recall in Chap. 1. We also assume that the reader is familiar with the terminology as well as results ddescribed in (Danilov (1988)).

Chapter 1
Homological Machinery

§ 1. Origins of Homological Concepts

1.1. The Idea of Homology. Homological concepts first appeared in the study of complex algebraic curves or, in the classical terminology, functions of one variable. The classical method consisted in investigating integrals of rational (or holomorphic) functions. Only the initial and final points were usually specified not the path of the integration. In fact, the integral remains unchanged under small deformations of the path. In more modern terms, we are dealing with 1-dimensional homology of the corresponding Riemann surface.

To deal with higher-dimensional manifolds, one needs a generalization of the notion of "path of integration". Poincaré suggested to consider k-dimensional strips contained in manifolds. A strip without a boundary or a closed strip is called a *cycle*. In general, a strip has a boundary which is a strip of dimension $k-1$. Two strips are said to be homologous if they bound (from different sides) a strip of one dimension greater. The classes of homologous strips are called the *homology classes* of the manifold in question. In simple examples, it is easy to see that the homology classes form a finitely generated object as in the case of Riemann surfaces.

1.2. Homology of Triangulated Spaces. The vague idea presented above requires a precise description. What do we mean by a strip? How to define a boundary? It is easy to define these concepts for manifolds or spaces that admit a *triangulation*, i. e., can be decomposed in a union of simplexes that intersect each other along their faces. For example, every algebraic variety over the field $\mathbb{C}$ (or $\mathbb{R}$) admits a triangulation and even a semi-algebraic triangulation (Chap. 3, Sect. 1.2). For such a *triangulation* T, we define a *k-chain* to be an algebraic sum of *k-dimensional simplexes* of the triangulation. The k-chains form an Abelian group $C_k(T)$. It is clear how to define the *boundary* of a k-dimensional simplex. It is a sum of all the faces of dimension $k-1$. We extend linearly the notion of boundary to all chains and obtain the boundary operator

$$\delta = \delta_k : C_k(T) \to C_{k-1}(T) .$$

The main property of δ is that $\delta \circ \delta = 0$ (every boundary is a cycle); of course, for that to hold we must employ the oriented simplexes. The collection of the groups $C_k(T)$ and the boundary homomorphisms δ form a *chain complex*

$$C_{\cdot}(T) = (\dots \xrightarrow{\delta_2} C_1(T) \xrightarrow{\delta_1} C_0(T) \to 0 \to \dots) .$$

By definition, the homology classes of the triangulation T (or the complex $C_{\cdot}(T)$) are cycles modulo boundaries:

$$H_k(T) = \operatorname{Ker} \delta_k / \operatorname{Im} \delta_{k+1} .$$

The following natural question may arise. Why do we pass to homology of $C_{\cdot}$ instead of considering the complex $C_{\cdot}(T)$ itself? The most important property of the homology groups is that they depend only on the space X and not on the choice of a triangulation. There are several ways to make this assertion precise. In the case of algebraic varieties, one may use a rather deep fact that every two semi-algebraic triangulation T' and T'' admit another finer semi-algebraic triangulation T. This means that each simplex of T is contained in a simplex of T' (and T''). Then each simplex of T' is a chain of finer simplexes of T, and we get natural homomorphisms

$$\varphi_k : C_k(T') \to C_k(T)$$

which commute with the corresponding boundary operators. A *morphism* from $C_{\cdot}(T')$ to $C_{\cdot}(T)$ is the collection (φ_k) of the homomorphisms φ_k; see Sect. 2.2. Although each φ_k is far from being an isomorphism, the induced map on the homology is, in fact, an isomorphism.

1.3. Singular Homology. There is another more intrinsic way to establish that the homology groups are topological invariants, which does not involve triangulations. Let

$$\Delta_k = \left\{ (t_0, \dots, t_k) \in \mathbb{R}_+^k \,\middle|\, \sum_{j=0}^{k} t_j = 1 \right\}$$

be a standard k-dimensional geometric simplex. A continuous mapping σ: $\Delta_k \to X$ is called a *singular k-dimensional simplex* of X. Let $\Sigma_k(X)$ denote the Abelian group generated by singular simplexes.

Next one defines a boundary operator δ: $\Sigma_k \to \Sigma_{k-1}$. For a singular simplex σ: $\Delta_k \to X$, we put

$$\delta\sigma = \sum_{j=0}^{k}(-1)^j \sigma \circ s_j$$

where $s_0, \ldots, s_k$: $\Delta_{k-1} \to \Delta_k$ are standard affine mappings from Δ_{k-1} onto the $(k-1)$-dimensional faces of Δ_k. Then $\delta \circ \delta = 0$ as before, and we get a singular chain complex $\Sigma.(X)$. The homology of $\Sigma.(X)$ are said to be the *singular homology* of X, and are denoted by $H.(X)$.

This notion is related to the previous one as follows. Suppose X is a space with a triangulation T. Then each (oriented) simplex of T is a singular simplex of X, and we again get a morphism of complexes $C.(T) \to \Sigma.(X)$. Although these complexes are far from being equal – there is a huge number of singular simplexes – they have the same homology groups. Employing various constructions and additional structures, one can built different complexes. However, those complexes are often equivalent, meaning they have the same homology groups. For the purpose of applications, we are interested in invariants of X, i. e. homology. To obtain deeper properties of the homology groups, we have to go back to complexes. In the next section, we will describe a few techniques for dealing with complexes.

1.4. Cohomology. One can employ the dual notion of cochain in place of chain. A *cochain* is a function on the set of simplexes with values in the group $\mathbb{Z}$. We obtain a cochain complex $\Sigma^{\cdot}(X)$ whose *coboundary operator* d (which is dual to the boundary operator) increases the dimension by one

$$d : \Sigma^k \to \Sigma^{k+1} .$$

The groups $H^*(X) = \operatorname{Ker} d / \operatorname{Im} d$ are called the cohomology groups. Unlike the homology groups, the cohomology groups are contravariant functors. Nothing new is gained by dealing with the cohomology groups, although $H^*(X)$ has an advantage of being a graded ring. In case X is a smooth variety, the multiplication in the ring $H^*(X)$ is dual to a more geometric operation of intersection of cycles or homology classes.

One can also define cohomology groups with coefficients in an arbitrary Abelian group A in place of $\mathbb{Z}$. The new groups, denoted by $H^*(X, A)$, provide very little new information. We obtain more interesting objects by employing "variable" systems of coefficients, i. e. sheaves.

1.5. Sheaves. The following two observations explain why we cannot restrict ourselves to the cohomology groups with fixed coefficients.

To better understand a variety or space, we often represent it as a fibering whose fibers have a smaller dimension. Let f: $X \to Y$ be such a fibering. It is clear that a close relation should exist between the cohomology of X, Y, and the fibers of f. However, very seldom f has no degenerations, i. e., all its fibers look alike. In general, some fibers acquire singularities that distinguish them from typical fibers. These special fibers play the most important role (see Morse theory and Picard-Lefschetz theory). The cohomology groups of $f^{-1}(y)$ do not form a constant family or even a continuous family. We encounter objects that had been named *sheaves* by Leray (see Sect. 3).

Secondly, various algebra-geometric objects can be described in the sheaf terminology (and those sheaves are usually coherent). As we have seen in (Danilov (1988)), questions concerning linear systems of divisors and therefore maps to projective spaces can be expressed in terms of global sections of invertible sheaves; differential forms are sections of Ω^p_X, etc. The deformations of a subvariety $Y \subset X$ are closely related to the conormal sheaf $\mathcal{N}_{Y/X}$. Interesting sheaves are often linked to already familiar sheaves by exact sequences. However, given an exact sequence of sheaves $0 \to A \to B \to C \to 0$, the corresponding sequence of global sections

$$0 \to \Gamma(A) \to \Gamma(B) \to \Gamma(C) \to 0$$

may not be exact in $\Gamma(C)$. Trying to understand why sections of C cannot be lifted to global sections of B, we obtain a certain group, depending on A and denoted by $H^1(A)$, that contains the obstructions to the lifting. In case A is a constant sheaf, this group coincides with $H^1(X, A)$. Further, one anticipates to be able to define the remaining groups, namely $H^*(A)$, $H^*(B)$, and so on, that fit into the following long exact sequence

$$0 \to \Gamma(A) \to \Gamma(B) \to \Gamma(C) \to H^1(A) \to H^1(B) \to H^1(C) \to H^2(A) \to \dots .$$

1.6. Cohomology of Sheaves. The preceding property suggests the following definition (or construction) of cohomology groups with sheaf coefficients. We assume that one can embed our sheaf A in an *acyclic sheaf* A^0, i. e., a sheaf with $H^q(A^0) = 0$ for every $q > 0$. Then it follows from the long exact sequence that $H^1(A) = \operatorname{Coker}(\Gamma(A^0) \to \Gamma(A^0/A))$. To describe $H^2(A)$, we embed A^0/A in an acyclic sheaf A^1, etc. This way we obtain a *resolution of the sheaf* A, i. e., a complex of sheaves $A^\cdot = (A^0 \to A^1 \to \dots)$. If the sheaves $A^0, A^1, \dots$ are acyclic, then $H^*(A)$ are the cohomology of the following complex of groups

$$\Gamma(A^\cdot) = (\Gamma(A^0) \to \Gamma(A^1) \to \dots).$$

An acyclic resolution is not unique. It is an auxiliary device whose role is similar to that of a triangulation of a space in the definition of homology groups. As above, we must either establish the independence of cohomology groups of the choice of resolutions (see Sect. 4.2) or employ a canonical resolution (see Sect. 4.1).

Sheaf cohomology are useful not only as a tool for calculating global sections. For sheaves canonically defined on varieties, they provide important invariants such as the Hodge numbers $h^{pq} = \dim H^q(\Omega^p)$. The 1-dimensional cohomology groups often admit interesting interpretations. For instance, $H^1(X, \mathcal{O}_X^*)$ coincides with the Picard group, $\operatorname{Pic} X$, of invertible sheaves (or line bundles); the 1-dimensional cohomology classes of the tangent sheaf of a variety X correspond to the infinitesimal deformations of X, etc.

1.7. Cohomology of Coherent Sheaves. Until the mid-1950s, homological methods were applied only to complex or real algebraic varieties which admit the usual classical topology. When Hirzebruch published his book on applications of sheaves to algebraic geometry in 1956, he had in mind exclusively complex varieties. However, it was clear that one could apply sheaves and cohomology to algebraic varieties with the Zariski topology over arbitrary fields. For constant sheaves (such as $\mathbb{Z}$), this theory provides nothing interesting. However, for coherent algebraic sheaves, the cohomology theory gives meaningful assertions, and it is accessible even more than the cohomology theory of coherent analytic sheaves. In Chap. 2, we describe its main results: the finiteness and semicontinuity theorems, the duality, and the Riemann-Roch theorem.

1.8. Cohomology of Etale Sheaves. The recipe for defining the cohomology groups, described in Sect. 1.6, works in a wider context and leads to a construction of the so-called derived functors. To utilize this recipe, we need the following data: a) an Abelian category $\mathcal{A}$ (in the previous case, it was the category of Abelian sheaves), b) a left exact additive functor F on $\mathcal{A}$ (the functor Γ above), and c) a sufficiently large class of "acyclic" objects where the functor F is exact.

In particular, one can utilize this recipe to define the cohomology of etale sheaves. More precisely, given an algebraic variety or a scheme, one can construct a certain category whose objects are sheaves in the so-called etale topology. Although the latter is not a topology, the definition of cohomology groups as derived functors makes sense. The etale topology is, in a sense, similar to the classical topology over $\mathbb{C}$. Moreover, it makes sense for varieties over arbitrary fields, and over $\mathbb{C}$, the etale results agree with the classical results. This enables us to extend the terminology and results familiar from the classical topology to abstract varieties.

§ 2. Complexes

2.1. Exact Sequences. Let A, B, C be Abelian groups. Let $u\colon A \to B$ and $v\colon B \to C$ be two homomorphisms. The sequence $A \to B \to C$ is *exact* (in B) if the kernel of v coincides with the image of u, i.e., $\operatorname{Ker} v = \operatorname{Im} u$. Then we get $v \circ u = 0$. In general, if $v \circ u = 0$, then the group $\operatorname{Ker} v / \operatorname{Im} u$ measures how far the sequence $A \to B \to C$ is from being exact.

It is easy to extend these definitions to modules, sheaves of Abelian groups, as well as arbitrary Abelian categories provided the notions of kernel, image, and cokernel make sense. The sequence $0 \to A \xrightarrow{u} B$ ($A \xrightarrow{u} B \to 0$) is exact if and only if u is monomorphism (epimorphism).

2.2. Complexes. A *complex of Abelian groups* is a collection $K^\cdot = (K^n, d^n)_{n\in\mathbb{Z}}$ of Abelian groups K^n and homomorphisms $d^n\colon K^n \to K^{n+1}$ such that $d^{n+1} \circ d^n = 0$ for all n. One can say that a *complex* is a diagram

$$K^\cdot = (\ldots \to K^{-1} \xrightarrow{d^{-1}} K^0 \xrightarrow{d^0} K^1 \xrightarrow{d_1} \ldots).$$

The homomorphisms d^n are called *differentials* or *boundary operators*; the index n is usually omitted. In the preceding section we have encounted several examples of complexes.

The *cohomology of the complex* K^n are defined to be the groups

$$H^n(K^\cdot) = \operatorname{Ker} d^n / \operatorname{Im} d^{n-1}.$$

It is the essential part of the complex. If $H^n(K^\cdot) = 0$ for all $n \in \mathbb{Z}$, then $K^\cdot$ is said to be *acyclic*.

The complexes defined above are said to be cochain complexes in order to stress that the differentials increase the dimension by 1. Inverting the arrows, we get a chain complex

$$K_. = (\ldots \leftarrow K_0 \leftarrow K_1 \leftarrow \ldots).$$

Since these two types of complexes differ only in terminology, we will restrict our attention mainly to cochain complexes. One may also consider complexes in an arbitrary Abelian category.

A *morphism of complexes*, $\varphi\colon K^\cdot \to L^\cdot$, is a set of morphisms $\varphi^n\colon K^n \to L^n$, $n \in \mathbb{Z}$, commuting with the differentials (i.e., $d_K \circ \varphi^n = \varphi^{n+1} \circ d_L$). Such a morphism induces a natural homomorphism in cohomology

$$H^n(\varphi) : H^n(K^\cdot) \to H^n(L^\cdot).$$

Thus cohomology depend functorially on complexes. We have encountered morphisms of complexes in Sect. 1.1 and 1.2. A morphism of complexes is said to be a *quasi-isomorphism* or homological equivalence if it induces an isomorphism in the cohomology. Given a category of complexes, we obtain the *derived category* by declaring that all quasi-isomorphisms are invertible. It is a key notion in the modern homological algebra. However, we are not going to use it, and will try to make our exposition more elementary with the least technical machinery.

Each Abelian group A defines a complex as follows: set $A^0 = A$ and augment it on the left and right by zeros. We denote such a complex by $A[0]$. A resolution of A is an exact sequence

$$0 \to A \xrightarrow{\varphi} K^0 \to K^1 \to \ldots.$$

In the above terminology, one may say that the morphism of complexes $A[0] \to K^{\cdot}$ is a quasi-isomorphism. By a *resolution of a complex* $L^{\cdot}$ we mean a complex $K^{\cdot}$ together with a quasi-isomorphism $L^{\cdot} \widetilde{\to} K^{\cdot}$.

2.3. A Long Exact Sequence. Now we come to the most important property of cohomology. Consider an *exact sequence of complexes*

$$0 \to K^{\cdot} \xrightarrow{\varphi} L^{\cdot} \xrightarrow{\psi} M^{\cdot} \to 0 \qquad (*)$$

(here φ and ψ are morphisms of complexes, and for each n, the corresponding sequence $0 \to K^n \to L^n \to M^n \to 0$ is exact). Then the exists a boundary or connecting homomorphism

$$\partial^n : H^n(M^{\cdot}) \to H^{n+1}(K^{\cdot}),$$

such that the following long exact sequence

$$\ldots \to H^n(L^{\cdot}) \xrightarrow{H^n(\psi)} H^n(M^{\cdot}) \xrightarrow{\partial^n} H^{n+1}(K^{\cdot}) \xrightarrow{H^{n+1}(\varphi)} H^{n+1}(L^{\cdot}) \to \ldots \qquad (**)$$

is exact. Furthermore, the ∂^n are functorial on short exact sequences $(*)$.

The homomorphisms ∂^n are constructed as follows. Assume $\mu \in H^n(M^{\cdot})$ is given by a cocycle $m \in M^n$. Since ψ is surjective, one can lift m to an element $l \in L^n$ such that $\psi(l) = m$; the element l is not necessary a cocyle, however $\psi(dl) = d\psi(l) = dm = 0$. In view of $(*)$, there exists $k \in K^{n+1}$ such that $dl = \varphi(k)$. Since $\varphi(dk) = d\varphi(k) = ddl = 0$ and φ is injective, we get $dk = 0$, i. e., k is a cocycle. The corresponding cohomology class in $H^{n+1}(K^{\cdot})$ is denoted by $\partial^n(\mu)$. One can easily verify that ∂ is well defined, functorial, and $(**)$ is exact.

Example. Let Y be a (usually closed) subset of a space X. We denote by $\Sigma^{\cdot}(X,Y)$ the kernel of the natural morphism

$$\Sigma^{\cdot}X \to \Sigma^{\cdot}Y$$

of singular cochain complexes. The cohomology of $\Sigma^{\cdot}(X,Y)$ are called the *cohomology of the pair* (X,Y). These cohomology are denoted by $H^*(X,Y)$, and they are intimately related to the cohomology of the space X/Y obtained from X by contracting Y to a point. The cohomology of X, Y, and (X,Y) are fit in a long exact sequence

$$\ldots \to H^{n-1}(Y) \to H^n(X,Y) \to H^n(X) \to H^n(Y) \to \ldots .$$

In some cases, this sequence allows us to determine cohomology of one term of the above triple, provided we know cohomology of the two remaining terms.

2.4. Filtered Complexes. The sequence $(**)$ relates the cohomology of $L^{\cdot}$ to the cohomology of $K^{\cdot} \subset L^{\cdot}$ and $L^{\cdot}/K^{\cdot} = M^{\cdot}$. We will generalize this as follows. A *filtered complex* $(K^{\cdot}, F)$ is a complex $K^{\cdot}$ equipped with a *filtration*, i. e., a descending family of subcomplexes of $F^pK^{\cdot}$:

$$\dots \supset F^{-1}K^{\cdot} \supset F^0K^{\cdot} \supset F^1K^{\cdot} \supset \dots .$$

We will assume that for every K^n, the filtration FK^n is finite, i. e., $F^pK^n = K^n$ for small p and $F^pK^n = 0$ for large p. In geometry such a filtration may arise from a filtration of a topological space, as in the previous example. For instance, a triangulated space has a filtration by skeletons of various dimensions. On the other hand, every complex $K^{\cdot}$ has a so-called *dummy filtration* σ:

$$\sigma^p(K^{\cdot}) = (\dots \to 0 \to K^p \to K^{p+1} \to \dots).$$

As with any filtered object, one can associate to $(K^{\cdot}, F)$ the corresponding graded object $\mathrm{Gr}^*_F(K^{\cdot})$, where $\mathrm{Gr}^p_F(K^{\cdot}) = F^{p+1}K^{\cdot}/F^pK^{\cdot}$.

2.5. Spectral Sequences. A filtration of a complex $K^{\cdot}$ allows us to calculate $H^*(K^{\cdot})$ using the sequence of approximations. In the nutshell, we are iterating the exact sequences of type (∗∗). This procedure is encoded in the notion of *spectral sequence.* Precisely, for each $r \geq 0$, we introduce the following objects

$$Z^{p,q}_r = \mathrm{Ker}(F^pK^{p+q} \xrightarrow{d} K^{p+q+1}/F^{p+r}K^{p+q+1}),$$
$$B^{p,q}_r = d(F^{p-r+1}K^{p+q-1}) + F^{p+1}K^{p+q},$$
$$E^{p,q}_r = (Z^{p,q}_r + B^{p,q}_r)/B^{p,q}_r .$$

For example,

$$E^{p,q}_0 = F^pK^{p,q}/F^{p+1}K^{p+q} = \mathrm{Gr}^p_F K^{p+q} .$$

One can verify that

$$d(Z^{p,q}_r) \subset Z^{p+r,q-r+1}_r , \quad d(B^{p,q}_r) \subset B^{p+r,q-r+1}_r ,$$

so that d induces a morphism d_r: $E^{p,q}_r \to E^{p+r,q-r+1}_r$. We get $d_r \circ d_r = 0$, and $E^{p,q}_{r+1}$ is identified with the cohomology of a short sequence

$$E^{p-r,q+r-1}_r \xrightarrow{d_r} E^{p,q}_r \xrightarrow{d_r} E^{p+r,q-r+1}_r .$$

This data is described by a sequence of tables $E^{p,q}_r$, one for each $r \geq 0$.

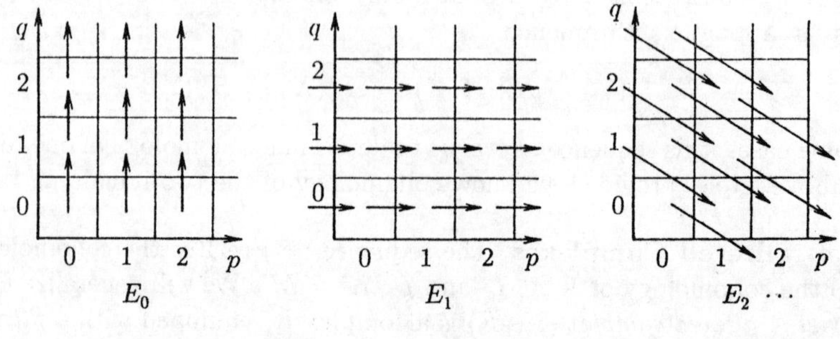

Furthermore, we get the so-called "final" table E_∞:

$$E_\infty^{p,q} = (Z_\infty^{p,q} + B_\infty^{p,q})/B_\infty^{p,q} ,$$

where

$$Z_\infty^{p,q} = \operatorname{Ker} d \bigcap F^p K^{p+q} = \bigcap_r Z_r^{p+q} ,$$

$$B_\infty^{p,q} = dK^{p+q-1} + F^{p+1}K^{p+q} = \bigcup_r B_r^{p,q} .$$

For large r, we get $E_r^{p,q} = E_\infty^{p,q}$. If for a certain r, $E_r = E_\infty$ (i. e., $d_r = d_{r+1} = \ldots = 0$), then we say that the spectral sequence *degenerates* at E_r.

The filtration F gives rise to a final filtration on $H(K^\cdot)$, i. e.

$$F^p H^n(K^\cdot) = \operatorname{Im}(H^n(F^p K^\cdot) \to H^n(K^\cdot)) .$$

The objects associated to the final filtration of $H(K^\cdot)$ are just the terms of E_∞, i. e. $E_\infty^{p,q} = \operatorname{Gr}_F^p(H^{p+q}(K^\cdot))$. Thus, a spectral sequence transforms an initial table

$$E_1^{p,q} = H^{p+q}(\operatorname{Gr}_F^p K^\cdot)$$

into a final table

$$E_\infty^{p,q} = \operatorname{Gr}_F^p H^{p+q}(K^\cdot) .$$

We express that symbolically as follows

$$E_1^{p,q} = H^{p+q}(\operatorname{Gr}_F^p K^\cdot) \Rightarrow H^{p+q}(K^\cdot) .$$

Spectral sequences are functorial. Given a morphism of filtered complexes

$$(K^\cdot, F) \to (\overline{K}^\cdot, \overline{F})$$

(i. e., $F^p K^\cdot$ goes to $\overline{F}^p \overline{K}^\cdot$), then there is a morphism of the corresponding sequences $E_r(K^\cdot) \to E_r(\overline{K}^\cdot)$, $0 \le r \le \infty$.

2.6. Bicomplexes. A resolution of a complex is often built from suitable resolutions of its terms. This gives rise to a bicomplex. A *bicomplex* consists of a bigraded object $K^{\cdot\cdot} = (K^{p,q})$, $p, q \in \mathbb{Z}$, and two commuting differentials

$$d' : K^{p,q} \to K^{p+1,q} , \quad d'' : K^{p,q} \to K^{p,q+1} .$$

For example, the tensor product $K^\cdot \otimes L^\cdot = (K^p \otimes L^q)_{p,q}$ of two complexes $K^\cdot$ and $L^\cdot$ is a bigraded complex with the differentials $d' = d_K \otimes 1_L$ and $d'' = 1_K \otimes d_L$.

Given a bicomplex $K^{\cdot\cdot}$, one can consider an ordinary (or *total*) *complex* $K^\cdot = \operatorname{tot}(K^{\cdot\cdot})$, where $K^n = \oplus_{q+q=n} K^{p,q}$, and the total differential d: $K^n \to K^{n+1}$ is given by the formula $d = d' + (-1)^p d''$ on $K^{p,q}$. The complex $\operatorname{tot}(K^{\cdot\cdot})$ admits two dummy filtrations, $'\Phi$ and $''\Phi$, where

$$'\Phi^p K^\cdot = \bigoplus_{i \ge p} K^{i,q} , \quad ''\Phi^p K^\cdot = \bigoplus_{j \ge q} K^{p,j} .$$

Clearly $\mathrm{Gr}^p_{'\Phi}(\mathrm{tot}(K^{\cdot\cdot}))$ is the complex $K^{p,\cdot}$ with shifted grading and the differential $(-1)^p d''$. The spectral sequence of the filtered complex $(\mathrm{tot}(K^{\cdot\cdot}), '\Phi)$ has the form

$$E_1^{p,q} = {}'H^p({}''H^q(K^{\cdot\cdot})) \Rightarrow H^{p+q}(\mathrm{tot}(K^{\cdot\cdot}))\,.$$

Here we denote the q-th cohomology of $K^{p,\cdot}$ by ${}''H^q$.

2.7. Mapping Cone. We would like to describe the preceding construction in one special case. Let $\varphi\colon K^\cdot \to L^\cdot$ be a morphism. Consider a bicomplex $A^{\cdot\cdot}$ such that $A^{\cdot,0} = K^\cdot$, $A^{\cdot,1} = L^\cdot$, and the remaining $A^{\cdot,q}$ are trivial. Let $d'' = \varphi$. We get a bicomplex of the form

$$A^{\cdot\cdot} = \left\{ \begin{array}{ccccccc} & 0 & & 0 & & 0 & \\ & \uparrow & & \uparrow & & \uparrow & \\ \ldots \to & L^{-1} & \xrightarrow{d} & L^0 & \xrightarrow{d} & L^1 & \to \ldots \\ & \uparrow\varphi & & \uparrow\varphi & & \uparrow\varphi & \\ \ldots \to & K^{-1} & \xrightarrow{d} & K^0 & \xrightarrow{d} & K^1 & \to \ldots \\ & \uparrow & & \uparrow & & \uparrow & \\ & 0 & & 0 & & 0 & \end{array} \right\}.$$

The corresponding ordinary complex $\mathrm{tot}(A^{\cdot\cdot})$ is called the *mapping cone of the morphism* φ, and is denoted by $\mathrm{Con}(\varphi)$ or $(K^\cdot \bmod L^\cdot)$. The cone of the trivial morphism $0 \to L^\cdot$ is denoted by $L^\cdot[-1]$. This complex is obtained from $L^\cdot$ by shifting the grading by one (i. e., $(L^\cdot[-1])^n = L^{n-1}$) and changing the sign of the differential.

We have an obvious exact sequence of complexes

$$0 \to L^\cdot[-1] \to \mathrm{Con}(\varphi) \to K^\cdot \to 0\,.$$

It gives rise to a long exact sequence

$$\begin{array}{ccccccccc} \ldots \to & H^n(L^\cdot[-1]) & \to & H^n(\mathrm{Con}(\varphi)) & \to & H^n(K^\cdot) & \xrightarrow{\partial} & H^{n+1}(L^\cdot[-1]) & \to \ldots\,. \\ & \| & & & & & & \| & \\ & H^{n-1}(L^\cdot) & & & & & & H^n(L^\cdot) & \end{array}$$

Note that the connecting homomorphism $\partial\colon H^n(K^\cdot) \to H^n(L^\cdot)$ coincides with $H^n(\varphi)$. So, there is no real difference between the connecting homomorphisms ∂ and the natural homomorphisms $H^n(\varphi)$. Clearly φ is a quasi-isomorphism if and only if the complex $\mathrm{Con}(\varphi)$ is acyclic.

2.8. Products. Let $K^\cdot$ and $L^\cdot$ be two complexes. One can form a complex $\mathrm{tot}(K^\cdot \otimes L^\cdot)$ (see Sect. 2.6). It is easy to verify that we have canonical homomorphism

$$H^p(K^\cdot) \otimes H^q(L^\cdot) \to H^{p+q}(\mathrm{tot}(K^\cdot \otimes L^\cdot))\,.$$

In particular, if $K^\cdot$ is a ring in the category of complexes, i. e., there exists a morphism $\mathrm{tot}(K^\cdot \otimes K^\cdot) \to K^\cdot$ with obvious properties, then $H^*(K^\cdot)$ admits a ring structure as well.

More explicit quantitative relationship between $H^*(K^\cdot \otimes L^\cdot)$ and $H^*(K^\cdot) \otimes H^*(L^\cdot)$ is called the *Künneth formula.* We will consider a typical example for complexes of modules over a commutative ring A.

Proposition. *We assume a complex $L^\cdot$ consists of flat A-modules. Then we have a spectral sequence*

$$E_2^{p,q} = \bigotimes_{i+j=q} \mathrm{Tor}^A_{-p}(H^i(K^\cdot), H^j(L^\cdot)) \Rightarrow H^{p+q}(\mathrm{tot}(K^\cdot \underset{A}{\otimes} L^\cdot)).$$

In particular, if A is a field, the formula takes the following simple form:

$$H^n(\mathrm{tot}(K^\cdot \underset{A}{\otimes} L^\cdot)) = \bigotimes_{i+j=n} (H^i(K^\cdot) \underset{A}{\otimes} H^j(L^\cdot)).$$

§ 3. Sheaves

3.1. Presheaves. Many geometric objects associated to a topological space or variety X posses a property that allows us to consider their "restrictions" to open subsets $U \subset X$. For instance, one may consider the restrictions to U of functions, vector fields, differential forms, vector bundles, etc. For an open subset U, we denote by $F(U)$ a set of objects of this kind. For $U' \subset U$, we have the restriction map $F(U) \to F(U')$ which preserves natural structures on $F(U)$ and $F(U')$. This leads to the definition of a presheaf.

Definition. A contravariant functor F from the category $O(X)$ of open subsets of a topological space X to the category of sets (groups, rings, etc.) is said to be a *presheaf* of sets (groups, rings, etc.) on X.

An element $s \in F(U)$ is called a *section* of F *over* U; in case $U = X$, we speak about *global sections.* The image of s under the map $F(U) \to F(U')$ is called the *restriction* of s to U'. We denote the restrictions by $s|_{U'}$.

Examples

a) For any set A, the constant functor $U \mapsto A$ is a presheaf of sets.

b) The functor that associates to an open subset $U \subset X$ its cohomology $H^q(U)$ is a presheaf of groups or rings.

c) We fix a space Y, and denote the set of all continuous maps from U to Y by $Y(U)$. As before, we get a presheaf of sets. The same can be done in the category of differential manifolds, complex spaces, or algebraic varieties. In particular, for an algebraic variety X, we denote by $\mathcal{O}_X$ the presheaf of regular maps to $\mathbb{A}^1$. It is a presheaf of rings. The presheaf of regular maps to $\mathbb{A}^1\backslash\{0\}$ is denoted by $\mathcal{O}^*_X$. It is presheaf of groups under multiplication.

In the following examples, X is an algebraic variety.

d) We associate to an open subset $U \subset X$ the ring of rational functions $K(U)$. We get a presheaf of rings denoted by $\mathcal{K}_X$.

e) We associate to an open subset $U \subset X$ the group $\operatorname{Div} U$ of Cartier divisors on U. We get a presheaf of groups denoted by $\mathcal{Div}_X$.

f) We associate to an open subset $U \subset K$ its Picard group $\operatorname{Pic} U$. We get a presheaf of groups denoted by $\mathcal{Pic}_X$.

3.2. Sheaves. A sheaf is a presheaf satisfying certain conditions. The sheaves describe objects that can be given locally. We say that sections s over U and s' over U' are *compatible* if their restrictions to $U \cap U'$ coincide.

Definition. A presheaf F is said to be a *sheaf* if it satisfies the following condition: For every collection of open subsets U_i $(i \in I)$ and sections $s_i \in F(U_i)$ that agree on the intersections, there is a unique section $s \in F(\cup_{i \in I} U_i)$ such that $s|_{U_i} = s_i$.

In short, one can say that a sheaf transforms direct limits (in the category $O(X)$) into inverse limits, i. e., a sheaf is "continuous" in a sense. The sheaf axiom essentially states that the following sequence is exact:

$$F(U \cup U') \to F(U) \times F(U') \rightrightarrows F(U \cap U').$$

For instance, the presheaves in Examples (c), (d), and (e) are, in fact, sheaves. On the other hand, very seldom, the presheaves in Examples (a), (b), and (f) are sheaves. We also recall that the definition of an algebraic variety utilizes the notion of sheaf: we are given affine variety structures on open sets U_i that agree on the intersections. Finally, we observe a similarly between the notions of sheaf and simplicial complex. The latter is obtained by glueing of simplexes Δ_n, while the former is obtained by glueing the "pieces" isomorphic to open sets $U \subset X$.

The sheaves, as well as presheaves, form a category, where by a morphism $F \to G$ we mean a morphism of functors, i. e., for every $U \subset X$, a unique map $F(U) \to G(U)$ and these maps agree on the intersections. For example, if we associate to a rational function f, $f \neq 0$, its Cartier divisor $\operatorname{div}(f)$, we get a morphism of the sheaf $\mathcal{K}_X^*$ to the sheaf $\mathcal{Div}_X$. Since every Cartier divisor is locally principle, this morphism is an epimorphism in the category of sheaves, however, not in the category of presheaves!

To every presheaf F one can associate a sheaf $\widetilde{F}$ by the following essentially tautological construction. Consider a "set" (in fact, a category) of sheaves G equipped with a morphism $F \to G$. We set $\widetilde{F} = \varprojlim G$, the inverse limit with respect to this "set". The presheaf $\varprojlim G$ is, in fact, a sheaf, because the corresponding inverse limits commute. This quite general construction is utilized each time one has to construct an adjoint functor.

For example, the sheaf associated to the presheaf $\mathcal{Pic}_X$ of Example (f) is a zero sheaf. The sheaf associated to the constant presheaf of Example (a) is called the *constant sheaf* with fiber A, and is denoted by A_X.

3.3. Direct and Inverse Images of Sheaves. Let $f\colon X \to Y$ be a continuous map, and F a sheaf on X. If we associate to every open subset $V \subset Y$ the set $F(f^{-1}(V))$, we get a sheaf on X (check!), denoted by f_*F, which is called the *direct image* of F under the map f.

The functor f_* has a left adjoint functor f^{-1}, the *inverse image*, which transforms the sheaves on Y into sheaves on X. The adjoint condition means, in our case, that for arbitrary sheaves F on X and G on Y, we have an equality (strictly speaking, a canonical bijection)

$$\mathrm{Hom}(G, f_*F) = \mathrm{Hom}(f^{-1}G, F).$$

This allows us to speak about morphisms of G to F over $f\colon X \to Y$, though those sheaves are defined on different spaces.

The following two special cases are particularly important.

Let i_x be the embedding of the point x in X. The sheaf $i_x^{-1}F$ (strictly speaking, the set $(i_x^{-1}F)(x)$) admits an explicit description as $\varinjlim F(U)$, where U runs over a system of open neighborhoods of x. This set is denoted by F_x and is called the *stalk of the sheaf F at x*.

Another special case is the inclusion $j\colon U \to X$ of an open subset. For an open subset $V \subset U$, we have $(j^{-1}F)(V) = F(V)$. We often denote the sheaf $j^{-1}F$ by $F|U$.

3.4. Abelian Sheaves. The sheaves of Abelian groups are said to be *Abelian*. The category of Abelian sheaves is an Abelian category. In particular, for any morphism of Abelian sheaves (homomorphism) $u\colon F \to G$, there exists a kernel $\mathrm{Ker}(u)$ and cokernel $\mathrm{Coker}(u)$. The kernel is given explicitly as follows:

$$U \mapsto \mathrm{Ker}(F(U) \to G(U)).$$

The case of cokernel is slightly more complicated: it is a sheaf associated to the presheaf

$$U \mapsto \mathrm{Coker}(F(U) \to G(U)).$$

So, it makes sense to speak about exact sequences. A sequences of Abelian sheaves

$$0 \to F \to G \to H \to 0$$

is exact if and only if for every $x \in X$, the corresponding sequence of stalks

$$0 \to F_x \to G_x \to H_x \to 0$$

is exact. On the other hand, the sequence of global sections

$$0 \to F(X) \to G(X) \to H(X) \to 0$$

is, in general, exact only in $F(X)$ and $G(X)$ but not in $H(X)$. Indeed, given an epimorphism $G \to H$, the sections of H can be lifted to the sections of G only locally. We will give several examples, since this phenomenon lead to the development of the cohomology theory.

Example 1. Let $Y \subset X$ be a closed subscheme of X defined by an ideal sheaf I. Then the sequence

$$0 \to I \to \mathcal{O}_X \to \mathcal{O}_Y \to 0$$

is exact. In particular, let $X = \mathbb{P}^1$ and Y consists of two distinct points on $\mathbb{P}^1$. Then the space $\mathcal{O}_Y(\mathbb{P}^1)$ has dimension two, while the space of global sections of $\mathcal{O}_{\mathbb{P}^1}$ contains only constants.

Example 2. For an arbitrary algebraic variety X, we have an exact sequence

$$1 \to \mathcal{O}_X^* \to \mathcal{K}_X^* \to \mathcal{D}iv_X \to 0 .$$

The quotient group $\operatorname{Div} X/K(X)^*$ is precisely the Picard group $\operatorname{Pic} X$ (as we shall see, isomorphic to $H^1(X, \mathcal{O}_X^*)$).

Example 3. Let X be a complex analytic variety. Let $\mathcal{O}_X$ and $\mathcal{O}_X^*$ be the sheaves of holomorphic maps from X to $\mathbb{C}$ and $\mathbb{C}\backslash\{0\}$. Then the sequence

$$0 \to \mathbb{Z}_X \to \mathcal{O}_X \xrightarrow{e} \mathcal{O}_X^* \to 1 .$$

is exact, where $e(s) = \exp(2\pi i s)$. The corresponding sequence of global sections is not exact, and the obstructions to lifting the sections of $\mathcal{O}_X^*$ lie in the cohomology group $H^1(X, \mathbb{Z})$.

In general, the obstructions to lifting of global sections of a sheaf H lie in the cohomology group $H^1(X, F)$, which is defined in the next section. However, if that cohomology group is trivial, then there are no obstructions and the homomorphism $G(X) \to H(X)$ is surjective.

Example 4. Let C and D be two curves in the projective plane $\mathbb{P}^2$ given by the equations $F = 0$ and $G = 0$. We assume that the third curve $\{H = 0\}$ is passing through all the intersections of C and D (for simplicity, we assume that C and D intersect transversely). Then $H = AF + BG$ for suitable forms A and B (*the $AF + BG$ theorem of M. Noether*).

To express a relationship between this theorem and cohomology we consider the ideal sheaf I of the subscheme $C \cap D$ of $\mathbb{P}^2$. The forms H of degree k that vanish at the points $C \cap D$ are global sections of the sheaf $I(k) = I \otimes_{\mathcal{O}} \mathcal{O}(k)$. This sheaf has a resolution

$$O \to \mathcal{O}(k-m-n) \xrightarrow{\alpha} \mathcal{O}(k-m) \oplus \mathcal{O}(k-n) \xrightarrow{\beta} I(k) \to 0 ,$$

where $m = \deg F$, $n = \deg G$, $\beta(a, b) = Fa + Gb$, and $\alpha(c) = (Gc, -Fc)$. Now, everything follows from the vanishing of $H^1(\mathbb{P}^2, \mathcal{O}(k-m-n))$ which will be established in Chap. 2, Sect. 3.

3.5. Flabby Sheaves. Before turning to the definition of the cohomology of sheaves, we would like to consider one class of sheaves. A sheaf F on X is said to be *flabby* if for every open set $U \subset X$, the restriction map

$F(X) \to F(U)$ is surjective. These sheaves are interesting in view of the following property:

Lemma. *Let* $0 \to F \to G \to H \to 0$ *be an exact sequence of Abelian sheaves on* X. *If* F *is flabby, then the sequence of global sections*

$$0 \to F(X) \to G(X) \to H(X) \to 0$$

is exact.

Indeed, consider open subsets $U, U' \subset X$ and a section $t \in H(X)$. Assume that t can be lifted to sections $s \in G(U)$ over U and $s' \in G(U')$ over U'. Then, on $U \cap U'$, those liftings differ by an element $r \in F(U \cap U')$. Since F is flabby, we can extend r to a section $\widetilde{r}$, and take $s' + \widetilde{r}$ in place of s', which is also a lifting of $t|U'$. Then s and s' coincide on $U \cap U'$, thus defining a lifting of t over $U \cup U'$. We conclude the proof by transfinite induction.

On the other hand, we have sufficiently many flabby sheaves. Let F be an arbitrary sheaf on X. Consider the sheaf $C^0(F)$ defined by the formula

$$C^0(F)(U) = \prod_{x \in U} F_x \,.$$

We get a canonical embedding $F \hookrightarrow C^0(F)$ by assigning to $s \in F(U)$ the family $(s(x)) \in \prod_{x \in U} F_x$. The sheaf $C^0(F)$ is always flabby. Furthermore, in the Abelian case, C^0 transforms exact sequences into exact sequences.

§ 4. Cohomology of Sheaves

4.1. Construction of Cohomology. From now on, we will consider only Abelian sheaves. Given a sheaf F, the functor C^0, from the preceding section, allows us to construct a *flabby Godement resolution* $F \rightrightarrows C^{\cdot}(F)$ which is functorial in F. We proceed as follows: $C^0(F)$ together with a morphism $F \hookrightarrow C^0(F)$ were constructed in the preceding section; set $C^1(F) = C^0(C^0(F)/F)$, and for $n \geq 1$, the sheaves $C^{n+1}(F)$ are defined by induction:

$$C^{n+1}(F) = C^0(C^n(F)/dC^{n-1}(F)) \,.$$

The differential d: $C^n(F) \to C^{n+1}(F)$ is defined as a composition

$$C^n(F) \to C^n(F)/d^{n-1}(F) \to C^0(C^n(F)/dC^{n-1}(F)) = C^{n+1}(F) \,.$$

Definition. The cohomology of the complex $C^{\cdot}(F)(X)$ of Abelian groups is said to be the *cohomology of the sheaf* F (or the *space* X *with coefficients in the sheaf* F). They are denoted by $H^n(X, F)$.

Clearly $H^0(X, F) = F(X)$ and $H^n = 0$ for $n < 0$. If $H^n(X, F) = 0$ for all $n \geq 1$, the sheaf F is said to be *acyclic*. The main property of the cohomology is that a short exact sequence

$$0 \to F \to G \to H \to 0$$

yields the following long exact sequence

$$0 \to H^0(X,F) \to H^0(X,G) \to H^0(X,H) \xrightarrow{\partial} H^1(X,F) \to H^1(X,G) \to \dots.$$

In fact, since C^0 is an exact functor, we get the exact sequence of sheaves

$$0 \to C^\cdot(F) \to C^\cdot(G) \to C^\cdot(H) \to 0.$$

Since $C^\cdot(F)$, $C^\cdot(G)$, and $C^\cdot(H)$ are flabby resolutions, we get an exact sequence of complexes of groups

$$0 \to C^\cdot(F)(X) \to C^\cdot(G)(X) \to C^\cdot(H)(X) \to 0,$$

by Lemma of Sect. 3.5. It remains to apply Sect. 2.3.

By induction and Lemma of Sect. 3.5, every flabby sheaf is acyclic. Now, let us take another look at the sequence of Example 2 of Sect. 3.4. That sequence is exact because $\mathcal{K}_X^*$ is a flabby sheaf. In general, we have an elementary assertion to the effect that every constant sheaf on an irreducible topological space is flabby.

4.2. Hypercohomology. It is possible to define cohomology groups not only for a single sheaf but for a complex of sheaves as well. We call them hypercohomology in order to distinguish them from the cohomology sheaves $H^*(F^\cdot)$. Given a complex of sheaves $F^\cdot$, we consider the corresponding bicomplex of sheaves $C^\cdot(F) = (C^p(F^q))$ $(p,q \in \mathbb{Z})$. The original complex can be embedded in the complex $K^\cdot = \mathrm{tot}(C^\cdot(F^\cdot))$. Moreover, this embedding is a quasi-isomorphism. The cohomology of the corresponding complex $K^\cdot(X) = \mathrm{tot}(C^\cdot(F^\cdot)(X))$ of global sections are said to be the *hypercohomology* of $F^\cdot$, and are denoted by $\mathbf{H}^*(X,F^\cdot)$.

The hypercohomology are functorial on the category of complexes, and every short exact sequence of complexes gives rise to an exact sequence as in Sect. 4.1. If $F^\cdot$ consists of a single sheaf F^0, the hypercohomology of $F^\cdot$ coincide with the cohomology of F^0.

As with any bicomplex, $C^\cdot(F^\cdot)$ gives rise to two spectral sequences whose limit is the hypercohomology of $F^\cdot$ (at least, if $F^\cdot$ is bounded from below – the only case needed in the sequel).

Consider the first spectral sequence

$$'E_2^{p,q} = H^p(X, H^q(F^\cdot)) \Rightarrow \mathbf{H}^n(X,F^\cdot).$$

In particular, if the complex $F^\cdot$ is acyclic, i. e., $H^q(F^\cdot) = 0$, then $\mathbf{H}^*(X,F^\cdot) = 0$. Utilizing the cone of the morphism (see Sect. 2.7), we see that a quasi-isomorphism of complexes $F^\cdot \xrightarrow{\approx} G^\cdot$ induces an isomorphism $\mathbf{H}^*(X,F^\cdot) \xrightarrow{\approx} \mathbf{H}^*(X,G^\cdot)$.

Now consider the second spectral sequence

$$''E_2^{p,q} = H^p(H^q(X, F^{\cdot})) \Rightarrow \mathbf{H}^n(X, F^{\cdot}) .$$

In particular, if $F^{\cdot}$ consists of acyclic sheaves, this spectral sequence degenerates, and we get isomorphisms

$$\mathbf{H}^n(X, F^{\cdot}) = H^n(H^0(X, F^{\cdot})) .$$

Comparing these two sequences, we obtain the following

Proposition. *Let $K^{\cdot}$ be a resolution of a sheaf F consisting of acyclic sheaves K^n. Then*

$$H^n(X, F) = H^n(K^{\cdot}(X))$$

for every n.

Thus, to calculate the cohomology of a sheaf F, one may employ an arbitrary acyclic resolution in place of the canonical resolution $C^{\cdot}(F)$.

Example 1. Let $X = [0, 1]$ be a segment of the real line $\mathbb{R}$. We will calculate the cohomology of the constant sheaf $\mathbb{Z}_X$. We denote by A and B the sheaves of germs of continuous functions on X with values in $\mathbb{R}$ and $\mathbb{R}/\mathbb{Z}$, respectively. We have a natural exact sequence

$$0 \to \mathbb{Z}_X \to A \to B \to 0 .$$

The sheaves A and B are acyclic, though they are not flabby (they are flasque (Godement (1958))). Since X is simply connected, the homomorphism $A(X) \to B(X)$ is surjective, hence the sheaf $\mathbb{Z}_X$ is acyclic.

Now, utilizing the Leray spectral sequence (see Sect. 4.3), we may obtain the same assertion for an arbitrary symplex Δ_n.

Example 2. Let X be a differential manifold, and Ω^p the sheaf of germs of differential p-forms on X. By the classical Poincaré lemma, the sequence

$$0 \to \mathbb{R}_X \to \Omega^0 \xrightarrow{d} \Omega^1 \xrightarrow{d} \dots$$

is exact. Furthermore, the sheaves Ω^p are acyclic because they are fine (Godement (1958)). It follows that the groups $H^n(X, \mathbb{R}_X)$ coincide with the cohomology of the complex $\Omega^{\cdot}(X)$ of global differential forms (*de Rham's theorem*). Later, we will discuss the analytic and algebraic versions of this result.

4.3. Higher Direct Images. The above construction allows us to define derived functors for many others functors. Let $\mathcal{A}$ and $\mathcal{B}$ be two Abelian categories, and $T: \mathcal{A} \to \mathcal{B}$ an additive left exact functor. To define derived functors on an object $A \in \mathcal{A}$, we take a "nice" resolution $A \xrightarrow{\sim} K^{\cdot}$, and let

$$(R^n T)(A) = H^n(T(K^{\cdot})) .$$

In case of sheaves and their global sections, a "nice" resolution consisted of flabby sheaves. In the general case, we employ injective or other suitable objects of $\mathcal{A}$, depending of the functor T. This procedure is used to define the functors Ext (the derived functor of Hom), the functor Tor (the derived functor of $\otimes$), the cohomology with compact support $H^n_c(X)$, the higher direct images $R^n f_*$, etc.

Now, we will discuss in detail the functors $R^n f_*$ which are immediate generalizations of cohomology. Let $f\colon X \to Y$ be a continuous map. The direct image functor f_*, from the category of Abelian sheaves on X to the category of Abelian sheaves on Y, is additive and left exact (but not necessarily exact). The corresponding derived functors, denoted by $R^n f_*$, are defined with the help of flabby resolutions. For a sheaf F on X, the sheaf $R^n f_*(F)$ is, in fact, a sheaf on Y associated to a presheaf

$$V \mapsto H^n(f^{-1}(V), F).$$

In particular, if f maps X to a point, then $R^n f_*(F)$ coincides essentially with the cohomology $H^n(X, F)$.

The functor $R^n f_*$ plays an important role in comparing cohomology of X and Y. Let F be a sheaf on X, and $K^{\cdot}$ a flabby resolution of F. Since $K^n(X) = (f_*K^n)(Y)$, one may view $H^n(X,F)$ as the cohomology of the complex of groups $f_*(K^{\cdot})(Y)$. The direct image of a flabby sheaf is flabby, so the latter cohomology coincide with $\mathbf{H}^*(Y, f_*K^{\cdot})$, and we obtain the spectral sequence

$$E_2^{p,q} = H^p(Y, H^q(f_*K^{\cdot})) \Rightarrow H^n(X,F).$$

By definition $H^q(f_*K^{\cdot}) = R^q f_*(F)$, and we get the so-called *Leray spectral sequence* for the map f:

$$E_2^{p,q} = H^p(Y, R^q f_*(F)) \Rightarrow H^n(X,F).$$

Similarly, we can establish a more general statement. For two continuous maps $f\colon X \to Y$ and $g\colon Y \to Z$, we have the spectral sequence

$$E_2^{p,q} = R^p g_*(R^q f_*(F)) \Rightarrow R^n(g \circ f)_*(F).$$

4.4. Cohomology of a Covering. We will describe a few additional devices helpful in calculating cohomology. Let X be a space covered by open sets U_0 and U_1. Let F be a sheaf on X. Then we have the sequence

$$0 \to F(X) \to F(U_0) \times F(U_1) \xrightarrow{\varphi} F(U_0 \cap U_1) \to 0.$$

where $\varphi(s_0, s_1) = s_0 - s_1$. By the definition of a sheaf, it is exact in the first two terms, and also in the third term provided F is flabby.

We will apply this remark to the flabby Godement resoltuion $C(F)$ of F. Since the restrictions of $C(F)$ to U_0, U_1 and $U_0 \cap U_1$ give flabby resoltuions of the corresponding restrictions of F, we obtain the exact *Mayer-Vietoris sequence*

$$\ldots \to H^{n-1}(U_0 \cap U_1, F) \to H^n(X, F) \to H^n(U_0, F) \times H^n(U_1, F) \to \\ \to H^n(U_0 \cap U_1, F) \to \ldots .$$

The above discussion can be generalized to an arbitrary open covering $\mathcal{U} = (U_i)_{i \in I}$ of X, and gives rise to a spectral sequence of the covering

$$E_1^{pq} = H^q(\mathcal{U}_p, F) \Rightarrow H^{p+q}(X, F) .$$

Here $H^q(\mathcal{U}_p, F)$ stands for $\prod H^q(U_{i_0} \cap \ldots \cap U_{i_p}, F)$, where the product is taken over all strictly increasing collections $i_0 < i_1 < \ldots < i_p$ of elements of I (the latter being a linearly ordered set). The differential d_1: $h^q(\mathcal{U}_p, F) \to H^q(\mathcal{U}_{p+1}, F)$ is given by a combinatorial (or simplicial) formula.

Definition. A covering (U_i) is said to be F-*acyclic* if

$$H^q(U_{i_0} \cap \ldots \cap U_{i_p}, F) = 0$$

for all $i_0, \ldots, i_p \in I$ and $q > 0$.

For an acyclic covering, $E_1^{p,q} = 0$ for $q > 0$, and the sequence degenerates into a single complex

$$H^0(\mathcal{U}_*, F) = \left(\prod_i F(U_i) \to \prod_{i<j} F(U_i \cap U_j) \to \ldots \right).$$

Now, for an arbitrary covering U, the preceding complex, denoted by $C^{\cdot}(\mathcal{U}, F)$, is called the *complex of a covering* $\mathcal{U}$ *with values in* F. The cohomology of $C^{\cdot}(\mathcal{U}, F)$, denoted by $H^*(\mathcal{U}, F)$, are called the *cohomology of the covering* $\mathcal{U}$ *with values in* F.

Proposition. *If the covering* $\mathcal{U}$ *is* F*-acyclic, then*

$$H^n(\mathcal{U}, F) = H^n(X, F) .$$

Another Version. The above definition of $C^{\cdot}(\mathcal{U}, F)$ requires I to be a linearly ordered set, although almost nothing depends on an ordering of I. From the theoretical point of view, it is more convenient to deal with the "unordered" version of the definition of a covering complex, where an arbitrary sequence $(i_0, \ldots, i_p)$ of I is called a p-simplex. Then the covering complex can be written in the form

$$F(X') \to F(X' \underset{X}{\times} X') \to F(X' \underset{X}{\times} X' \underset{X}{\times} X') \to \ldots ,$$

where $X' = \coprod_{i \in I} U_i$. In this form, the notion can be generalized to an arbitrary "covering" $X' \to X$, and even an arbitrary "simplicical space"

$$X' \leftleftarrows X'' \overset{\leftarrow}{\underset{\leftarrow}{\leftarrow}} X''' \ldots .$$

We get a similar spectral sequence also for a closed covering of X, provided that covering is locally finite. In particular, it follows that for a locally finite polyhedron X, the cohomology of the constant sheaf, denoted by $H^*(X, \mathbb{Z}_X)$, coincide with the cohomology of a triangulation, as well the singular cohomology $H^*(X, \mathbb{Z}_X)$.

4.5. The Acyclicity Criterion for Coverings. The following theorem of H. Cartan allows us to establish the acyclicity of some coverings.

Theorem. *Let $\mathcal{A}$ be a class of open subsets of X that satisfies the following two conditions:*

a) *$\mathcal{A}$ is closed under finite intersections, and*
b) *$\mathcal{A}$ contains sufficiently small open subsets.*

We assume that for every $U \in \mathcal{A}$ and an arbitrary $\mathcal{A}$-covering $\mathcal{U} = (U_i)$ of U (which means that every $U_i \in \mathcal{A}$), we get $H^q(\mathcal{U}, F) = 0$ for $q > 0$. Then every $\mathcal{A}$-covering is acyclic.

In particular, for every $\mathcal{A}$-covering of X, we get an isomorphism

$$H^*(\mathcal{U}, F) = H^*(X, F).$$

In view of (a), it will suffice to verify that $H^q(U, F) = 0$ for every $U \in \mathcal{A}$ and $q > 0$. We proceed by induction. Assume the assertion for $q < n$.

We assume $\alpha \in H^n(U, F)$ is represented by a cocycle $a \in C^n(F)(U)$. Since $d\alpha = 0$, a is locally a coboundary. So, by (b), one can find an $\mathcal{A}$-covering (U_i) of U such that the image of $a|U_i$ in $H^n(U_i, F)$ is trivial for every i. By induction hypothesis, we get

$$E_2^{p,q} = 0 \qquad (p \geq 0, \quad 0 < q < n)$$

in the spectral sequence of our covering (see Sect. 4.4). This yields an exact sequence

$$0 \to H^n((U_i), F) \to H^n(U, F) \to E_2^{0,n} \subset \prod_i H^n(U_i, F).$$

Since the image of α in $H^n(U_i, F)$ is trivial for every i, the class α lies in the group $H^n((U_i), F)$, which is trivial by the assumption of the theorem.

Chapter 2
Cohomology of Coherent Sheaves

Since every algebraic variety is endowed with an algebraically defined Zariski topology, it makes sense to consider sheaves and their cohomology on algebraic varieties. However, it is not clear if we obtain something of interest. In the classical setting, many important invariants of variety appear as cohomology of constant sheaves like $\mathbb{Z}$. We cannot expect similar assertions when dealing with the Zariski topology.

Indeed, interesting algebraic varieties are irreducible, and on such spaces, every constant sheaf is flabby, so its cohomology are trivial. Similarly, there are no nontrivial locally constant sheaves. Of course, for reducible varieties, we obtain some information employing even constant sheaves, whose cohomology reflect the combinatorial structure of a variety as a union of its irreducible components. However, it is more interesting to consider the "pliable" coherent sheaves, more so that many geometric problems can be formulated in terms of coherent sheaves.

This chapter is devoted to an exposition of the cohomology of coherent sheaves on schemes.

§ 1. Cohomology of Quasi-Coherent Sheaves

1.1. Quasi-Coherent Sheaves. Let X be a scheme (or, if the reader prefers, an algebraic variety) with a structure sheaf $\mathcal{O}_X$. The quasi-coherent sheaves on X are sheaves of $\mathcal{O}_X$-modules that have a specific local structure. Therefore, we first describe the quasi-coherent sheaves on affine schemes.

Let $X = \operatorname{Spec} A$ be an affine scheme, where A is a commutative ring with unity element 1. Given an A-module M, we consider a sheaf $\widetilde{M}$, which associates to an open subset $U \subset X$ the module $M \otimes_A \mathcal{O}_X(U)$. In particular, for a principal open subset $D(f) = \{x \in X \,|\, f(x) \neq 0\}$, where $f \in A$,

$$\widetilde{M}(D(f)) = M_f = M \underset{A}{\otimes} A_f ,$$

Here $A_f = A[f^{-1}]$ is the ring of fractions of the form a/f^m. The sheaves of the form $\widetilde{M}$ are said to be *quasi-coherent*.

Now, for an arbitrary scheme X, a sheaf of $\mathcal{O}_X$-modules F is said to be *quasi-coherent* if for every open affine subscheme $U \subset X$, the restriction $F|_U$ is *quasi-coherent* on U. The quasi-coherent sheaves play an important role in algebraic geometry because many geometric objects associated to algebraic varieties can be described in terms of such sheaves.

Example 1. The sections of *invertible sheaves* (i. e., sheaves locally isomorphic to $\mathcal{O}_X$) determine effective divisors on X. For instance, the sections of $\mathcal{O}(m)$ on the projective space $\mathbb{P}$ determine hypersurfaces of degree m in $\mathbb{P}$. The divisors passing through given points or sets in $\mathbb{P}$ are described by sections of subsheaves of $\mathcal{O}(m)$. So, the questions concerning the existence and "number" of such divisors are reduced to calculations of the dimensions of the spaces of global sections of sheaves F (*Riemann's problem*). As we shall see, it is easier to calculate the *Euler characteristic* of a sheaf, which includes, in addition to $H^0(X, F)$, cohomology of F.

Example 2. Let $Y \subset X$ be a closed subscheme. The question about deformations of Y inside X is closely related to the normal sheaf $\mathcal{N}_{Y/X}$ of Y in X. Roughly speaking, the infinitesimal deformations of Y are described by elements of $H^0(\mathcal{N})$, while the obstructions lie in $H^1(\mathcal{N})$.

The deformations of a variety X provide a more complicated example. Those deformations are closely related to cohomology of the tangent sheaf of X (Palamodov (1986)).

Example 3. Cohomology of sheaves intimately related to a variety (such as the sheaf Ω_X^p) provide important invariants of the variety. For instance, the *genus of a curve* X can be defined as the dimension of the space $H^0(\Omega_X^1)$ or $H^1(\mathcal{O}_X)$.

1.2. Serre's Theorem. The following theorem established in (Serre (1955)) is a corner-stone in the cohomology theory of coherent sheaves. It is an algebraic analog of Cartan's theorem (Theorem B).

Theorem. *Let F be a quasi-coherent sheaf on an affine variety X. Then $H^q(X, F) = 0$ for all $q > 0$.*

One can show that converse is also true in the Noetherian case: if the cohomology of every quasi-coherent sheaf on a scheme X are trivial, then X is an affine scheme (Hartshorne (1977)). In particular, a scheme is affine if and only if its components are affine.

Since Serre's theorem plays an important role, we will sketch its proof. Let $\mathcal{A}$ denote a class of open subsets of the form $D(f)$ of an affine scheme $X = \operatorname{Spec} A$, where $f \in A$. It satisfies conditions (a) and (b) of the acyclicity criterion for coverings (Chap. 1, Sect. 4.5). Therefore, it will surface to establish the vanishing of cohomology of any $\mathcal{A}$-covering of X. Let $U_i = D(f_i)$, $i \in I$, be such a covering; we may assume that I is a finite set. The corresponding complex has the form ($F = \widetilde{M}$):

$$\otimes_i M_{f_i} \to \bigotimes_{i<j} M_{f_i f_j} \to \dots .$$

It is natural to augment that complex from the left by M; we get the sequence

$$0 \to M \to \bigotimes_i M_{f_i} \to \bigotimes_{i<j} M_{f_i f_j} \to \dots . \qquad (*)$$

We claim that $(*)$ is exact. This will imply the theorem. Moreover, the exactness in the terms M and $\otimes M_{f_i}$ shows once again that $\widetilde{M}$ is indeed a sheaf.

To verify the exactness of $(*)$, we need only to show that the sequence remains exact after tensoring with any ring A_{f_i}, $j \in I$. The sequence $(*) \otimes_A A_{f_i}$ corresponds to the covering

$$(U_i \cap D(f_j)), \qquad i \in I,$$

of the set $D(f_j)$. This covering is, however, trivial in the sense that one of its elements coincides with $D(f_j)$, and for such trivial coverings, the (augmented) covering complex is acyclic.

Remark. The triviality of the cohomology of the covering $(D(f_i))$ of $\operatorname{Spec} A$ admits the following generalization. Let B be a strictly flat A-algebra (in the previous case $B = \oplus A_{f_i}$). Then the following natural sequence

$$0 \to M \to M \underset{A}{\otimes} B \to M \underset{A}{\otimes} B \underset{A}{\otimes} B \to \dots$$

is exact; compare Chap. 1, Sect. 4.4, *Another Version.* This result of Grothendieck forms the foundation of his descent theory.

1.3. The Koszul Complex. Complexes of modules, like (*), arise in many problems. We will recall a few facts; for details, see (Fulton-Lang (1985), Griffiths-Harris (1978), Grothendieck (1968b)). For simplicity, we restrict ourselves to the case $M = A$. Then the complex (*) is a tensor product of the "elementary" two term complexes

$$K^{\cdot}(f_i^{\infty}) = (A \to A_{f_i}),$$

concentrated in dimensions 0 and 1.

In general, given an element $f \in A$ and an integer $n \geq 0$, we denote by $K^{\cdot}(f^n)$ the two term complex

$$\left(A \to \frac{1}{f^n} A\right) \simeq \left(A \xrightarrow{f^n} A\right).$$

Then $K^{\cdot}(f^{\infty})$ is the inductive limit of $K^{\cdot}(f^n)$ as $n \to \infty$. Since cohomology commute with $\varinjlim$, it is helpful to examine the complex $K^{\cdot}(f^n)$.

The *Koszul complex* of a sequence $(f_1, \dots, f_n)$ of elements of A is defined as the tensor product of complexes:

$$K^{\cdot}(f_1, \dots, f_n) = K^{\cdot}(f_1) \underset{A}{\otimes} \dots \underset{A}{\otimes} K^{\cdot}(f_n).$$

It is natural to study its cohomology by induction on n. We will utilize the following lemma, which follows from the Künneth formula (Chap. 1, Sect. 2.8) or can be derived directly.

Lemma. *Let $K^{\cdot} = (K^0 \to K^1)$ be a two term complex of A-modules, where the A-modules K^0 and K^1 are flat. Then for an arbitrary complex of A-modules $L^{\cdot}$, the long sequence*

$$\dots \to H^{q-1}(L^{\cdot}) \underset{A}{\otimes} K^0 \xrightarrow{u^{q-1}} H^{q-1}(L^{\cdot}) \underset{A}{\otimes} K^1 \to H^q(L^{\cdot} \underset{A}{\otimes} K^{\cdot}) \to$$
$$\to H^q(L^{\cdot}) \underset{A}{\otimes} K^0 \xrightarrow{u^q} H^q(L^{\cdot}) \underset{A}{\otimes} K^1 \to \dots$$

is exact.

Here the homomorphisms u^q are induced by the differential $d\colon K^0 \to K^1$. We will describe two cases when this lemma allows us to calculate the cohomology of the Koszul complex.

First, we assume that $f_1, \dots, f_n$ generate the unit ideal in A, i.e., the open sets cover $\operatorname{Spec} A$. We observe that if in the lemma, $d\colon K^0 \to K^1$ is an isomorphism, then each u^q is an isomorphism, hence $L^{\cdot} \otimes_A K^{\cdot}$ is an acyclic complex for an arbitrary $L^{\cdot}$. Since f_i is invertible over $D(f_i)$, the complex $K^{\cdot}(f_1, \dots, f_n)$ is acyclic over each $D(f_i)$ hence acyclic everywhere. This gives yet another proof that $(*)$ is exact.

The second case is even more interesting. Recall that a sequence $f_1, \dots, f_n$ of elements of A is *regular* if for each i, f_i is not a zero divisor in $A/(f_1, \dots, f_{i-1})$. Utilizing the lemma, it is easy to verify by induction that for any regular sequence $f_1, \dots f_n$, we get

$$H^q(K^{\cdot}(f_1 \dots f_n)) = \begin{cases} 0 & \text{for } q \neq n\,, \\ A/(f_1, \dots f_n) & \text{for } q = n\,. \end{cases}$$

If A is a local ring, the length of the maximal regular sequence is said to be the *depth* of A, and is denoted by $\operatorname{depth} A$. It is an important numerical invariant of A. In general $\operatorname{depth} A \leq \dim A$. If we have an equality, than A is said to be a *Cohen-Macaulay ring.* If all the local rings of a scheme X are Cohen-Macaulay, then X is said to be a *Cohen-Macaulay scheme.* For the geometric meaning of this notion, see (Danilov (1988), Chap. 2, Sect. 6).

1.4. A Theorem on Affine Coverings. The most important consequence of Serre's theorem is that it enables us to calculate the cohomology of quasi-coherent sheaves with the help of arbitrary affine coverings.

Theorem. *Let X be a separated scheme, $\mathcal{U} = (U_i)$ an open affine covering of X, and F a quasi-coherent sheaf on X. Then*

$$H^*(X, F) = H^*(\mathcal{U}, F)\,.$$

Indeed, since X is separated, all the intersections

$$U_{i_0} \cap \dots \cap U_{i_p}$$

are affine. By Serre's theorem, the covering $\mathcal{U}$ is F-acyclic, and the theorem follows from (Chap. 1, Sect. 4.4, Proposition).

In fact, we replace the sheaf F by a complex $C^{\cdot}(\mathcal{U}, F)$ "equivalent" to F. In a sense, the true cohomological invariant of F is this complex as an object in the derived category. The reason we are interested only in its cohomology is more psychological than anything else.

Henceforth, all schemes are separated and all sheaves are quasi-coherent unless stated otherwise. We may ignore derived functors and flabby resolutions, and regard the cohomology just as the cohomology of an arbitrary affine covering. So, why to bother with all those general notions? There are two reasons. First, we get the independence of the choice of a covering. Second, the general definition of cohomology relates quasi-coherent sheaves with arbitrary Abelian sheaves, for example, $\mathcal{O}_X^*$.

Example. We will calculate the cohomology of $\mathcal{O}_X$ for the simplest non-affine variety $X = \mathbf{A}^2 \backslash \{0\}$. Let T_1, T_2 be coordinates in $\mathbf{A}^2$. Then X is covered by two affine charts: $U_i = D(T_i)$, $i = 1, 2$. Consider the complex of this covering:

$$K\left[T_1, T_2, T_1^{-1}\right] \oplus K\left[T_1, T_2, T_2^{-1}\right] \xrightarrow{d} K\left[T_1, T_2, T_1^{-1}, T_2^{-1}\right],$$

where $d(f_1, f_2) = f_1 - f_2$. Clearly $H^0(X, \mathcal{O}_X) = \operatorname{Ker} d = K[T_1, T_2]$ (i.e., a function on X can be extended to a regular function on $\mathbf{A}^2$), while $H^1(X, \mathcal{O}_X) = \operatorname{Coker} d$ is generated by the monomials $T_1^{m_1} T_2^{m_2}$ with $m_1 < 0$ and $m_2 < 0$. Thus $H^1(X, \mathcal{O}_X)$ is not trivial and even infinite-dimensional.

1.5. Cohomological Dimension. A consequence of the theorem on affine coverings is the vanishing of the cohomology H^q for large q. Precisely, if a scheme X can be covered by n open affine charts, then $H^q(X, F) = 0$ for $q \geq n$.

In particular, the cohomology H^q of an arbitrary sheaf on $\mathbb{P}^n$ are trivial for $q > n$. On the other hand, in the next section, we will see that there are sheaves F on $\mathbb{P}^n$ with $H^n(\mathbb{P}^n, F) \neq 0$. Since an arbitrary n-dimensional projective variety X admits a covering by $n+1$ affine charts, $H^q(X, F) = 0$ for $q > n = \dim X$. This is, in fact, true for every (Noetherian) scheme X and every (Abelian) sheaf F: $H^q(X, F) = 0$ for $q > \dim X$ (Grothendieck (1957), Godement (1958)).

1.6. Higher Direct Images. Let $f\colon X \to Y$ be a morphism of schemes. Given a sheaf G on Y, one can define its inverse image

$$f^*G = f^{-1}G \underset{f^{-1}\mathcal{O}_Y}{\otimes} \mathcal{O}_X,$$

which is obviously a quasi-coherent sheaf. With the help of the theorem on affine coverings, one can easily verify that the higher direct images are also quasi-coherent.

Proposition. *Let $f\colon X \to Y$ be a quasi-compact morphism, and F a quasi-coherent sheaf on X. Then the sheaves $R^q f_* F$ are quasi-coherent. Furthermore*

$$(R^q f_* F)(V) = H^q(f^{-1}(V), F).$$

for open affine subsets $V \subset Y$.

Corollary. *If $X \to Y$ is an affine morphism, then $R^q f_* F = 0$ for $q > 0$. In particular $H^*(X, F) = H^*(Y, f_* F)$.*

The latter follows from the Leray spectral sequence, and the former follows from Serre's theorem. The corollary is very often applied to a closed embedding $i\colon X \to Y$; the sheaf F on X is often identified with $i_* F$ on Y.

1.7. The Künneth Formula. Let X and Y be S-schemes, and F on X and G on Y be sheaves. We form the fiber product and consider the sheaf

$$F \underset{S}{\boxtimes} G = \bar{g}^* F \otimes \bar{f}^* G$$

on $X \times_S Y$. What is the relation between its cohomology and the cohomology of F and G?

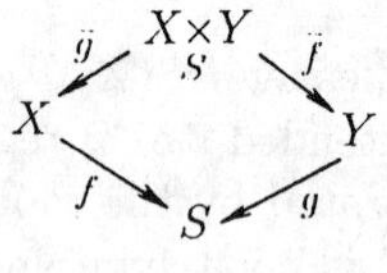

We restrict ourselves to the case when the base is affine, i. e., $S = \operatorname{Spec} A$. Let $\mathcal{U} = (U_i)$ and $\mathcal{V} = (V_j)$ be affine coverings of X and Y, respectively. Then $U_i \times_S V_j$ are affine and cover $X \times_S Y$. We get an isomorphism of covering complexes

$$C^{\cdot}(\mathcal{U} \times \mathcal{V}, F \underset{A}{\boxtimes} G) = C^{\cdot}(\mathcal{U}, F) \underset{A}{\otimes} C^{\cdot}(\mathcal{V}, G) .$$

So, we can apply the methods of Chap. 1, Sect. 2.8. Recall that a sheaf G on an S-scheme Y is said to be *flat* over S if the module G_y is flat over A for every point $y \in Y$. Then $C^{\cdot}(\mathcal{V}, G)$ consists of flat A-modules, and we may apply the spectral sequence from Chap. 1, Sect. 2.8. If, in addition, all the cohomology $H^i(X, F)$ (or all $H^j(Y, G)$) are flat A-modules, we obtain the *the Künneth formula*

$$H^q(X \underset{S}{\times} Y, F \underset{S}{\boxtimes} G) = \bigoplus_{i+j=q} (H^i(X, F) \underset{A}{\otimes} H^j(Y, G)) .$$

For instance, this will be the case if A is a field. Here is another special case: if $Y = \operatorname{Spec} B$ and Y is flat over $\operatorname{Spec} A$, then

$$H^q(X, F) \underset{A}{\otimes} B \simeq H^q(X \underset{A}{\otimes} B, F \underset{A}{\otimes} B) .$$

1.8. Cohomology of Open Inclusions. This chapter is devoted, for the most part, to the cohomology of complete varieties and proper morphisms. We will now briefly look at the opposite case that of open inclusions. Let $X = \operatorname{Spec} A$ be an affine scheme, and $j \colon U \to X$ an open inclusions. What can one say about the sheaf $R^q j_*(\mathcal{O}_U)$ or, what is essentially the same, the A-module $H^q(U, \mathcal{O}_U)$?

We cover U by open sets $U_i = D(f_i)$, where $f_i \in A$. Now, we compare $C^{\cdot}(\mathcal{U}, \mathcal{O}_U)$ with the Koszul complex

$$K^{\cdot}(f_1^{\infty}, \ldots, f_n^{\infty})$$

from Sect. 1.3. We get a natural morphism of complexes

$$\begin{array}{ccccccc} 0 \to & 0 & \to & C^0(\mathcal{U}, \mathcal{O}_U) & \to & C^1(\mathcal{U}, \mathcal{O}_U) & \to \dots \\ & \downarrow & & \downarrow \wr & & \downarrow \wr & \\ 0 \to & K^0(f^\infty) & \to & K^1(f^\infty) & \to & K^2(f^\infty) & \to \dots \end{array}$$

This diagram yields an exact sequence

$$0 \to H^0(K^{\cdot}(f^\infty)) \to A \to H^0(U, \mathcal{O}_U) \to H^1(K^{\cdot}(f^\infty)) \to 0$$

and, for $q \geq 2$, an isomorphism

$$H^q(U, \mathcal{O}_U) \simeq H^{q+1}(K^{\cdot}(f_1^\infty, \dots, f_n^\infty)) .$$

In particular, assume that $f_1, \dots, f_n$ form a regular sequence (and $n \geq 2$). Then

$$H^q(U, \mathcal{O}_U) = \begin{cases} A & \text{for } q = 0 , \\ 0 & \text{for } 0 < q < n-1 , \\ (A_{f_1}/A) \underset{A}{\otimes} \dots \underset{A}{\otimes} (A_{f_n}/A) & \text{for } q = n-1 . \end{cases}$$

The sheaves $R^q j_*(\mathcal{O}_U)$ have a similar structure for any open inclusion j. They are trivial for $0 < q < \operatorname{depth} X - 1$, and usually nontrivial and even infinite dimensional provided

$$\operatorname{depth} X - 1 \leq q < \dim X .$$

For more details, see (Grothendieck (1968b)).

§ 2. Cohomology of Projective Space

2.1. Sheaves on $\mathbb{P}^n$ and Graded Modules. Let $\mathbb{P} = \mathbb{P}^n$ be a projective space over a field K. Precisely, our $\mathbb{P}$ is the space of lines in a vector space V, and $\pi \colon V \backslash \{0\} \to \mathbb{P}$ denotes the natural projection.

Let F be a sheaf on $\mathbb{P}$. One can associate to F the module $H^0(V \backslash \{0\}, \pi^* F)$ over the polynomial ring $K[T_0, \dots, T_n]$ (or, in an invariant form, the symmetric algebra $\operatorname{Sym}(V^*)$). It has a natural grading of type $\mathbb{Z}$. Indeed, the group K^* acts on $V \backslash \{0\}$ as well as the sections on $\pi^* F$. The component of weight m of the module consists of the global sections s of $\pi^* F$ such that $s(tx) = t^m s(x)$ for $t \in K^*$ and $x \in V \backslash \{0\}$. The sections of F over $\mathbb{P}$ are in a one-to-one correspondence with the elements of weight 0. Similarly, the elements of weight m are in a one-to-one correspondence with the sections of a twisted sheaf

$$F(m) = F \underset{\mathcal{O}_{\mathbb{P}}}{\otimes} \mathcal{O}(m) ,$$

where $\mathcal{O}_{\mathbb{P}}(m)$ (or $\mathcal{O}(m)$) is the m-th tensor power of the tautological sheaf $O(1)$ on $\mathbb{P}$.

The same is true for higher cohomology. For each q, the space $H^q(\mathbb{P}, F(m))$ can be identified with the component of weight m of the graded module

$H^q(V\backslash\{0\}, \pi^*F)$ (see the preceding paragraph). This follows from the fact that the complex of the standard covering (U_i) of $\mathbb{P}$ coincides, as a graded module, with the complex of the covering $(\pi^{-1}(U_i))$ of $V\backslash\{0\}$. Thus we have an isomorphism

$$H^q(\mathbb{P}, F(*)) = \bigoplus_{m\in\mathbb{Z}} H^q(\mathbb{P}, F(m)) \simeq H^q(V\backslash\{0\}, \pi^*F),$$

2.2. Applications to Invertible Sheaves. We will give the summary of calculations for $F = \mathcal{O}_{\mathbb{P}}$. We utilize the description of $H^q(V\backslash\{0\}, \mathcal{O}_V)$ obtained in Sect. 1.8 (see also Example in Sect. 1.4). Let $T_0, \ldots, T_n$ be homogeneous coordinates in $\mathbb{P}$. They form a regular sequence in $K[T_0 \ldots, T_n]$. Hence

$$H^q(\mathbb{P}^n, \mathcal{O}(*)) = \begin{cases} K[T_0, \ldots, T_n] & \text{for } q = 0, \\ 0 & \text{for } 0 < q < n, \\ \bigotimes_{i=0}^{n} (K[T_i, T_i^{-1}]/K[T_i]) & \text{for } q = n. \end{cases}$$

The latter space is generated by the monomials $T_0^{m_0} \cdot \ldots \cdot T_n^{m_n}$, where all $m_i < 0$. The right-hand side is graded by the usual degree.

This can be expressed in the following invariant form:

$$H^0(\mathbb{P}^n, \mathcal{O}(m)) = \mathrm{Sym}^m(V^*) \quad \text{for } m \geq 0,$$
$$H^n(\mathbb{P}^n, \mathcal{O}(-n-1-k)) \simeq \mathrm{Sym}^k(V) \quad \text{for } k \geq 0;$$

The remaining $H^q(\mathbb{P}^n, \mathcal{O}(m))$ are equal to zero. We collect the data in a table.

q \ m		$-n-2$	$-n-1$	$-n$		-1	0	1	
n	S^2V	V	K	0	$\ldots$	0	0	0	0
1	0	0	0	$0 \vdots 0$	0	$0 \vdots 0$	0	0	0
0	0	0	0	0	$\ldots$	0	K	V^*	S^2V^*

The symmetry of the table hints for a possibility of the existence of a duality. In Sect. 5, we will discuss it in a more general setting. For the same reason, we have identified $H^n(\mathbb{P}^n, \mathcal{O}(-n-1-k))$ with $\mathrm{Sym}^k(V)$. In fact, we have a natural pairing

$$H^0(\mathcal{O}(k)) \otimes H^n(\mathcal{O}(-n-1-k)) \to H^n(\mathcal{O}(-n-1)) \simeq K,$$

given by multiplication in the cohomology. It is easily seen from explicit formulas that the pairing is perfect, and allows us to identify $H^n(\mathcal{O}(-n-1-k))$ with the dual of $H^0(\mathcal{O}(k)) = \mathrm{Sym}^k(V^*)$.

If a sheaf can be expressed in terms of sheaves of the form $\mathcal{O}(m)$, one is often able to calculate its cohomology.

Example 1. On $\mathbb{P} = \mathbb{P}^n$, we have an important *Euler sequence*

$$0 \to \Omega^1_{\mathbb{P}} \to V^* \otimes \mathcal{O}_{\mathbb{P}}(-1) \to \mathcal{O}_{\mathbb{P}} \to 0\,,$$

where Ω^1 is the cotangent sheaf. Passing to the cohomology, we derive from the previous formulas that

$$H^q(\mathbb{P}, \Omega^1_{\mathbb{P}}) = \begin{cases} 0 & \text{for } q \neq 1\,, \\ K & \text{for } q = 1\,. \end{cases}$$

Taking the p-th exterior power of the Euler sequence, we get an exact sequence

$$0 \to \Omega^p_{\mathbb{P}} \to (\wedge^p V^*) \otimes \mathcal{O}_{\mathbb{P}}(-p) \to \Omega^{p-1}_{\mathbb{P}} \to 0\,.$$

It follows by induction on p (for $p \leq \dim \mathbb{P}$) that

$$H^q(\mathbb{P}, \Omega^p_{\mathbb{P}}) = \begin{cases} 0 & \text{for } q \neq p\,, \\ K & \text{for } q = p\,. \end{cases}$$

Glueing the above short exact sequences, we get a long exact sequence of sheaves on $\mathbb{P}^n$:

$$0 \to \wedge^{n+1} V^* \otimes \mathcal{O}(-n-1) \to \ldots \to V^* \otimes \mathcal{O}(-1) \to \mathcal{O} \to 0\,.$$

It begins with a canonical sheaf

$$\omega_{\mathbb{P}} = \Omega^n_{\mathbb{P}} = \wedge^{n+1} V^* \otimes \mathcal{O}(-n-1) \simeq \mathcal{O}(-n-1)$$

followed by acyclic sheaves. So it is an acyclic resolution of $\omega_{\mathbb{P}}$.

Example 2. Let X be a hypersurface of degree d in $\mathbb{P} = \mathbb{P}^n$. For $\mathcal{O}_X$, as a sheaf on $\mathbb{P}$, we get a resolution

$$0 \to \mathcal{O}_{\mathbb{P}}(-X) \to \mathcal{O}_{\mathbb{P}} \to \mathcal{O}_X \to 0\,,$$

moreover $\mathcal{O}_{\mathbb{P}}(-X) \simeq \mathcal{O}(-d)$ hence $H^0(X, \mathcal{O}_X) = K$. The latter also follows at once from the connectedness of X. Furthermore, $H^q(X, \mathcal{O}_X) = 0$ for $0 < q < \dim X$, and

$$H^{n-1}(X, \mathcal{O}_X) \simeq H^n(\mathbb{P}, \mathcal{O}(-d))$$

and its dimension equals $(n!)^{-1}(d-1)(d-2)\ldots(d-n)$. In a similar manner one can calculate the cohomology of $\mathcal{O}_X(m)$ (Serre (1955)).

2.3. Applications to Coherent Sheaves. Recall that a quasi-coherent sheaf on a Noetherian scheme is said to be *coherent* if it is generated locally by finitely many local sections (Danilov (1988)). In other words, it is locally of the form $\widetilde{M}$, where M is a module of finite type. Thus, in addition to the restrictions on the local structure, the coherent sheaves satisfy a finiteness condition.

A coherent sheaf on $\mathbb{P}$ is not necessary generated by its global sections. For example, the sheaf $\mathcal{O}(-1)$ has only a zero global section. However, one can easily show that after a suitable twisting by $\mathcal{O}(m)$, we get a sheaf $F(m)$ that has sufficiently many global sections, i. e., there exists an epimorphism

$$\mathcal{O}^N \to F(m) .$$

By twisting these sheaves, we see that any coherent sheaf F on $\mathbb{P}$ is a quotient sheaf of a sheaf of the form $\mathcal{O}(-m)^N$.

This fact is similar to a representation of a module as a quotient module of a free module, and it plays a similar role. Suppose we want to prove a general assertion concerning coherent sheaves. We verify it, first, for sheaves of the form $\mathcal{O}(m)$, next, for direct sums of such sheaves, and finally, using a resolution consisting of such sheaves, we extend the assertion to arbitrary coherent sheaves. We will demonstrate this principle in the following theorem of Serre.

Theorem. *Let F be a coherent sheaf on $\mathbb{P}$. Then*
a) *the space $H^q(\mathbb{P}, F)$ has finite dimension for every q; and*
b) *$H^q(\mathbb{P}, F(m)) = 0$ if $q > 0$ and m is sufficiently large.*

It follows from the explicit calculations of Sect. 2.2 that the theorem holds for all $\mathcal{O}(m)$ hence direct sums of the $\mathcal{O}(m)$'s. Furthermore, for $q > \dim \mathbb{P}$, the theorem follows from Sect. 1.5. We proceed by decreasing induction on q. Consider an exact sequence

$$0 \to G \to E \to F \to 0 ,$$

where E is a sum of sheaves of the form $\mathcal{O}(m)$. In the exact sequence

$$H^q(E) \to H^q(F) \to H^{q+1}(G)$$

the left-hand term and the right-hand term are finite-dimensional spaces. Hence the middle term has also a finite dimension, thus proving (a). A similar argument proves (b) as well.

2.4. Regular Sheaves. The Serre theorem implies that sufficiently ample sheaves are acyclic. To what extend the converse is true? For example, is it true that an acyclic sheaf is generated by its global sections? The sheaf $\mathcal{O}(-1)$ gives a counterexample. To get a handle on the problem, one has to look at $H^q(\mathbb{P}, F(-q))$ in place of $H^q(\mathbb{P}, F)$.

Definition. A coherent sheaf F on $\mathbb{P}$ is said to be *regular* (in the sense of Castelnuovo-Mumford) if $H^q(\mathbb{P}, F(-q)) = 0$ for all $q \geq 1$.

For example, the sheaf $\mathcal{O}(m)$ is regular for $m \geq 0$.

Proposition. *Let F be a regular sheaf on $\mathbb{P}$. Then F is generated by global sections, and $F(1)$ is regular. In particular, F is acyclic.*

We will verify the proposition assuming, for simplicity, that F is flat. Let H be a hyperplane in $\mathbb{P}$. Tensoring the exact sequence

$$0 \to \mathcal{O}(-H) \to \mathcal{O} \to \mathcal{O}_H \to 0$$

with F, we get the exact sequence

$$0 \to F(-1) \to F \to F_H \to 0,$$

since F is flat. We claim that F_H is also regular. Indeed, the cohomology $H^q(F_H(-q))$ is squeezed between $H^q(F(-q))$ and $H^{q+1}(F(-q-1))$, which are trivial by assumption. By induction hypothesis, F_H is generated by global sections. Since the sequence

$$H^0(\mathbb{P}, F) \to H^0(H, F_H) \to H^1(\mathbb{P}, F(-1)) = 0$$

is exact, F is also generated by global sections at the points of H. Since H is an arbitrary hyperplane, F is generated by global sections. Further, by induction, $F_H(1)$ is regular on H. It follows from the exact sequence

$$0 \to F(-H) \to F \to F_H \to 0$$

that $H^q(F(1-q))$ is squeezed between $H^q(F(-q)) = 0$ and $H^q(F_H(1-q)) = 0$. Therefore $F(1)$ is also regular.

2.5. The Euler Characteristic. It is often convenient to bundle the information about all the cohomology of F in a single integer

$$\chi(\mathbb{P}, F) = \sum_q (-1)^q \dim_K H^q(\mathbb{P}, F),$$

called the *Euler characteristic* of F. Here we use the finiteness of the cohomology of coherent sheaves. The main property of the Euler characteristic is its additivity. Let

$$0 \to F \to G \to H \to 0$$

be an exact sequence of coherent sheaves. Then

$$\chi(G) = \chi(F) + \chi(H).$$

This often makes the calculation of $\chi(.)$ an easy task (Sect. 4). Using the table from Sect. 2.2, one can verify that the Euler characteristic of $\mathcal{O}(m)$ is given by the formula

$$\chi(\mathbb{P}^n, \mathcal{O}(m)) = \frac{(m+n)\cdot \ldots \cdot (m+1)}{n!},$$

which is a polynomial in m of degree $n = \dim \mathbb{P}$. The latter is true in a more general case.

Proposition. *Let F be a coherent sheaf on $\mathbb{P}$. Then there exists a polynomial* (Hilbert polynomial) $P_F(t) \in \mathbb{Q}[t]$ *such that* $\chi(\mathbb{P}, F(m)) = P_F(m)$ *for all integers* m.

One proves the proposition by comparing F with $F(-1)$. Note that the degree of the Hilbert polynomial P_F is equal to the dimension of the support of F. Recall that a similar fact holds for a graded module $M = \oplus_\nu M_\nu$ of finite type (Danilov (1988), Chap. 3). There, however, the Hilbert function gives the dimension of M_ν for large ν only. Taking into account the cohomology, we arrive at the statement valid for all $\nu \in \mathbb{Z}$.

2.6. Relative Case. Almost all facts described in this section can be extended to the case when the ground field K is replaced by an arbitrary Noetherian scheme S, and the projective space $\mathbb{P}$ is replaced by a relative projective space $\mathbb{P}_S$ or an arbitrary projective bundle $f\colon \mathbb{P}(E) \to S$. One should, now, replace the cohomology space $H^q(\mathbb{P}, F)$ by the higher direct image sheaf $R^q f_*(F)$. In particular, the sheaves $R^q f_*(\mathcal{O}(m))$ are locally free and have correct ranks, as in Sect. 2.2. The Serre theorem from Sect. 2.3 can be generalized as follows: the sheaves $R^q f_*(F)$ are always coherent, and $R^q f_*(F(m)) = 0$ for $q > 0$ and $m \gg 0$.

§ 3. Cohomology of Proper Morphisms

3.1. The Finiteness Theorem. Many basic results on the cohomology of sheaves on projective spaces or projective bundles hold for arbitrary proper morphisms. Henceforth, we assume all schemes to be Noetherian.

Theorem (Grothendieck). *Let $f\colon X \to Y$ be a proper morphism, and F a coherent sheaf on X. Then the sheaves $R^q f_* F$ on Y are coherent for every q.*

Corollary. *Let F be a coherent sheaf on a complete algebraic variety X. Then $H^q(X, F)$ is a finite dimensional space.*

We will explain the strategy of proof of this and similar results. Since the assertion is local on Y, we may assume that Y is affine. Given a projective morphism $f\colon X \to Y$, one can decompose it into a closed embedding $X \hookrightarrow \mathbb{P}^n_Y$ and the projection $\mathbb{P}^n_Y \to Y$. The coherence, now, follows from Sect. 2.6.

The case of an arbitrary proper morphism is reduced to a projective morphism by the following trick. According to the Chow lemma, there is a projective Y-scheme $\overline{f}\colon \overline{X} \to Y$ and a birational projective morphism $\pi\colon \overline{X} \to X$. By induction on the dimension of the support of F, we may assume that the theorem holds for any sheaf on X whose support is not the whole X. Now, we consider the sheaf $\overline{F} = \pi^* F$ on $\overline{X}$ in place of the sheaf F on X. These sheaves are related by a Leray spectral sequence

$$E_2^{p,q} = R^p f_*(R^q \pi_* \overline{F}) \Rightarrow R^{p+q} \overline{f}_*(\overline{F}).$$

We observe that $E_2^{p,q}$ are coherent provided $q \neq 0$. Indeed, since π is birational, the support of the sheaf $R^q \pi_* \overline{F}$ is distinct from X provided $q \neq 0$; moreover, it is coherent because π is projective. The limit sheaves $R^{q+q} \overline{f}_*(\overline{F})$ are also coherent because $\overline{f}\colon \overline{X} \to Y$ is a projective morphism. It follows that $E_2^{p,0} = R^p f_*(\pi_* \overline{F})$ are also coherent. But $\pi_* \overline{F}$ differ from F by sheaves with smaller supports, so $R^p f_* F$ are also coherent.

3.2. The Comparison Theorem. We now consider the following question. Let $f\colon X \to Y$ be a proper morphism, and F a coherent sheaf on X. How to describe the fiber of $R^q f_* F$ at a point y? Intuitively, this fiber should be intimately related to the cohomology of $f^{-1}(y)$ (the fiber of f). Is it possible to make this relationship more precise?

A systematic approach to such questions is via the base change homomorphism described in a general setting in Sect. 1.7. Let $\mathfrak{m}$ be the maximal ideal of the local ring $A = \mathcal{O}_{Y,y}$. For an integer $n \geq 0$, we denote by X_n the n-th infinitesimal neighborhood of $f^{-1}(y)$, i. e.,

$$X \underset{Y}{\times} \operatorname{Spec}(A/\mathfrak{m}^{n+1}).$$

As a topological space, X_n coincides with $X_0 = f^{-1}(y)$. However, it has a larger sheaf of functions. Set

$$F_n = F \underset{\mathcal{O}_X}{\otimes} \mathcal{O}_{X_n}.$$

The sections of F_n take into account not only the restrictions of the sections of F to X_0 but the first n partial derivatives as well. Then the base change homomorphism takes the form

$$\varphi_n : (R^q f_* F)_y \otimes A/\mathfrak{m}^{n+1} \to H^q(X_n, F_n).$$

In the general case, one can say nothing about φ_n.

We will discuss the case when F is flat below. Now, however, we return to the general case. It turns out that in the limit the homomorphisms become isomorphisms.

Theorem (Grothendieck). *With the above notation and assumptions, the limit homomorphisms*

$$\varphi_\infty : \varprojlim_n (R^q f_* F) \otimes A/\mathfrak{m}^{n+1} \xrightarrow{\sim} \varprojlim_n H^q(X_n, F_n)$$

are isomorphisms for each q.

3.3. Sketch of the Proof. On the left-hand side, we have the completion of the A-module $(R^q f_* F)_y$ in the $\mathfrak{m}$-adic topology. Replacing the scheme Y

by Spec $\hat{A}$, we assume that Y is the spectrum of a complete local ring A, and prove that

$$H^q(X,F) = \varprojlim H^q(X_n, F_n).$$

As in Sect. 3.1, first we can reduce everything to projective morphisms, and then to $X = \mathbb{P}_A^N$.

Now we can employ the principle from Sect. 2.3. First, the theorem is true for $\mathcal{O}(m)$. In that case, the φ_n's are already isomorphism. Second, we proceed by descending induction on q. For $q > N$, all the cohomology vanish. So, we suppose that the theorem is true for any q' greater than q. Consider an exact sequence

$$0 \to G \to E \to F \to 0, \tag{$*$}$$

where E is a direct sum of sheaves of the form $\mathcal{O}(m)$. Let $J = \mathfrak{m}\mathcal{O}_X$ be the sheaf of ideals of X_0 inside X. We assume, for a moment, that the sequences

$$0 \to G/J^nG \to E/J^nE \to F/J^nF \to 0, \tag{$*_n$}$$

derived from $(*)$ by tensoring with $\mathcal{O}/J^n$ are exact. Then we get a commutative diagram

$$\begin{array}{ccccccc}
H^q(G) & \longrightarrow & H^q(E) & \longrightarrow & H^q(F) & \longrightarrow \\
\downarrow\varphi_G & & \downarrow\varphi_E & & \downarrow\varphi_F & \\
\varprojlim H^q(G/J^nG) & \to & \varprojlim H^q(E/J^nE) & \to & \varprojlim H^q(F/J^nF) & \to
\end{array}$$

$$\begin{array}{ccccc}
\longrightarrow & H^{q+1}(G) & \longrightarrow & H^{q+1}(E) \\
 & \downarrow\varphi'_G & & \downarrow\varphi'_E \\
\to & \varprojlim H^{q+1}(G/J^nG) & \to & \varprojlim H^{q+1}(E/J^nE).
\end{array}$$

The top row is exact as the long cohomology sequence for the short exact sequence $(*)$. The bottom row comes from the exact cohomology sequences for the short exact sequence $(*_n)$ by taking inverse limits. The $\varprojlim$ does not preserve exact sequences in general. However, in our case, the bottom row is also exact since the corresponding modules are Artin modules. We already know that φ_E and φ'_E are isomorphisms, and φ'_G is an isomorphism by induction hypothesis. By a diagram chase, we first derive that φ_F is surjective. Since this is true for every module, φ_G is surjective. Now, one can easily deduce that φ_F is surjective.

Now, we turn to the general case. Of course, in general $(*_n)$ is not exact, however, the following sequence is always exact

$$0 \to G/G \cap J^nE \to E/J^nE \to F/J^nF \to 0.$$

So, it will suffice to show that

$$\varprojlim H^*(G/G \cap J^nE) \xrightarrow{\sim} \varprojlim H^*(G/J^nG).$$

The latter easily follows from the fact that the filtrations J^nG and $G \cap J^nE$ on G are equivalent, i. e.

$$G \cap J^m E \subset J^n G \subset H \cap J^n E$$

for $m \gg n$. The latter inclusion is a special case of a more precise Artin-Rees lemma from commutative algebra (Grothendieck-Dieudonné (1962–1963), Hartshorne (1977)).

3.4. The Theorem on Formal Functions. The most famous applications of the comparison theorem utilize only the 0-dimensional cohomology of $\mathcal{O}_X$ (though the proof makes an essential use of all the cohomology). In that case, the comparison theorem gives an isomorphism

$$(\widetilde{f_*\mathcal{O}_X})_y \tilde{\to} \varprojlim H^0(\mathcal{O}_X/\mathfrak{m}^n\mathcal{O}_X)\,.$$

An element of the right-hand side ring is a compatible system (s_n) of infinitesimal germs of functions along the fiber $f^{-1}(y)$, i.e., a *formal function* on a formal neighborhood of $f^{-1}(y)$. The theorem asserts that for every n, there exists a genuine (Zariski) neighborhood $U \subset X$ of $f^{-1}(y)$ and a regular function s on U that coincides with s_n up to the n-th order. In other words, a formal function admits a sufficiently close approximation by regular functions.

In (Danilov (1988)), we have already presented an application of that fact to the Zariski connectedness theorem. Other applications are similar – something, defined in a formal neighborhood, can be extended to a genuine Zariski neighborhood. Applications of this technique to the Lefschetz-type theorems are given in the remarkable work (Grothendieck (1968b)).

3.5. Continuous Families of Sheaves. One may view an arbitrary morphism of finite type, $f\colon X \to Y$, as a family of algebraic schemes $X_y = f^{-1}(y)$, $y \in Y$, each defined over its own field $k(y)$. It is not "continuous" in general. For instance, the dimension of special fibers can be larger than the dimension of a typical general fiber. An intuitive concept of "continuity" of a family $(X_y)_{y\in Y}$ can be formalized by requiring the morphism $f\colon X \to Y$ to be flat (Danilov (1988), Chap. 4).

One can introduce a similar notion for families of sheaves on families of schemes (X_y). Let $f\colon X \to Y$ be a morphism, and F a coherent sheaf on X. It induces on the fiber

$$X_y = X \underset{Y}{\times} \operatorname{Spec} k(y)$$

the sheaf

$$F_y = F \underset{\mathcal{O}_Y}{\otimes} k(y)$$

(or $F_Y \boxtimes k(y)$, in the notation of Sect. 1.7). The family of sheaves (F_y), $y \in Y$, on the family of schemes (X_y) is said to be "continuous" if F is flat over Y. We will try to convince that this definition indeed corresponds to the intuitive notion of continuity.

3.6. The Semicontinuity Theorem. Throughout the rest of the section, $f\colon X \to Y$ is a proper morphism, and F a coherent sheaf on X that is flat over Y. Given a point $y \in Y$, we denote by $h^q(y, F)$ the dimension of the vector space $H^q(X_y, F_y)$ over the field $k(y)$. We would like to understand how the $h^q(y, F)$ vary as a function of $y \in Y$. We begin with two examples.

Example 1. let C be an elliptic curve. We consider a family of sheaves $\mathcal{O}_C(P - Q)$ on C, depending on the points P, $Q \in C$. If $P \neq Q$ then $H^0(C, \mathcal{O}(P - Q)) = 0$. Indeed, otherwise we would get a morphism $C \to \mathbb{P}^1$ of degree 1. But if $P = Q$ then

$$H^0(C, \mathcal{O}(P - P)) = H^0(C, \mathcal{O}) = K\,.$$

Thus, $h^q(y)$ can jump under a specialization.

Example 2. Let E be a locally free sheaf of rank 2 on $\mathbb{P}^1$. One can show that E is a direct sum $\mathcal{O}(m) \oplus \mathcal{O}(m')$; the pair (m, m') is called the *type* of E. It is possible to construct a family (E_t) of two-dimensional sheaves on $\mathbb{P}^1$ such that E_0 has type $(-2, 0)$, while each E_t has type $(-1, -1)$ for $t \neq 0$. Then, according to Sect. 2.2,

$$H^0(\mathbb{P}^1, E_t) = H^1(\mathbb{P}^1, E_t) = 0 \qquad (t \neq 0)\,,$$

while $H^0(\mathbb{P}^1, E_0) = K$ and $H^1(\mathbb{P}^1, E_0) \simeq K$. So, for $t = 0$, we again get a jump for h^0 as well as h^1. The following theorem shows that this is a general phenomenon.

Theorem. *If F is a flat sheaf, then the $h^q(., F)$ is an upper semicontinuous function on Y.*

In particular, if $h^q(X_y, F_y) = 0$ for a point y, then is also vanishes in a neighborhood of y. By the comparison theorem of Sec. 3.2, it follows that $R^q f_*$ is a zero sheaf in that neighborhood.

The statement of the theorem is local on Y, so we can assume that $Y = \operatorname{Spec} A$. The proof of the semicontinuity theorem and other similar fact is based on the following trick, which seems rather technical.

3.7. The Lemma on Equivalent Complex. Under the above assumptions, the lemma asserts the *existence of a complex $K^{\cdot}$ of A-modules with the following properties:*

a) *each K^q is a flat A-module of finite type, and*
b) *for every A-module M*

$$H^q(X, F \underset{A}{\otimes} M) \simeq H^q(K^{\cdot} \underset{A}{\otimes} M)\,.$$

Thus, we may deal with a complex of sheaves on Y, i. e. a complex of A-modules, instead of the sheaf F on X. Of course, this gives nothing new – we have already seen that the complex $C^{\cdot}(\mathcal{U}, F) = C^{\cdot}$ of an affine covering of X can be used in place of F, and it consists of flat modules provided F is flat. The only new ingredient is the finiteness of K^q. To construct such a complex (see Mumford (1970)), one uses the finiteness of the cohomology $H^*(C^\cdot) = H^*(X, F)$ established in Sect. 3.1. Now we will explain how this lemma works.

The complex $K^\cdot$ is built of homomorphisms

$$d^q : K^q \to K^{q+1}$$

of two A-modules, which we may assume to be free of finite rank. So, d^k is a matrix with coefficients in A. The specialization to a point $y \in Y$ (i.e., the homomorphism $A \to A/\mathfrak{m}_y = k(y)$) gives a matrix d^q_y over $k(y)$. Clearly the rank of d^q_y is a lower semicontinuous function of y, while the corank, i.e., the dimension of the kernel of d^q_y, is an upper semicontinuous function of y. Since

$$H^q(X, F \otimes_A k(y)) = H^q(K^\cdot \otimes k(y)) = \operatorname{Ker} d^q_y / \operatorname{Im} d^{q-1}_y ,$$

we deduce the semicontinuity theorem of Sect. 3.6.

Now, we assume, in addition, that the $h^q(., F)$ is not only upper semicontinuous but continuous (i.e., locally constant). Then the ranks of d^q_y and d^{q-1}_y are (locally) constant. If, in addition, Y is reduced, then the modules $\operatorname{Ker} d^q$ and $\operatorname{Im} d^{q-1}$ are direct summands in K^q. Since

$$\operatorname{Ker} d^q / \operatorname{Im} d^{q-1} = H^q(K^\cdot) = H^q(X, F) ,$$

we deduce the following

Proposition. *Let Y be a reduced scheme, and F a sheaf that is flat over Y. If the function $h^q(., F)$ is locally constant on Y, then the sheaf $R^q f_* F$ is locally constant, and for every point $y \in Y$, the base change homomorphism*

$$(R^q f_* F) \otimes_A k(y) \to H^q(X_y, F_y)$$

is an isomorphism.

3.8. The Constancy of Euler Characteristic. Another corollary of the existence of $K^\cdot$ is the constancy of the Euler characteristic of the sheaves F_y in a continuous family.

Proposition. *If F is flat over Y, then $\chi(X_y, F_y)$ is locally constant as a function of $y \in Y$.*

Indeed

$$\chi(X_y, F_y) = \sum_q (-1)^q h^q(y, F) = \sum_q (-1)^q \dim(K^q \otimes k(y))$$
$$= \sum_q (-1)^q \operatorname{rk} K^q .$$

Thus, it comes as no surprise that both h^0 and h^1, in Example 2, have jumped up simultaneously and cancel each other. If $X = \mathbb{P}_Y$, the Hilbert polynomial P_{F_y} is also constant. One can show that the converse is also true: If the base Y is reduced and the Hilbert polynomial P_{F_y} of the family of sheaves (F_y) is independent of $y \in Y$, the family is continuous (Hartshorne (1977)). For additional facts on the base change, see (Grothendieck-Dieudonné (1962–1963), § 7).

§ 4. The Riemann-Roch Theorem

4.1. The Riemann-Roch Theorem for Curves. To begin with we consider a classical problem. Let X be a smooth projective curve over an algebraically closed field, and $D = \sum_{P\in X} n_P[P]$ a divisor on X. The Riemann problem is to describe $H^0(X, \mathcal{O}_X(D))$. The elements of this space are rational functions f on X with restrictions on the order of zeros and poles, namely:

$\mathrm{ord}_P(f) \geq -n_P$ for every point $P \in X$. As we know from the preceding section, the space $H^0(X, \mathcal{O}(D))$ and its dimension depend considerably on the divisor D. However, the situation is much simpler if we are kepping track of the Euler characteristic

$$\chi(X, \mathcal{O}_X(D)) = \dim H^0(X, \mathcal{O}(D)) - \dim H^1(X, \mathcal{O}(D))\,.$$

We know from Sect. 3.8 that $\chi(\mathcal{O}(D))$ depends only on discrete invariants of D. A discrete invariant that comes to mind at once is the degree of D, $\deg D = \sum_P n_P$. Indeed, one can easily show, by comparing D and $D + [P]$, that

$$\chi(X, \mathcal{O}_X(D)) = \deg D + \chi(X, \mathcal{O}_X)\,.$$

This fact is called the Riemann-Roch theorem. In fact, Riemann proved the inequality

$$h^0(D) \geq \deg D + 1 - g\,,$$

which follows from the formula (here g is the genus of the curve); Roch then gave the interpretation of $h^1(\mathcal{O}(D))$ as $h^0(\mathcal{O}(K - D))$ (the latter is discussed in Sect. 5).

4.2. The General Riemann Problem. Let X be an arbitrary complete variety, and F a coherent sheaf on X. As we know, the Euler characteristic

$$\chi(X, F) = \sum_q (-1)^q \dim H^q(X, F)$$

is invariant under continuous variations of X and F. Therefore, we expect the existence of a formula that represents $\chi(X, F)$ in terms of "discrete" invariants of X and F. For curves, such invariants were the genus of X and the degree of D. In the general case, we should have been able to associate to F some symbols, perform algebraic manipulations with those symbols, and obtain the number $\chi(X, F)$ as a result.

It seems that the Hilbert polynomials P_F could be convenient symbols in the projective case. However, the theory developed along a different path, without appealing to an embedding in a projective space. Given a variety X, we consider a ring $A(X)$ together with a linear functional deg: $A(X) \to \mathbb{Q}$ and a distinguished element $\mathrm{Td}(X)$. Each sheaf F on X determines an element $\mathrm{ch}(F) \in A(X)$. Then the required formula for $\chi(X, F)$ takes the form

$$\chi(X, F) = \deg(\mathrm{ch}(F) \cdot \mathrm{Td}(X)).$$

In any case, Hirzebruch obtained such a formula for complex projective manifolds. He employed the ring of the singular cohomology $H^*(X, \mathbb{Q})$ as $A(X)$. The functional deg arises from a natural identification of $H^n(X, \mathbb{Q})$ with $\mathbb{Q}$, where $n = \dim X$. Using Chern classes, one defines elements $\mathrm{ch}(F)$ and $\mathrm{Td}(X) = \mathrm{td}(T_X)$, where T_X is the tangent sheaf of X.

In the abstract case, one can employ the Chow ring $A(X)$ of algebraic cycles on X (Danilov (1988), Chap. 3) in place of the singular cohomology ring. We will briefly explain how to construct Chern classes in the abstract case.

4.3. Chern Classes. Let E be a locally free sheaf of rank r on a smooth algebraic variety X. We consider the corresponding projective bundle π: $\mathbb{P}(E) \to X$, and the tautological invertible sheaf $\mathcal{O}(1)$ on $\mathbb{P}(E)$. Let ξ denote the divisor class in $A^1(\mathbb{P}(E))$ corresponding to $\mathcal{O}(1)$. It is known that $A(\mathbb{P}(E))$ is freely generated as an $A(X)$-module by $1, \xi, \ldots, \xi^{r-1}$ (Danilov (1988), Chap. 3). So, we get the following expression for ξ^r in terms of that basis:

$$c_0\xi^r - c_1\xi^{r-1} + \ldots + (-1)^r c_r \xi^0 = 0, \qquad c_0 = 1, \quad c_i \in A^i(X).$$

In this decomposition, the coefficients $c_i = c_i(E)$ are said to be *Chern classes of the sheaf* E. The *total Chern class* is the element

$$c(E) = c_0(E) + c_1(E) + \ldots + c_r(E) \in A(X).$$

The Chern classes satisfy the usual formal properties:

1. *Functoriality.* For every morphism $f\colon Y \to X$

$$c(f^*E) = f^*c(E).$$

2. *The Whitney Condition.* For every short exact sequence

$$0 \to E' \to E \to E'' \to 0,$$

we have

$$c(E) = c(E') \cdot c(E'').$$

3. *Normalization.* For a divisor D on X

$$c(\mathcal{O}_X(D)) = 1 + D.$$

In particular, if E is isomorphic to a direct sum of invertible sheaves $\mathcal{O}(D_1), \ldots, \mathcal{O}(D_r)$, then $c(E) = \prod_i (1 + D_i)$. It is convenient to imagine that every sheaf E admits a decomposition $E = \oplus_{i=1}^r \alpha_i$ into "quarks" α_i. As a matter of fact, in the final formula the "quarks" α_i appear only in symmetric expressions like $\alpha_1 + \ldots + \alpha_r, \ldots, \alpha_1 \ldots \alpha_r$ that equal $c_1(E), \ldots, c_r(E)$.

Example 1. It follows from the Euler exact sequence on $\mathbb{P}^n$

$$0 \to \Omega^1_{\mathbb{P}^n} \to \mathcal{O}_{\mathbb{P}^n}(-1)^{n+1} \to \mathcal{O}_{\mathbb{P}^n} \to 0$$

that $c(\Omega^1_{\mathbb{P}^n}) = c(\mathcal{O}(-1)^{n+1}) = (1-H)^{n+1}$, where H is the class of a hyperplane in $\mathbb{P}^n$. Similarly $c(T_{\mathbb{P}^n}) = (1 + H)^{1+n}$.

Example 2. Let X be an n-dimensional Abelian variety embedded in $\mathbb{P}^m$. Then $m \geq 2n$.

Indeed, consider an exact sequence of sheaves on X

$$0 \to T_X \to T_{\mathbb{P}^m}|_X \to \mathcal{N} \to 0,$$

where T_X is the tangent sheaf of X, and $\mathcal{N}_{X/\mathbb{P}}$ the normal sheaf of X in $\mathbb{P}^m$. Since T_X is trivial,

$$c(\mathcal{N}) = c(T_{\mathbb{P}|X}) = i^* c(T_{\mathbb{P}}) = (1 + h)^{m+1},$$

where $i \colon X \hookrightarrow \mathbb{P}^m$ is the embedding, and $h = i^* H$. In particular, $c_n(\mathcal{N}) = \binom{m+1}{n} h^n$ is not trivial since $\deg h^n = \deg X$. So, the rank of the sheaf $\mathcal{N}$, which is equal to $m - n$, is at least n whence $m \geq 2n$.

Now we assume that $m = 2n$. One can show that $(X \cdot X)_{\mathbb{P}^m} = \deg c_n(\mathcal{N})$. Since $[X] = \deg X \cdot [H]^n$, we get $\deg X = \binom{2n+1}{n}$. For example, every Abelian surface in $\mathbb{P}^4$ should have degree 10. Such a surface was constructed by Horrocks and Mumford (Horrocks-Mumford (1973)).

Now we return to Chern classes. Given a locally free sheaf E, its *Chern character* is defined by the following expression

$$\operatorname{ch}(E) = \sum_{i=1}^r e^{\alpha_i},$$

where $c(E) = \prod_{i=1}^r (1 + \alpha_i)$, or in terms of the Chern classes of E,

$$\operatorname{ch}(E) = \operatorname{rk} E + c_1(E) + \tfrac{1}{2}(c_1(E)^2 - 2c_2(E)) + \ldots .$$

The Chern character translates a direct sum of sheaves into a sum, and a tensor product into a product.

The *Todd class* of E, $\operatorname{td}(E)$, is given by the following formal expression

$$\operatorname{td}(E) = \prod_{i=1}^r \frac{\alpha_i}{1 - e^{-\alpha_i}} = \prod_{i=1}^r \left(\sum_{k=1}^\infty (-1)^k \frac{\alpha_i^{k-1}}{k!} \right),$$

where the α_i's have the same meaning. In terms of Chern classes

$$\mathrm{td}(E) = 1 + \tfrac{1}{2}c_1(E) + \tfrac{1}{12}(c_1(E)^2 + c_2(E)) + \dots .$$

Since the expressions for ch and td contain denominators, they are elements of the ring $A(X)_{\mathbb{Q}} = A(X) \otimes \mathbb{Q}$.

4.4. The Riemann-Roch-Hirzebruch Theorem. We denote by deg the homomorphism $A(X)_{\mathbb{Q}} \to \mathbb{Q}$ that assigns to a cycle α the degree of its 0-dimensional component.

Theorem (Hirzebruch). *Let X be a nonsingular projective variety, and E a locally free sheaf on X. Then*

$$\chi(X, E) = \deg(\mathrm{ch}(E) \cdot \mathrm{td}(T_X)) .$$

We will mention several special cases.

Example 1. Let X be a curve with a canonical class K. For a locally free sheaf E on X, we get

$$\chi(X, E) = -\frac{\mathrm{rk}\, E}{2} \cdot \deg K + \deg c_1(E) .$$

In particular, $\deg K = -2\chi(X, \mathcal{O}_X)$ is even and

$$\chi(X, \mathcal{O}(D)) = \deg D + \chi(X, \mathcal{O}_X) .$$

Example 2. Let X be a surface, and $c_i = c_i(T_X)$. Then $c_0 = 1$, $c_1 = -K$ where K is the canonical divisor on X, but c_2 is a new invariant of X. Roughly speaking, c_2 is the number of zeros of a "general" vector field on X. We get

$$\chi(X, \mathcal{O}_X) = \tfrac{1}{12}(K^2 + c_2) ,$$

so $K^2 + c_2$ is divisible by 12. For every divisor D on X

$$\chi(X, \mathcal{O}_X(D)) = \tfrac{1}{2}D(D - K) + \chi(X, \mathcal{O}_X) .$$

For an arbitrary locally free sheaf E, the formula gives

$$\chi(X, E) = \mathrm{rk}\, E \cdot \chi(X, \mathcal{O}_X) - \tfrac{1}{2}K \cdot c_1(E) + \tfrac{1}{2}(c_1(E)^2 - 2c_2(E)) .$$

Example 3. For an arbitrary manifold X, $\chi(X, \mathcal{O}_X) = \deg \mathrm{td}(T_X)$ is an integer. This implies the existence of some divisibility relations between the Chern classes c_i of X. In particular, let X be an n-dimensional Abelian variety. Since the tangent sheaf T_X is trivial, $\mathrm{td}(T_X) = 1$, hence for every divisor D on X

$$\chi(X, \mathcal{O}(D)) = \frac{1}{n!}(D \cdot \ldots \cdot D)_X ,$$

so $(D^n)_X$ is divisible by $n!$. This topic is discussed in Schwarzenberger's appendix in the book (Hirzebruch (1966)).

Example 4. For $\mathbb{P}^n$, we get $\operatorname{td}(T_{\mathbb{P}^n}) = (\xi/(1-e^{-\xi}))^{n+1}$. Hence for the sheaf $\mathcal{O}(m)$, we get the formula

$$\binom{n+m}{n} = \deg\left(e^{m\xi}\cdot\left(\frac{\xi}{1-e^{-\xi}}\right)^{n+1}\right).$$

One can also verify it by hand. The right-hand side is the coefficient at z^n in the formal power series $z^{n+1}e^{mz}(1-e^{-z})^{-n-1}$, i.e., the residue of the differential $e^{mz}(1-e^{-z})^{-n+1}dz$ at the point $z=0$. Now, making a substitution $u = 1 - e^{-2}$, we get the differential $u^{-n-1}(1-u)^{-m-1}du$.

4.5. The Riemann-Roch-Grothendieck Theorem. We will now discuss Grothendieck's generalization of the Hirzebruch formula. Grothendieck observed that since the Euler characteristic and the Chern character are additive, one may consider arbitrary coherent sheaves F in place of locally free sheaves E. One can show that an arbitrary coherent sheaf F on a smooth variety admits a locally free resolution

$$0 \to E_r \to \ldots \to E_1 \to E_0 \to F \to 0.$$

We then set $\operatorname{ch}(F) = \sum_{i=0}^{r}(-1)^i \operatorname{ch}(E_i)$.

As a matter of fact, the formula makes sense not only for "genuine" sheaves but for the "differences" $F - G$ as well. Let $K(X)$ be the Grothendieck ring of coherent or locally free (which is the same provided X is smooth) sheaves on X. One may substitute the elements of $K(X)$ in place of E and F.

Next, we observe that the Euler characteristic $\chi(X, F)$ can be thought of as a Chern character ch, namely, the Chern character of the element $\sum(-1)^q H^q(X, F)$ in the ring $K(Y)$, where Y is a point. Indeed, for a space consisting of a point, ch is just the dimension of a vector space. Note that deg can also be thought of as a direct image homomorphism f_*: $A(X)_{\mathbb{Q}} \to A(Y)_{\mathbb{Q}} = \mathbb{Q}$. This leads to the following more general statement of the Riemann-Roch theorem.

Let $f\colon X \to Y$ be a proper morphism of smooth varieties. For a sheaf F on X, we set $f_k(F) = \sum(-1)^q R^q f_*(F)$. We get an (additive) homomorphism $f_k\colon K(X) \to K(Y)$. Now, the Riemann-Roch formula describes the extend to which the Chern character ch falls to commute with direct images.

Theorem (Grothendieck). *For every* $\alpha \in K(X)$

$$\operatorname{ch}(f_k\alpha)\cdot\operatorname{td}(T_Y) = f_*(\operatorname{ch}(\alpha)\cdot\operatorname{td}(T_X)).$$

4.6. Principle of the Proof. As it often happens, it is easier to prove a general formula than its special case – the formula helps itself. In the case in question, we are dealing with a morphism instead of a fixed variety. Clearly, if Grothendieck's formula is valid for morphisms $f\colon X \to Y$ and $g\colon Y \to Z$, then it also valid for the composition $g \circ f\colon X \to Z$. This follows essentially from the Leray spectral sequence, which gives $(g \circ f)_k = g_k \circ f_k$.

Now, every morphism f is a composition of a closed embedding $i\colon X \to \mathbb{P}^n_Y$ and a projection $g\colon \mathbb{P}^n_Y \to Y$. Those two cases are treated separately.

Projections. As in the case of the Chow ring, one can verify that $K(\mathbb{P}^n \times Y)$ is generated over $K(Y)$ by the classes of sheaves $\mathcal{O}(m)$. (By the way, the long exact sequence of Example 1 from Sect. 2.2 yields the relation among the $\mathcal{O}(m)$'s in $K(\mathbb{P}^n)$). So, it is suffice to verify the formula for $\mathcal{O}(m)$, which can be done by explicit calculations, as in Example 4 of Sect. 4.4.

Embeddings. By the deformation to the normal bundle, everything is reduced to the case when X is embedded as the zero section of a vector bundle $V(E)$. Using the splitting principle, one may assume that the E is an invertible sheaf on X. We then conclude the proof by explicit calculation.

(For further generalizations of the Riemann-Roch theorem to schemes with singularities, and the relations with the higher K-theory, see (Fulton (1984), Fulton-Lang (1985), Grothendieck et al. (1971), Manin (1969)).)

§ 5. Duality

5.1. Heuristic Remarks. The Riemann-Roch formula is symmetric with respect to one half of the canonical class K. For instance, on a curve

$$\chi(\mathcal{O}(D)) = \deg D - \tfrac{1}{2}\deg K$$

whence

$$\chi(\mathcal{O}(K - D)) = -\chi(\mathcal{O}(D)).$$

On a surface we get

$$\chi(\mathcal{O}(K - D)) = \chi(\mathcal{O}(D)),$$

etc. This suggests that $H^i(\mathcal{O}(D))$ and $H^{n-i}(\mathcal{O}(K - D))$ have the same dimension. The duality theory confirms that presemption, namely, for a smooth n-dimensional variety X, we have natural isomorphisms

$$H^{n-q}(X, \Omega^n_X(-D)) \xrightarrow{\sim} H^q(X, \mathcal{O}_X(D))^*.$$

These isomorphisms arise from the multiplication in the cohomology

$$H^q(X, \mathcal{O}_X(D)) \times H^{n-q}(X\Omega^n_X(-D)) \to H^q(X, \Omega^n_X)$$

and the trace isomorphism

$$H^n(X, \Omega^n_X) \xrightarrow{\sim} K.$$

5.2. Duality for Curves. We will explain where the trace isomorphism and the duality come from in the case of a smooth projective curve X. In this case the duality is essentially due to Roch.

Let S be an effective divisor on X without multiple components, i.e., S is a collection of several distinct points. We can view $\Omega^1_X(S) = \Omega^1_X \otimes_{\mathcal{O}_X} \mathcal{O}(S)$

as the sheaf of differentials with simple (or logarithmic) poles at the points of S. For each $P \in S$, one can define the residue of such a form. We can write a form locally as $at^{-1}dt$, where t is a local parameter at P. Then its residue equals $a(P)$, the value of the function a at P. Clearly the residue gives an exact sequence of sheaves

$$0 \to \Omega^1_X \to \Omega^1_X(S) \xrightarrow{\text{res}} \mathcal{O}_S \to 0\,.$$

If S is sufficiently large, then $\Omega(S)$ is acyclic by Theorem of Sect. 2.3 (see condition (b)). So, we get an exact sequence

$$H^0(X, \Omega^1_X(S)) \xrightarrow{\text{res}} H^0(X, \mathcal{O}_S) \to H^1(X, \Omega^1_X) \to 0\,.$$

Now, we consider a linear functional on the space $H^0(X, \mathcal{O}_S) = K^S$ given by a sum of coordinates. We claim that this functional vanishes on $H^0(X, \Omega^1(S))$, hence it determines a functional

$$H^1(X, \Omega^1_X) \to K$$

(trace map). The claim is just the classical

Lemma on Residues. *For a form $\omega \in H^0(X, \Omega^1(S))$*

$$\sum_{P \in S} \operatorname{res}_P(\omega) = 0\,.$$

One proves the lemma as follows. Take a projection $X \to \mathbb{P}^1$ that maps S to the point $\infty \in \mathbb{P}^1$. The trace homomorphism

$$\text{tr:}\ H^0(X, \Omega^1_X(S)) \to H^0(\mathbb{P}^1, \Omega^1_{\mathbb{P}^1}(\infty))$$

preserves the sums of residues. On the other hand $H^0(\mathbb{P}^1, \Omega^1(\infty)) = 0$.

Thus, we get a (nontrivial) functional which is called the trace:

$$H^1(X, \Omega^1_X) \to K\,.$$

Similarly, for every divisor D, we get

$$H^0(\mathcal{O}_X(D)) \hookrightarrow H^1(\Omega^1_X(-D))^*\,.$$

It follows from the Riemann-Roch formula that this inclusion is an isomorphism.

Corollary. *For a connected curve X, the space $H^1(X, \Omega^1_X)$ is 1-dimensional.*

5.3. The Serre Duality. Similarly one can establish the duality for (smooth, projective, and irreducible) n-dimensional varieties. We can find a sufficiently ample and smooth (by Bertini's theorem) divisor $Y \subset X$. Using the Poincaré residue, we get an exact sequence of sheaves

$$0 \to \Omega_X^n \to \Omega_X^n(Y) \to \Omega_Y^{n-1} \to 0 .$$

Since $\Omega_X^n(Y)$ is acyclic, we obtain the trace isomorphism

$$H^n(X, \Omega_X^n) = H^{n-1}(Y, \Omega_Y^{n-1}) \simeq K$$

by induction. The multiplication in the cohomology

$$H^q(X, E) \times H^{n-q}(X, \Omega_X^n \otimes E^*) \to H^n(X, \Omega_X^n)$$

gives a duality homomorphism for every E:

$$H^{n-q}(X, \Omega_X^n \otimes E^*) \to H^q(X, E)^* .$$

Theorem. *For a locally free sheaf E, those homomorphisms are isomorphisms.*

We will explain how to prove the theorem in case $X = \mathbb{P}^n$. If E is invertible, the duality follows by explicit calculations. To prove the theorem for a general E, we employ the following standard argument: we represent E as a quotient sheaf of a sum of $\mathcal{O}(m)$'s, and then apply a decreasing induction on the dimension of cohomology.

See Sect. 5.6 below for a proof for an arbitrary X. We observe that the duality theorem is also true for smooth complete varieties (Hartshorne (1966)).

Corollary 1. $H^q(X, E(-m)) = 0$ *for* $q < n$ *and* $m \gg 0$.

Corollary 2. $H^q(X, \Omega_X^p) \simeq H^{n-q}(X, \Omega_X^{n-p})$.

5.4. The Hodge Index Theorem. We will give an application of the Riemann-Roch theorem and the duality to projective surfaces.

Theorem. *Let H be an arbitrary hyperplane sections of a surface X, and D a divisor on X such that the intersection number $(D.H)$ equals 0. Then $(D.D) \leq 0$, and if $(D.D) = 0$ then D is numerically equivalent to zero (i. e., $(D.C) = 0$ for every $C \subset X$).*

We will give another equivalent statement of the theorem, which also explains its name. Let $N(X)$ denote the quotient of the divisor group $\operatorname{Div} X$ by the subgroup of divisors numerically equivalent to zero. It is an Abelian group with a nondegenerate intersection pairing $N(X) \times N(X) \to \mathbb{Z}$. The index theorem states that this pairing is negative on the orthogonal complement to H.

First, we establish the following fact.

Lemma. *Let C be a divisor on X with $(C.H) > 0$ and $(C.C) > 0$. Then nC is linearly equivalent to an effective divisor provided n is large.*

To show this observe that $h^0(K - nC) = 0$ for large n. Indeed, otherwise, the divisor class $K - nC$ would contain an effective cycle, hence $(K - nC.H) \geq 0$ and $(K.H) \geq n(C.H) > 0$, which is absurd.

Now, by the Serre duality, $h^2(nC) = 0$ for $n \gg 0$. Applying the Riemann-Roch theorem, we get

$$h^0(nC) + h^2(nC) \geq \tfrac{1}{2}(C.C)n^2 - \tfrac{1}{2}(C.K)n + \chi(X, \mathcal{O}_x),$$

whence, in view of $(C.C) > 0$, $h^0(nC) > 0$ for $n \gg 0$.

Returning to the index theorem, suppose to the contrary that there exists a divisor D with $(D.H) = 0$ and $(D.H) > 0$. Consider a divisor $H' = D + mH$. It is also an ample divisor for large m. Since

$$(D.H') = (D.D + mH) = (D.D) > 0,$$

nD is equivalent to an effective divisor by the lemma. Therefore $(nD.H) > 0$ contrary to the assumption $(D.H) = 0$.

5.5. General Duality. One can rewrite the Serre duality in the form

$$H^{n-q}(X, \mathcal{H}om(E, \Omega_X^n)) = \mathrm{Hom}(H^q(X, E), K),$$

where K is the ground field. In this form, it looks like a commutativity rule between the cohomology and the functor Hom. This suggests a possibility of the existence of a general duality theorem for arbitrary morphisms (instead of varieties) and arbitrary sheaves (instead of locally free sheaves). In its most naive form, a duality for a morphism $f\colon X \to Y$ would mean the existence of a functor $f^!$, right adjoint to the functor f_*, that gives an isomorphism

$$\mathrm{Hom}(f_*F, G) \simeq \mathrm{Hom}(F, f^!G)$$

for arbitrary sheaves F on X and G on Y.

The main obstacle to a realization of that naive idea is that f_* is not exact. In fact, for a functor to have an adjoint it is necessary (and essentially sufficient) that it commutes with limits (in our additive case, this is equivalent to the functor being right exact). However, f_* is not a right exact functor, which motivated the development of the cohomology theory. To make f_* exact, one has to pass to derived categories, which we would like to avoid. Then $f^!G$ is constructed as an object of a derived category. (This topic is discussed in (Hartshorne (1966)) as well as (Gelfand-Manin (1988)).)

There is, however, one special case where all the difficulties disappear and the above naive idea can be realized. Assume that $f\colon X \to Y$ is a finite morphism. Then the functor f_* is exact by Sect. 1.6, and we may set

$$f^!G = \mathcal{H}om_{\mathcal{O}_Y}(f_*\mathcal{O}_X, G)$$

and view the right-hand side as a sheaf of modules on X. We get a canonical isomorphism

$$f_*\,\mathcal{H}om_{\mathcal{O}_X}(F, f^!G) \simeq \mathcal{H}om_{\mathcal{O}_Y}(f_*F, G). \qquad (*)$$

5.6. Duality on Cohen-Macaulay Schemes. Let X be a projective Cohen-Macaulay scheme of pure dimension n (see Sect. 1.3). There exists a finite morphism $f\colon X \to \mathbb{P}^n$, and every such morphism is flat (Danilov (1988)). The sheaf $\omega_X = f^!\omega_{\mathbb{P}^n}$ is said to be the *dualizing sheaf* on X, where $\omega_{\mathbb{P}^n} = \Omega^n_{\mathbb{P}^n}$ is the canonical sheaf on $\mathbb{P}^n$. One can show that ω_X depends only on X and not on the choice of the projection f. Moreover, ω_X coincides with Ω^n_X provided X is smooth.

Theorem. *With the above notation, let E be a locally free sheaf on X. Then*

$$H^{n-q}(X, \mathcal{H}om_{\mathcal{O}_X}(E, \omega_X)) = H^q(X, E)^*.$$

Indeed, f_*E is a locally free sheaf on $\mathbb{P}^n$ so, in view of $(*)$ from Sect. 5.5,

$$\begin{aligned} H^{n-q}(X, \mathcal{H}om(E, \omega_X)) &= H^{n-q}(\mathbb{P}^n, f_*\,\mathcal{H}om(E, \omega_X)) \\ &= H^{n-q}(\mathbb{P}^n, \mathcal{H}om(f_*E, \omega_{\mathbb{P}^n}))\,. \end{aligned}$$

By the duality on $\mathbb{P}^n$ (see Sect. 5.3), the latter term is isomorphic to

$$H^q(\mathbb{P}^n, f_*E)^* = H^q(X, E)^*.$$

Remark. If E is not assumed to be locally free, then we have to replace $H^{n-q}(X, \mathcal{H}om(E, \omega_X))$ by a more general expression $\operatorname{Ext}^{n-q}(X; E, \omega_X)$. In particular, for any sheaf E on X, the space $\operatorname{Ext}^0(X; E, \omega_X) = \operatorname{Hom}(E, \omega_X)$ is dual to $H^n(X, E)$. This implies the uniqueness of the dualizing sheaf ω_X up to an isomorphism. Note that ω_X has rank 1, though it is not invertible in general.

We will present another geometric application of the duality; compare (Danilov (1988), Chap. 3).

Corollary. *Let X be an irreducible projective variety of dimension at least 2. Then any hyperplane section of X is connected.*

We will prove the corollary for surfaces. Replacing X by its normalization, we may assume that X is a normal surface hence a Cohen-Macaulay scheme. Let H be a hyperplane section of X. We consider the following exact sequence on X:

$$0 \to \mathcal{O}_X(-mH) \to \mathcal{O}_X \to \mathcal{O}_{mH} \to 0\,.$$

By the duality, $H^1(X, \mathcal{O}(-mH)) = 0$ for large m (compare Sect. 5.3). It follows that the map $K = H^0(\mathcal{O}_X) \to H^0(\mathcal{O}_{mH})$ is surjective, hence mH (which is equal to H as a space) is connected.

(For other applications of the duality to geometric problems, see (Griffiths-Harris (1978), Chap. 5).)

§ 6. The de Rham Cohomology

6.1. Definition. In Chap. 1, Sect. 4.2, we have already mentioned the classical de Rham theory, which enables to represent cohomology of differential manifolds by differential forms. It has a beautiful algebraic analog. To motivate the following definition, we will first consider a complex analytic manifold M. Then the (analytic) de Rham complex $\Omega^{\cdot}_M$, which consists of sheaves of germs of holomorphic forms, is a resolution of the sheaf $\mathbb{C}_M$ (holomorphic Poincaré lemma). Thus, the cohomology $H^*(M, \mathbb{C})$ are isomorphic to the hypercohomology $\mathbf{H}^*(M, \Omega^{\cdot}_M)$ of the de Rham complex $\Omega^{\cdot}_M$. In the algebraic case, the algebraic de Rham complex of an algebraic variety is not a resolution of any sheaf. Nevertheless, it makes sense to consider its hypercohomology.

Definition. Let X be an algebraic scheme over the field K. The hypercohomology of the de Rham complex are said to be the *de Rham cohomology* of X, namely:

$$H^*_{\mathrm{DR}}(X \,|\, K) = \mathbf{H}^*(X, \Omega^{\cdot}_X)\,.$$

It is a graded cocommutative K-algebra which is contravariant in X. The definition makes sense for arbitrary schemes, but one should expect it to satisfy nice properties only for smooth complete varieties, which is tacitly assumed in the sequel. If $K = \mathbb{C}$, then $H^*_{\mathrm{DR}}(X \,|\, \mathbb{C})$ coincides with the classical cohomology $H^*(X(\mathbb{C}), \mathbb{C})$; see Chap. 3, Sect. 2.

As in case of any hypercohomology (see Chap. 1, Sect. 4.2), we have a spectral sequence for the de Rham cohomology:

$$E_1^{p,q} = H^q(X, \Omega^p_X) \Rightarrow H^{p+q}_{\mathrm{DR}}(X \,|\, K)\,,$$

called the *Hodge-de Rham spectral sequence.* The final filtration in H^*_{DR} is called the *Hodge filtration.* The numbers $h^{pq}(X) = \dim H^q(X, \Omega^p_X)$ are called the *Hodge numbers.*

Clearly the dimension of $H^k_{\mathrm{DR}}(X)$ is finite ($\leq \sum_{p+q=k} h^{pq}$) and $H^k_{\mathrm{DR}}(X) = 0$ for $k > 2 \dim X$. We will see that already $H^{2n}_{\mathrm{DR}}(X)$ does not vanish.

6.2. A Degeneration Theorem. These and many other facts depend on a deeper study of the Hodge-de Rham spectral sequence. To begin with consider the simplest case that of a curve. Then the term E_1 has the form

q			
	0	0	0
1	$H^1(\mathcal{O}_X)$ $\longrightarrow$	$H^1(\Omega^1_X)$	0
0	$H^0(\mathcal{O}_X)$ $\longrightarrow$	$H^0(\Omega^1_X)$	0
	0	1	p

while all the remaining entries are trivial. The bottom differential

$$d_1 : H^0(\mathcal{O}_X) \to H^0(\Omega^1_X)$$

equals zero, since the global regular functions are constants. The top differential

$$d_1 : H^1(\mathcal{O}_X) \to H^1(\Omega^1_X)$$

is dual to the bottom one, and also equals zero. Hence $E_1 = E_2$. Clearly the higher differentials d_2, d_3, etc. equal zero. Therefore, the Hodge-de Rham spectral sequence degenerates at E_1, and we can fit H^1_{DR} in the following exact sequence:

$$0 \to H^0(X, \Omega^1_X) \to H^1_{\mathrm{DR}}(X|K) \to H^1(X, \mathcal{O}_X) \to 0\,.$$

In particular $\dim H^1_{\mathrm{DR}}(X) = 2g$, where $g = h^{10} = h^{01}$ is the genus of the curve X. Clearly $\dim H^0_{\mathrm{DR}}(X) = \dim H^2_{\mathrm{DR}}(X) = 1$.

The situation is more complicated for surfaces; there are examples with non-closed global forms. This is, however, more a pathological phenomenon than a rule, as the following theorem illustrates.

Theorem. *The Hodge-de Rham spectral sequence degenerates at E_1 in the following two cases:*

a) $\operatorname{char} K = 0$;

b) $\dim X \leq \operatorname{char} K$ *and X admits a lifting to characteristic zero.*

The assertion (a), in case $K = \mathbb{C}$, is one of the basic results in the Hodge theory; see (Deligne (1971)). It was proved by Hodge utilizing harmonic analysis, and until recently only a transcendental proof was available. Recently Deligne and Illusie have proved (b), and that make it possible to give an algebraic and "elementary" proof of (a) (Deligne-Illusie (1987)).

The main ingredient is the case when K is a finite field or even $K = \mathbb{Z}/p\mathbb{Z}$, where p is a prime integer. In that case, it is easy to explain the meaning of the condition of lifting to characteristic 0. It means the existence of a flat $\mathbb{Z}$-scheme $\widetilde{X}$ such that $X = \widetilde{X} \otimes_{\mathbb{Z}} (\mathbb{Z}/p\mathbb{Z})$ is the reduction modulo p. As a matter of fact, it will suffice to have a flat lifting of X to $\operatorname{Spec} \mathbb{Z}/p^2\mathbb{Z}$.

6.3. Reduction to Finite Fields. We shall see more than once that many geometric statements over fields like $\mathbb{C}$ can be deduced from similar statements over finite fields. Geometry over finite fields is often a key to geometry over arbitrary fields. We will now illustrate this technique in case of the previous theorem.

The general approach is as follows. Let X be a scheme over K. It is glued from a finite number of affine charts, and only finitely many elements of K are involved in defining each chart and in glueing. Let A denote a subring of K generated by those elements. Then there is an A-scheme $\widetilde{X}$ of finite type such that $X = \widetilde{X} \otimes_A K$. Removing degenerate fibers of the morphism $\widetilde{X} \to \operatorname{Spec} A$, we may assume that the morphism is smooth an proper (provided X was a smooth and proper scheme). The ground scheme $\operatorname{Spec} A$ is of finite type over

Spec $\mathbb{Z}$, hence it has many closed points s with finite residue fields $k(s)$. Given some facts about the variety $\widetilde{X}_s = \widetilde{X} \otimes_A k(X)$ over $k(X)$, one can derive similar facts for $X = \widetilde{X} \otimes_A K$.

Now, we will explain how it works in our case. We can construct a similar spectral sequence for the relative de Rham complex $\Omega_{\widetilde{X}/A}$, namely:

$$\widetilde{E}_1^{p,q} = H^q(\widetilde{X}, \Omega^p_{\widetilde{X}/A}) \Rightarrow \mathbf{H}^{p+q}(\widetilde{X}, \Omega_{\widetilde{X}/A}) .$$

The original spectral sequence is obtained from $\widetilde{E}$ by multiplying $\widetilde{E}$ with $\otimes_A K$. Therefore, it will suffice to establish that $\widetilde{E}$ degenerates (perhaps, after a localization of A).

Localizing A, we can assume that all the A-modules that form $\widetilde{E}$ are free. Thus, the degeneration of $\widetilde{E}$ is reduced to the equality of ranks:

$$\sum_{p+q=k} \operatorname{rk}_A H^q(X, \Omega^p_X) = \operatorname{rk}_A \mathbf{H}^k(\widetilde{X}) .$$

Next, we compare $\widetilde{E}$ with a similar spectral sequence $\widetilde{E}(s)$ for the variety $\widetilde{X}_s$ over the point $s \in \operatorname{Spec} A$. By the base change theorem (see Proposition of Sect. 3.7), taking the spectral sequence $\widetilde{E}$ commutes with $\otimes_A k(X)$, so suffice to verify the degeneration of $\widetilde{E}(s) = \widetilde{E} \otimes_A k(s)$. Finally, if char $K = 0$, then Spec A is flat over Spec $\mathbb{Z}$ and contains points s of an arbitrary large characteristic.

6.4. The Finite Field Case. It appears that varieties over finite fields are list acessible to geometric intuition and could pose the greatest obstacle for us. However, they posses an invaluable property, namely, they have only a finite number of points. This opens up additional venues for studying them. In Chap. 4, we will discuss this phenomenon in detail. For varieties in positive characteristic, one can also define the so-called *Frobenius morphism.*

We fix a prime integer p, and assume that all varieties are defined over $K = \mathbb{Z}/p\mathbb{Z}$. Such a variety X admits the *Frobenius endomorphism*

$$\Phi : X \to X .$$

It is an identity on the set of points, i. e., it leaves the points fixed; however, it acts nontrivially on functions, namely, as the p-th power map: $\Phi^*(a) = a^p$. This is a ring homomorphism since $p = 0$ in X! Obviously, the morphism Φ differ from everything we used to deal with over the complex numbers. Its differential

$$d_x : T_x X \to T_x X$$

is a zero map at every point $x \in X$.

The subsheaf $\Phi^*(\mathcal{O}_X) \subset \mathcal{O}_X$, of the p-th powers, has vanishing derivatives ($da^p = pa^{p-1}da = 0$), so it is convenient to regard the de Rham complex $\Omega^{\cdot}_X$ as a complex of modules over the subring $\Phi^*(\mathcal{O}_X)$. Equivalently, we pass to the complex of sheaves $\Phi_*(\Omega^\cdot_X)$ over $\mathcal{O}_X$. The differential of this complex is linear over $\mathcal{O}_X$. Now, we are going to explain how to calculate the cohomology sheaves $H^*(\Phi_*\Omega^\cdot_X)$ of the complex $\Phi_*\Omega^\cdot_X$.

6.5. The Cartier Operators. To begin with consider the simplest case. Let $X = \mathbb{A}^1 = \operatorname{Spec} K[X]$. Then the complex $\Omega^\cdot_X$ takes the form

$$K[T] \xrightarrow{d} K[T]\,dT\,.$$

The kernel of d coincides with the subring $K[T^p]$ of the p-th powers, while the cokernel is generated (over $K[T]$) by the form $T^{p-1}dT$.

Similarly, in general, one can define a homomorphism of $\mathcal{O}_X$-algebras

$$c_X : \Omega^\cdot_X \to H^*(\Phi_*\Omega^\cdot_X)$$

(*Cartier operator*). Locally, for a "function" a on X, the cohomology class $c_X(da)$ is represented by a closed 1-form $a^{p-1}da$, and this can be extended to other forms by multiplication. Note that in general, the form $(a+b)^{p-1}d(a+b)$ does not equal to $a^{p-1}da + b^{p-1}db$, only homologically equivalent.

Proposition (Cartier). *For a smooth scheme, the Cartier homomorphism c_X is an isomorphism.*

The proof is rather straightforward. Using local coordinates, the assertion is reduced to the case $X = \mathbb{A}^n$ where it can be verified by hand, as we have done above for $n = 1$.

We now return to the degeneration theorem. It turns out that if X can be lifted to characteristic 0, the homomorphism c_X can be realized not only on the level of cohomology but on the level of forms as well. Precisely, consider first a rather ideal situation when it is possible to lift not only the scheme X but the Frobenius morphism as well, which means there is a morphism $\widetilde{\Phi}$: $\widetilde{X} \to \widetilde{X}$ whose reduction modulo p is the original morphism Φ: $X \to X$. Then there exists a homomorphism of graded $\mathcal{O}_X$-algebras

$$c_{\widetilde{\Phi}} : \Omega^\cdot_X \to \Phi_*(\Omega^\cdot_X)\,,$$

such that the form $c_{\widetilde{\Phi}}(\omega)$ is closed and homologically equivalent to $c_X(\omega)$. We will define it only on Ω^1_X and for the forms of type $\omega = da$, where a is a local section of $\mathcal{O}_X$.

First, we lift a to a section $\widetilde{a}$ on $\widetilde{X}$ and consider $d\widetilde{\Phi}^*(\widetilde{a})$.

This form equals zero modulo p, so we can divide it by p. Finally, $c_{\widetilde{\Phi}}(da)$ is the reduction modulo p of the form $p^{-1}d\widetilde{\Phi}^*(\widetilde{a})$. One can easily verify that the construction is independent of the lifting of a, and the form $c_{\widetilde{\Phi}}(da)$ is homologically equivalent to $a^{p-1}da$.

Thus, in the ideal case, the de Rham complex $(\Phi_*\Omega^\cdot_X, d)$ is quasi-isomorphic to the complex $(\Omega^\cdot_X, 0)$ with zero differential! Now the degeneration follows at once. Indeed

$$\mathbf{H}^k(X, \Omega^{\cdot}_X) = \mathbf{H}^k(X, \Phi_* \Omega^{\cdot}_X) \simeq \mathbf{H}^k(X, (\Omega^{\cdot}_X, 0)) = \bigoplus_{i+j=m} H^j(X, \Omega^i_X).$$

For complete varieties, the ideal situation – when the Frobenius morphism can be lifted to characteristic zero – is rather exceptional (although it happens for projective spaces and arbitrary toric varieties). However, one can always lift the Frobenius locally. Analyzing different liftings of Φ, we obtain a morphism from Ω^1_X to $\Phi_* \Omega^{\cdot}_X$, which is, however, not a morphism of complexes but objects in the derived category. If $p > \dim X$, one can extend this morphism by multiplication to a quasi-isomorphism from $(\Omega^{\cdot}_X, 0)$ to $(\Phi_* \Omega^{\cdot}_X, d)$. Then the degeneration follows as above. (For precise statements and details, see (Deligne-Illusie (1987)).)

6.6. Vanishing Theorems. One can employ the above method to deduce vanishing theorems.

It is often helpful to know that some cohomology vanish. For instance, the Riemann-Roch formula calculates $\chi(X, F)$; if it is known that F is acyclic, we obtain the dimension of $H^0(X, F)$. We have already encountered two results of this kind, namely, the vanishing of H^q for $q > \dim X$ (Sect. 1.5), and the Serre theorem: $H^q(X, F(m)) = 0$ for $q > 0$ and $m \gg 0$ (Sect. 2.3). The latter means that a sufficiently ample sheaf is acyclic. A result due to Kodaira, which we are going to describe below, is much more precise: an invertible sheaf is acyclic provided it is more ample than the canonical sheaf ω_X.

Let X be a smooth projective variety, and L an ample sheaf on X. By duality (Sect. 5.3), the following statements are equivalent:

(i) $H^i(X, \Omega^j_X \otimes L) = 0$ for $i + j > \dim X$;
(ii) $H^i(X, \Omega^j_X \otimes L^{-1}) = 0$ for $i + j < \dim X$.

Each of the statements is called the *Kodaira-Nakano vanishing theorem.*

Theorem. *We assume that L is ample. Then the Kodaira-Nakano vanishing theorem holds in the following cases:*

a) $\operatorname{char} K = 0$;
b) $\dim X \leq \operatorname{char} K$ *and* X *admits a lifting to characteristic zero.*

As in Sect. 6.3, assertion (a) is reduced to finite fields and assertion (b). To prove (b), we observe that $\Phi_* \Omega^{\cdot}_X$ and $\Omega^{\cdot}_X$ are quasi-isomorphic by Sect. 6.5. We get the following inequality valid for an arbitrary invertible sheaf M:

$$\sum_{i+j=k} \dim H^j(X, \Omega^i_X \otimes M) \leq \sum_{i+j=k} \dim H^j(X, \Omega^i_X \otimes M^{\otimes p}).$$

(here $p = \operatorname{char} K$). Indeed, since the complex $(\Phi_* \Omega^{\cdot}_X) \otimes M$ is quasi-isomorphic to the complex $\Omega^{\cdot}_X \otimes M$ with zero differential,

$$\sum_{i+j=k} \dim H^j(X, \Omega_X^i \otimes M) = \dim \mathbf{H}^k(X, \Omega_X^* \otimes M)$$

$$= \dim \mathbf{H}^k(X, \Phi_* \Omega_X^{\cdot} \otimes M) \leq \sum_{i+j=k} \dim H^j(X, \Phi_* \Omega_X^i \otimes M).$$

By the projection formula

$$\Phi_*(\Omega_X^i) \otimes M = \Phi_*(\Omega_X^i \otimes \Phi^* M) = \Phi_*(\Omega_X^i \otimes M^{\otimes p}),$$

whence $H^j(X, \Phi_*(\Omega_X^i) \otimes M) = H^j(X, \Omega_X^i \otimes M^{\otimes p})$.

Now, we apply those inequalities to the vanishing theorem. Since L is ample, $H^j(X, \Omega_X^i \otimes L^{\otimes p^m}) = 0$ for $j > 0$ and m large by the Serre theorem. Together with previous inequalities, this shows that $H^j(X, \Omega_X^i \otimes L) = 0$ for $i + j > \dim X$.

(For other generalizations of the Kodaira vanishing theorem and applications, see (Esnault-Vieweg (1986), Kollár (1986), Shiffman-Sommese (1985)).)

6.7. Properties of the de Rham Cohomology. For smooth complete varieties, the de Rham cohomology provide a purely algebraic cohomology theory that has many traits of the theory of singular cohomology $H^*(X, \mathbb{C})$ over $\mathbb{C}$. We will list the main properties:

I. *Poincaré Duality*: If $n = \dim X$, then $H_{\mathrm{DR}}^{2n}(X|K) = H^n(X, \Omega_X^n) \simeq K$ and the pairing

$$H_{\mathrm{DR}}^k(X|K) \times H_{\mathrm{DR}}^{2n-k}(X|K) \to H_{\mathrm{DR}}^{2n}(X|K) \simeq K$$

is nondegenerate. This follows from the selfduality of $\Omega_X^{\cdot}$.

II. *Künneth's Formula*: $H_{\mathrm{DR}}^*(X \times Y|K) = H_{\mathrm{DR}}^*(X|K) \otimes_K H_{\mathrm{DR}}^*(Y|K)$.

III. *Cycle Map*: There is a ring homomorphism of the Chow ring $A^*(X)$ to the ring $H_{\mathrm{DR}}^{2*}(X \times Y|K)$. For a smooth subvariety $Y \subset X$ of codimension k, its class is defined to be the class Poincaré dual to the homomorphism

$$H_{\mathrm{DR}}^{2n-2k}(X|K) \to H_{\mathrm{DR}}^{2n-2k}(Y|K) = K.$$

In particular, we have the first Chern class homomorphism:

$$c\colon \operatorname{Pic} X \to H_{\mathrm{DR}}^2(X|K).$$

It can be given more explicitly by a sheaf homomorphism

$$d\log\colon \Omega_X^* \to \Omega_X^1, \quad d\log a = da/a.$$

6.8. Crystalline Cohomology. The de Rham cohomology were introduced by Grothendieck in (Grothendieck (1966)). This cohomology theory is a "correct" one provided $\operatorname{char} K = 0$, at least in the sense that it coincides with the theory of $H^*(X, \mathbb{C})$ if $K = \mathbb{C}$. However, over a field of positive characteristic, this theory not always gives the "correct" Betti numbers – sometimes

$h^{0,1} \neq h^{1,0}$, etc. These shortcomings can be removed, if the variety X can be lifted to characteristic 0 to a scheme $\widetilde{X}$ over the ring $W = W(k)$ of Witt vectors. In that case, the hypercohomology $\mathbf{H}^*(\widetilde{X}, \Omega^*_{\widetilde{X}/W})$ are independent of $\widetilde{X}$ and give the "correct" Betti numbers. The range of applications of such a theory, however, is restricted to varieties that can be lifted to characteristic zero. To obtain a good cohomology theory with coefficients in $W(K)$, Grothendieck invented the *crystalline cohomology* $H^*_{\mathrm{cris}}(X, W)$ which utilizes local liftings to characteristic 0. This theory has the properties described in Sect. 6.7. We refer the reader to (Berthelot-Ogus (1987)) for a detailed treatment, mentioning here only the following three points.

i) Linkage to the de Rham cohomology. Roughly speaking, the latter is the reduction modulo p of the crystalline cohomology. Precisely, we have an exact sequence of "universal coefficients"

$$0 \to H^k_{\mathrm{cris}}(X) \underset{W}{\otimes} K \to K^k_{\mathrm{DR}}(X) \to \mathrm{Tor}_1^W(H^{k+1}_{\mathrm{cris}}(X), K) \to 0\,.$$

In particular, the first (crystalline) Betti number b^1, i.e. the rank of H^1_{cris}, satisfies the inequality

$$b^1 \leq \dim H^1_{\mathrm{cris}} \otimes K \leq \dim H^1_{\mathrm{DR}} \leq h^{0,1} + h^{1,0}\,.$$

ii) In (Illusie (1979)), Illusie constructed a complex $W\Omega^{\cdot}_X$, which generalizes the de Rham complex and

$$\mathbf{H}^*(X, W\Omega^{\cdot}_X) = H^*_{\mathrm{cris}}(X/W)\,.$$

The corresponding "slope" spectral sequence

$$E_1^{ij} = H^j(X, W\Omega^i_X) \Rightarrow H^{i+j}_{\mathrm{cris}}(X/W)$$

degenerates at E_1 (at least modulo torsion).

iii) The Frobenius endomorphism $\Phi\colon X \to X$ induces an action on $H^*_{\mathrm{cris}}(X/W)$ as well as on the "slope" spectral sequence. Furthermore, the p-adic estimates of the *Frobenius action* on the space $H^j(X, W\Omega^i_X)$ lie in the inverval $[i, j+1)$, which explains the degeneration of the spectral sequence. In the final analysis, these facts provide a logical framework for the Warning-Chevalley type theorems (see Chap. 4, Sect. 1; and (Mazur (1975))).

Chapter 3
Cohomology of Complex Varieties

Algebraic schemes defined over the field $\mathbb{C}$ are endowed, in addition to the Zariski topology, with a more conventional and stronger classical topology. This topology is induced locally by the Euclidean metric on $\mathbb{C}^n$. For some time, this topology made it possible to apply analysis and algebraic topology to complex algebraic varieties. Later, many results obtained by transcendental methods were expressed in the algebraic form and generalized to abstract algebraic varieties defined over an arbitrary field; some results still remain transcendental. In any case, complex varieties provided typical examples and set the direction for correct abstract statements (although we also encounter examples of this influence in the opposite direction).

We should mention that the most delicate questions in the topology of complex algebraic varieties, which involve the Hodge theory, are considered in the forthcoming survey on complex algebraic varieties. We touch only briefly upon this subject in Sect. 3.

Throughout this chapter, all schemes are separated and algebraic $\mathbb{C}$-schemes.

§ 1. Complex Varieties as Topological Spaces

1.1. Classical Topology. Given a $\mathbb{C}$-scheme X, one may consider the complex analytic space X^{an} consisting of the set $X(\mathbb{C})$ (with the classical topology) and the sheaf of rings $\mathcal{O}_{X^{\mathrm{an}}}$. First we will treat affine schemes. Assume an affine scheme X is given as a closed subscheme of an affine space $\mathbb{A}^n_{\mathbb{C}}$ by equations $f_i = 0$, where each f_i is a polynomial in $T_1, \dots, T_n$. Regarding those polynomials as analytic functions on $\mathbb{C}^n$, we get an analytic subspace in $\mathbb{C}^n$, denoted by X^{an}. As a set, it is identified with $X(\mathbb{C})$, the set of $\mathbb{C}$-valued points of X. The classical topology on $X(\mathbb{C})$ is induced by the Euclidean metric on $\mathbb{C}^n$. The sheaf $\mathcal{O}_{X^{\mathrm{an}}}$ is the sheaf of analytic functions on $X(\mathbb{C})$. This construction is clearly functorial, so it can be extended to arbitrary schemes. We get the analytification functor $X \mapsto X^{\mathrm{an}}$ from the category of schemes to the category of analytic spaces. We will now discuss the topology on $X(\mathbb{C})$, leaving a more delicate analytic aspect of the theory for the next section.

Example. The analytification of $\mathbb{P}^1$ gives the usual complex projective space $\mathbb{P}^1(\mathbb{C})$. In particular, $\mathbb{P}^1$ is homeomorphic to the Riemann sphere S^2. For an elliptic curve X, the space $X(\mathbb{C})$ is homeomorphic to the 2-dimensional torus $S^1 \times S^1$. A puntured line $\mathbb{A}^1 \backslash \{0\}$ is homotopy equivalent to the circle S^1.

As as set, $X(\mathbb{C})$ consists of $\mathbb{C}$-valued points of X; the word "point" is used henceforth only in that sense. The topology of $X(\mathbb{C})$ is indeed stronger than the restriction of the Zariski topology to $X(\mathbb{C})$. Finally, we observe that the topological space $(X \times Y)(\mathbb{C})$ coincides with the product of $X(\mathbb{C})$ and $Y(\mathbb{C})$.

1.2. Properties of the Classical Topology. The spaces of the form $X(\mathbb{C})$ have several nice properties that distinguish them among all topological spaces. The following proposition lists the simplest properties.

Proposition. a) *For every scheme X, the space $X(\mathbb{C})$ is separated, locally compact, and with a countable base.*

b) *The scheme X is complete if and only if $X(\mathbb{C})$ is compact.*

In general, the morphism $f\colon X \to Y$ is proper if and only if the map $f(\mathbb{C})$: $X(\mathbb{C}) \to Y(\mathbb{C})$ is proper, i. e., the inverse image of a compact is a compact.

The space $X(\mathbb{C})$ is triangulizable. In other words, $X(\mathbb{C})$ is homeomorphic to a polyhedron of a simplicial set, which is, in addition, locally finite and at most countable; this imposes a strong restriction on the local structure of $X(\mathbb{C})$.

We will explain the principle of constructing such a triangulation in case of an algebraic curve. Projecting our curve to $\mathbb{P}^1$, we obtain a finite covering $X \to \mathbb{P}^1$ ramified at a finite number of points: $p_1, \ldots, p_n \in \mathbb{P}^1$. Next, we choose a triangulation of the Riemann sphere $\mathbb{P}^1(\mathbb{C})$ that contains the points p_i as vertexes. We then lift each simplex of the triangulation of $\mathbb{P}^1(\mathbb{C})$ to $X(\mathbb{C})$. For a very schematic picture, see Fig. 1.

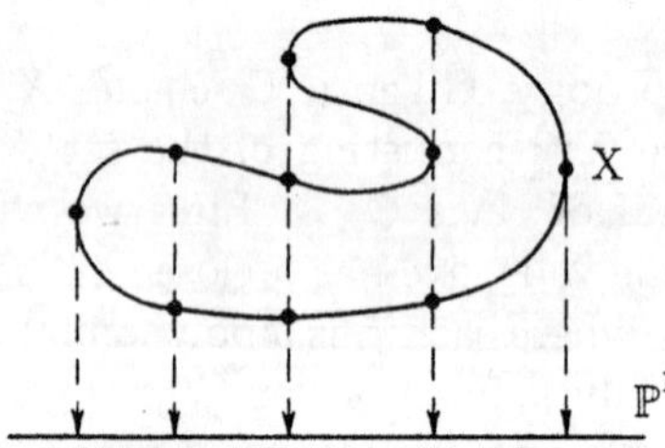

Fig. 1

In the general case, we also employ general projections, although, from the technical point of view, everything becomes much more complicated and requires a more precise inductive procedure. We get, however, a more precise statement. First, one can make each simplex to be semi-algebraic (in the real sense; i. e., it is given by inequalities $g_i \geq 0$, where each g_i is a real-algebraic function on $X(\mathbb{C})$). Second, if a part of $X(\mathbb{C})$ is already semi-algebraically triangulated, one can find a triangulation of $X(\mathbb{C})$ compatible with the partial triangulation. (For details, see (Hironaka (1975)).)

An immediate consequence of triangulation is local connectedness. Indeed, a star of each vertex can be contracted to that vertex. We will discuss global consequences of triangulation in Sect. 1.3.

Finally, $X(\mathbb{C})$ is even-dimensional and orientable. However, since it can be singular, we will postpone the precise statement till Sect. 1.4.

1.3. Singular (Co)homology. Given a topological space $X(\mathbb{C})$, it makes sense to consider its *singular homology* $H_k(X) = H_k(X(\mathbb{C}), \mathbb{Z})$ and *cohomology* $H^k(X) = H^k(X(\mathbb{C}), \mathbb{Z})$ (we will often take $\mathbb{Q}$ or $\mathbb{C}$ as coefficients in place of $\mathbb{Z}$). Since $X(\mathbb{C})$ is triangulizable, its singular cohomology coincide with the cohomology of an arbitrary triangulation, as well as the cohomology $H^k(X) = H^k(X(\mathbb{C}), \mathbb{Z}_{X(\mathbb{C})})$ of the constant sheaf $Z_{X(\mathbb{C})}$.

Proposition. *For an arbitrary scheme X, the groups $H_k(X)$ and $H^k(X)$ are finitely generated.*

To prove this, we embed X in a complete variety $\overline{X}$. We set $Y = \overline{X} \backslash X$, and consider a triangulation of $\overline{X}(\mathbb{C})$ extending a triangulation of $Y(\mathbb{C})$. Let U be a star-like neighborhood of $Y(\mathbb{C})$. Then $X(\mathbb{C})$ is a star-like neighborhood of $\overline{X}(\mathbb{C}) \backslash U$, so $\overline{X}(\mathbb{C}) \backslash U$ is a deformation retract of $X(\mathbb{C})$ (see Fig. 2), and they have the same (co)homology. However, $\overline{X}(\mathbb{C}) \backslash U$ is a finite polyhedron.

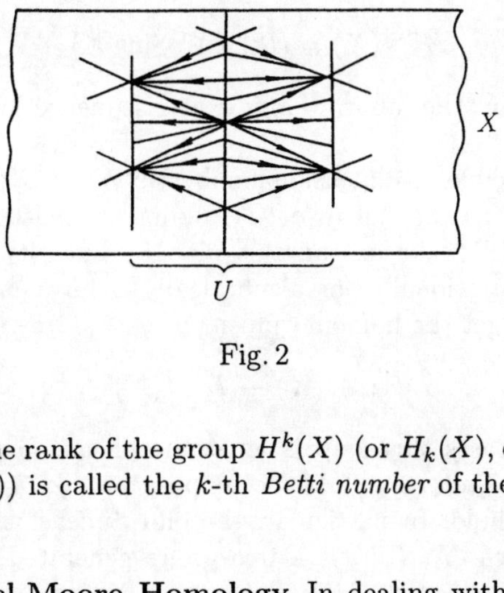

Fig. 2

Definition. The rank of the group $H^k(X)$ (or $H_k(X)$, or dimension of the space $H_k(X(\mathbb{C}), \mathbb{Q})$) is called the k-th *Betti number* of the scheme X, and is denoted by $b^k(X)$.

1.4. The Borel-Moore Homology. In dealing with noncompact varieties, it is sometimes convenient to employ homology constructed with the help of locally finite chains. Utilizing a triangulation of $X(\mathbb{C})$, that can be done in an effective fashion.

We will consider another approach. Let $X \subset \overline{X}$ be an algebraic compactification of $\overline{X}$ (i. e., an open inclusion in a complete variety $\overline{X}$) and $D = \overline{X} \backslash X$. The relative homology $H_k(\overline{X}(\mathbb{C}), D(\mathbb{C}); \mathbb{Z})$ of the pair $(\overline{X}(\mathbb{C}), D(\mathbb{C}))$ is said to be the *Borel-Moore homology*, and is denoted by $H_k^{\mathrm{BM}}(X)$.

One can show that this definition is independent of the choice of the compactification $X \hookrightarrow \overline{X}$, and those groups have finite type. The Borel-Moore homology are covariant with respect to proper morphisms. If X is a complete variety, then they obviously coincide with $H_k(X)$. This is not the case in general. For example

$$H_k^{\mathrm{BM}}(\mathbb{A}^n) = \begin{cases} 0 & \text{for } k \neq 2n\,, \\ \mathbb{Z} & \text{for } k = 2n\,. \end{cases}$$

Given a closed subscheme Y of X, we get a long exact sequence

$$\ldots \to H_k^{\mathrm{BM}}(Y) \to H_k^{\mathrm{BM}}(X) \to H_k^{\mathrm{BM}}(X \backslash Y) \to H_{k-1}^{\mathrm{BM}}(Y) \to \ldots\,.$$

The Borel-Moore homology are interesting because they contain the so-called fundamental class. First, let us assume that X is a smooth variety of dimension n. Then $X(\mathbb{C})$ is a topological manifold of dimension $2n$. By choosing an orientation of $\mathbb{C}$ (i. e., by choosing, essentially, one of the roots $\sqrt{-1}$), we get a canonical orientation of $X(\mathbb{C})$. In terms of triangulations, we may express this as follows: each $2n$-dimensional simplex has an orientation such that the orientations of nearby simplexes are compatible. We get a $2n$-dimensional chain which is, in fact, a cycle of $H_{2n}^{\mathrm{BM}}(X)$, called the *fundamental class* of X, and is denoted by μ_X. In general, when X has singularities, one defines the fundamental class μ_X using the following equality

$$H_{2n}^{\mathrm{BM}}(X) = H_{2n}^{\mathrm{BM}}(X \backslash \operatorname{Sing} X)\,,$$

which follows from the above long exact sequence and the inequality $\dim \operatorname{Sing} X < n$.

Given an irreducible m-dimensional subvariety $Y \subset X$, we define its class $\mathrm{cl}(Y)$ in $H_{2m}^{\mathrm{BM}}(X)$ as the image of μ_X under the natural homomorphism $H_{2m}^{\mathrm{BM}}(Y) \to H_{2m}^{\mathrm{BM}}(X)$. We extend this map by linearity to arbitrary algebraic cycles. Since rationally (or algebraically) equivalent cycles determine the same class, we get the homomorphism

$$\mathrm{cl}\colon A_*(X) \to H_{2*}^{\mathrm{BM}}(X)\,.$$

In several simple cases this map is an isomorphism. For instance, cl is an isomorphism if X admits a cellular decomposition, like a projective space or a Grassmann manifold. In particular, the odd dimensional homology of $\mathbb{P}^n$ are trivial, and each $H_{2m}(\mathbb{P}^n)$ is a free group generated by the class of an m-dimensional subspace $\mathbb{P}^m$ of $\mathbb{P}^n$.

1.5. The Intersection Theory. For an arbitrary scheme X, we have a topological $\cap$-product

$$H^k(X) \otimes H_i^{\mathrm{BM}}(X) \to H_{i-k}^{\mathrm{BM}}(X)\,.$$

In particular, if X is a purely n-dimensional scheme, the multiplication by $\mu_X \in H_{2n}^{\mathrm{BM}}(X)$ defines a homomorphism

$$H^k(X) \to H_{2n-k}^{\mathrm{BM}}(X)\,.$$

Theorem (Poincaré-Lefschetz duality). *For smooth varieties, this homomorphism is an isomorphism.*

A proof can be found in (Dold (1972)). One can define the inverse map geometrically using the dual cellular decomposition of a triangulation of $X(\mathbb{C})$ (Griffiths-Harris (1978)). By the duality, the product in cohomology defines the intersection product in homology

$$H_k^{\mathrm{BM}}(X) \otimes H_m^{\mathrm{BM}} \to H_{2n-k-m}^{\mathrm{BM}}(X).$$

The intersection of cycles is defined geometrically as follows. Replacing the cycles by equivalent, we may assume that they intersect transversely. Then the simplexes in the intersection have the sign $\pm$ depending on the orientation. The intersection products in the rings A_* and H_*^{BM} commute with the map cl.

Corollary. *Let X be a smooth complete purely n-dimensional scheme. Then the intersection pairing*

$$H_k(X) \otimes H_{2n-k}(X) \to H_0(X) = \mathbb{Z}$$

is unimodular, i. e., the corresponding map

$$H_k(X) \to \mathrm{Hom}(H_{2n-k}(X), \mathbb{Z})$$

is surjective and its kernel coincides with the torsion in $H_k(X)$.

In particular, the middle (co)homology $H_n(X)$ of a smooth complete variety admits a nondegenerate intersection form, which is symmetric for n even and skew-symmetric for n odd. It follows that the dimension of $H_n(X)$ is even provided n is odd. The same is true for arbitrary odd dimensional homology by the Hodge theory.

If a variety has singularities, the spaces H_k and H_{2n-k} do not necessary have the same dimension. Goresky and MacPherson proposed an interesting version of "homology", intermediate between H_* and H_*^{BM}, that always satisfy the duality (Goresky-MacPherson (1980)).

1.6. The Lefschetz Formula. Let $f\colon X \to Y$ be an endomorphism of a complete smooth scheme X. The Lefschetz formula gives an expression for the number of fixed points of f in terms of the action of f on the cohomology

$$H^k(f)\colon H^k(X) \to H^k(X).$$

Of course, we count the fixed points with their multiplicities. Let $\Gamma_f \subset X \times X$ be the graph of f. We identify the fixed points of f with $\Gamma_f \cap \Delta$, where $\Delta \subset X \times X$ is the diagonal. Thus, it is natural to call the intersection number of the cycles Γ_f and Δ in the ring $A^*(X \times X)$ (or $H^*(X \times X)$) the *number of fixed points of f*. The Lefschetz formula

$$(\Gamma_f . \Delta) = \sum_{k\geq 0}(-1)^k \operatorname{Tr} H^k(f)$$

is a formal consequence of the Poincaré duality and the Künneth formula

$$H^*(X \times Y, \mathbb{Q}) = H^*(X, \mathbb{Q}) \otimes H^*(Y, \mathbb{Q}).$$

Indeed, let (e_i^r) be a basis of $H^r(X)$ and (ε_i^{2n-r}) the corresponding dual basis of $H^{2n-r}(X)$. Clearly

$$\operatorname{cl}(\Gamma_f) = \sum_{i,r} f^*(e_i^r) \otimes \varepsilon_i^{2n-r}.$$

Similarly

$$\operatorname{cl}(\Delta) = \sum_{i,r} e_i^r \otimes \varepsilon_i^{2n-r} = \sum_{i,r}(-1)^r \varepsilon_i^{2n-r} \otimes e_i^r.$$

Multiplying them in the ring $H^*(X \times X)$, we get

$$(\Gamma_f . \Delta) = \sum_r (-1)^r \sum_i (f^*(e_i^r) \cdot \varepsilon_i^{2n-r}) = \sum_r (-1)^r \operatorname{Tr} H^r(f).$$

In particular, for the identity morphism, we get a formula

$$(\Delta . \Delta) = \sum_k (-1)^k b^k(X) = \chi(X).$$

We observe that $(\Delta . \Delta)$ is also the top Chern class, $c_n(T_X)$, of the tangent sheaf of X.

The Lefschetz formula is one of the profound reasons for the introduction of cohomological methods into algebraic geometry. Its left-hand side is purely algebraic, while the right-hand side involves all the cohomology.

Example. Let X denote an elliptic curve viewed as the quotient space of $\mathbb{C}$ by a lattice $\mathbb{Z} + \mathbb{Z} \cdot \tau$, $\operatorname{Im} \tau > 0$. Let f be the reflection with respect to zero: $z \mapsto -z$. Then $H^0(f)$ is an identity, $H^1(f)$ is given by the matrix

$$\begin{pmatrix} -1 & 0 \\ 0 & -1 \end{pmatrix},$$

while $H^2(f)$ is a multiplication by a power of f so an identity too. Thus, by the Lefschetz formula, the endomorphism f has $1 - (-2) + 1 = 4$ fixed points. Clearly those are the following points of order two: 0, $1/2$, $\tau/2$, $1/2 + \tau/2$.

§ 2. Cohomology of Coherent Sheaves

2.1. The Analytification Functor. In Sect. 1.1, we associated to a scheme X an analytic space X^{an} with the underlying topological space $X(\mathbb{C})$ and the sheaf of rings $\mathcal{O}_{X^{\mathrm{an}}}$. We also have a natural morphism of ringed spaces

$$\varphi \colon X^{\mathrm{an}} \to X .$$

Therefore, given a sheaf F of $\mathcal{O}_{X^{\mathrm{an}}}$-modules, we can define its *analytification*

$$F^{\mathrm{an}} = \varphi^* F = \varphi^{-1} F \underset{\varphi^{-1}\mathcal{O}_X}{\otimes} \mathcal{O}_{X^{\mathrm{an}}} .$$

Obviously, it is a functor, which is exact and faithful (i. e., if $F^{\mathrm{an}} = 0$ then $F = 0$ too). For a coherent sheaf F on X, the sheaf F^{an} is coherent and analytic on X^{an}. Now, we will study the connection between the cohomology of F in the Zariski topology on X and the cohomology of F^{an} in the classical topology.

Theorem. *Let X be an affine scheme. Then $H^q(X(\mathbb{C}), F^{\mathrm{an}}) = 0$ for $q > 0$ and every coherent sheaf F on X.*

This is a special case of a theorem of H. Cartan (Gunning-Rossi (1965), Theorem B). Note the analogy with Serre's theorem from Chap. 2, Sect. 1.2. In particular, one may calculate, as before, the cohomology of F^{an} using affine coverings of X. In fact, the cohomology theory of coherent sheaves on algebraic varieties and analytic spaces are similar. One may even tempt to say that the classical topology provides no new information for coherent sheaves. Strictly speaking, this is obviously not true – there are much more entire functions on $\mathbb{C}^n$ than polynomials. Presumably, the theories will indeed coincide if we impose growth conditions at infinity.

If a scheme is complete, there is no need to impose growth conditions. The results of these kind are refered to as GAGA, because they were established by Serre in the projective case in (Serre (1956)).

2.2. The Comparison Theorem. *Let $f \colon X \to Y$ be a proper morphism of schemes, and F a coherent sheaf on X. Then a natural base change homomorphism*

$$(R^q f_* F)^{\mathrm{an}} \to (R^q f_*^{\mathrm{an}})(F^{\mathrm{an}})$$

is an isomorphism.

Applying standard reductions (Chap. 2, Sect. 3), we may assume that $X = \mathbb{P}^n_Y$ and $F = \mathcal{O}(m)$. We cover $\mathbb{P}^n$ by standard affine charts $U_i \simeq \mathbb{A}^n$. Then we can calculate both sides using that covering. The only difference is that on the left-hand side we are dealing with Laurent polynomials, while on the right-hand side we are dealing with Laurent series. This, however, does not affect the cohomology.

In particular, the sheaves $(R^q f_*^{\mathrm{an}})(F^{\mathrm{an}})$ are coherent. More generally, this fact is true for an arbitrary proper morphisms of analytic spaces (*Grauert's theorem*). Here is a more elementary fact: the cohomology of coherent analytic sheaves on compact analytic spaces are finite-dimensional (Gunning-Rossi (1965)).

Corollary. *Let F be a coherent sheaf on a complete scheme X. Then $H^q(X, F) \simeq H^q(X(\mathbb{C}), F^{\mathrm{an}})$.*

This corollary has numerous applications.

2.3. Applications to the de Rham Cohomology. As we have already mentioned in Chap. 2, Sect. 6, the analytic de Rham complex $\Omega^{\cdot}_{X^{\mathrm{an}}} = (\Omega^{\cdot}_X)^{\mathrm{an}}$ is a resolution of the constant sheaf $\mathbb{C}_{X(\mathbb{C})}$ on the space $X(\mathbb{C})$. Therefore, for a smooth X, there is an isomorphism:

$$H^*(X(\mathbb{C}), \mathbb{C}) \simeq \mathbf{H}^*(X(\mathbb{C}), \Omega^{\cdot\,\mathrm{an}}_X).$$

On the other hand, we have the following

Theorem (Grothendieck). *For a smooth scheme X*

$$\mathbf{H}^*(X, \Omega^{\cdot}_X) \simeq \mathbf{H}^*(X(\mathbb{C}), \Omega^{\cdot\,\mathrm{an}}_X).$$

We will show this for complete schemes only (the argument is more involved in the general case (Grothendieck (1966))). Both expressions are limits of spectral sequences whose initial terms, $H^j(X, \Omega^i_X)$ and $H^j(X(\mathbb{C}), \Omega^{i\,\mathrm{an}}_X)$, are isomorphic by GAGA.

Thus, for smooth $\mathbb{C}$-schemes, the singular cohomology $H^*(X, \mathbb{C})$ coincide with the de Rham cohomology $H^*_{\mathrm{DR}}(X/\mathbb{C})$.

Corollary. *Let X be a smooth affine scheme. Then $H^m(X(\mathbb{C}), \mathbb{C}) = 0$ provided $m > \dim X$.*

Indeed, since the sheaves $\Omega^{i\,\mathrm{an}}_X$ are acyclic (see Theorem of Sect. 2.1), the space $H^m(X(\mathbb{C}), \mathbb{C})$ coincides with the cohomology of the complex $\Omega^{\cdot\,\mathrm{an}}_X(X(\mathbb{C}))$ (or $\Omega^{\cdot}_X(X)$) of length $n = \dim X$.

Remark. In fact, any Stein manifold M is homotopy equivalent to a finite cell complex of dimension $\leq \dim M$ (Onishchik (1986)).

2.4. The Weak Lefschetz Theorem. *Let X be a smooth projective variety of pure dimension n, and $Y \subset X$ a hyperplane section (or ample divisor). Then the homomorphisms*

$$H_m(Y) \to H_m(X)$$

are isomorphisms for $m < n-1$ and epimorphisms for $m = n-1$.

In view of the long exact sequence of Sect. 1.4, it will suffice to verify that $H^{\mathrm{BM}}_m(X \backslash Y) = 0$ for $m < n$ or, by Poincaré-Lefschetz duality, $H^m(X \backslash Y) = 0$ for $m > n$. The latter, however, follows from Corollary in Sect. 2.3, since $X \backslash Y$ is affine.

Similarly, the map $H^m(X) \to H^m(Y)$ is an isomorphism for $m < n$ and monomorphism for $m = n-1$. So, given an n-dimensional projective variety X, the really new cohomology of X (with respect to the cohomology of varieties of smaller dimensions) lie in the middle dimension n.

2.5. The Algebraization Theorem. On a complete scheme X, not only the cohomology of the sheaves F and F^{an} coincide but even any coherent analytic sheaf is algebraizable. Precisely, we have the following

Theorem. *Let X be a complete scheme. Then the analytification functor induces an equivalence of the category of coherent sheaves on X and the category of coherent analytic sheaves on X^{an}.*

First, we observe that no new morphisms arise under analytification. Indeed, for sheaves F and G on X

$$\begin{aligned}\mathrm{Hom}(F,G) &= H^0(X, \mathcal{H}om(F,G)) = H^0(X^{\mathrm{an}}, \mathcal{H}om(F,G)^{\mathrm{an}}) \\ &= H^0(X^{\mathrm{an}}, \mathcal{H}om(F^{\mathrm{an}}, G^{\mathrm{an}})) = \mathrm{Hom}(F^{\mathrm{an}}, G^{\mathrm{an}}).\end{aligned}$$

Now, we will verify that any analytic sheaf is algebraizable. Using standard reductions (Grothendieck (1971)), we may restrict ourselves to the case $X = \mathbb{P}^n$.

Let $\widetilde{F}$ be a coherent analytic sheaf on X^{an}. Taking into account that the cohomology are finite-dimensional (see Sect. 2.2), we can easily deduce, by induction on dimension, that a twisted sheaf $\widetilde{F} \otimes \mathcal{O}(m)^{\mathrm{an}}$ is generated by global sections (Griffiths-Harris (1978)). So, we can represent $\widetilde{F}$ as a cokernel of a sheaf homomorphism

$$\widetilde{\alpha} : \bigoplus_i \mathcal{O}(m_i)^{\mathrm{an}} \to \bigoplus_j \mathcal{O}(m_j)^{\mathrm{an}}.$$

We already know taht $\widetilde{\alpha} = \alpha^{\mathrm{an}}$ for a homomorphism α: $\oplus_i \mathcal{O}(m_i) \to \oplus_j \mathcal{O}(m_j)$. Hence $\widetilde{F} = \mathrm{Coker}(\alpha)^{\mathrm{an}}$.

Corollary 1. *Any analytic subspace of a complete scheme is an algebraic subscheme.*

This was established for the subspaces of $\mathbb{P}^n$ by Chow.

Corollary 2. *Any analytic morphism of two complete schemes is algebraizable.*

Indeed, its graph is a closed analytic subspace of $X \times Y$. In particular, complete schemes that are isomorphic as analytic spaces are isomorphic as schemes as well. This fails for non-complete schemes, see an example in (Hartshorne (1977)).

2.6. The Connectedness Theorem. *If a scheme X is connected, then the space $X(\mathbb{C})$ is also connected.*

The converse is obviously true and easy to show. We will sketch the proof of the theorem. First, we can assume that X is reduced and irreducible. Next, using Hironaka's resoltuion of singularities, we can assume that X is smooth. Now, we take a smooth compactification $X \hookrightarrow \overline{X}$; since $\overline{X}\backslash X$ has a smaller dimension, $X(\mathbb{C})$ and $\overline{X}(\mathbb{C})$ are connected at the same time. So we can assume that X is a complete scheme. Finally, we apply GAGA: $H^0(X(\mathbb{C}), \mathcal{O}_{X^{an}}) = H^0(X, \mathcal{O}_X) = \mathbb{C}$. Hence $X(\mathbb{C})$ is connected.

It seems that the connectedness theorem is only a fine clue relating the topological properties of X and $X(\mathbb{C})$. However, in view of its universal nature – it holds for an arbitrary scheme – this clue allows us to derive, by pure algebra, almost the whole classical topology (see Sect. 4).

2.7. The Riemann Existence Theorem. The question of which compact analytic spaces are algebraizable is rather delicate and still very much open. Several aspects of the problem are described in (Hartshorne (1977), Shafarevich (1988)). Here we will consider the simplest relative version of the problem.

Let Y be a complete scheme, $\widetilde{X}$ an analytic space, and

$$f : \widetilde{X} \to Y^{an}$$

a proper holomorphic map with finite fibers. Applying the algebraization theorem of Sect. 2.5 to the coherent sheaf of algebras $f_* \mathcal{O}_{\widetilde{X}}$, we deduce the existence of a finite Y-scheme X such that $\widetilde{X} = X^{an}$.

Interestingly enough, in the previous statement we can drop the assumption that Y is complete assuming instead that Y and $\widetilde{X}$ are normal. The following version is even more important (Riemann, Grauert-Remmert, Grothendieck).

Theorem. *Let X be a scheme. Then every finite unramified covering of the analytic space X^{an} is algebraizable.*

In view of rigidity of etale coverings, we can shrink X and assume it to be affine and smooth. Next, we take a smooth compactification $X \hookrightarrow \overline{X}$ such that the divisor $D = \overline{X}\backslash X$ has normal crossings. Utilizing the explicit local descriptions of etale coverings ramified along D, we can extend the covering $\widetilde{Y} \to X^{an}$ to a finite covering $\widetilde{\overline{Y}} \to \overline{X}^{an}$. We then apply the Riemann existence theorem, stated above, to a complete scheme $\overline{X}$.

2.8. The Exponential Sequence. It provides a good example of relationship between topology, geometry, and analysis. Let X be a scheme. Then we have an exact sequence of Abelian groups on $X(\mathbb{C})$, called the *exponential sequence*

$$0 \to \mathbb{Z}_{X(\mathbb{C})} \to \mathcal{O}_{X^{an}} \to \mathcal{O}^*_{X^{an}} \to 1$$

(see Example 3 of Chap. 1, Sect. 3.4). It yields a cohomology sequence

$$H^1(X(\mathbb{C}), \mathbb{Z}) \to H^1(X(\mathbb{C}), \mathcal{O}_{X^{an}}) \\ \to H^1(X(\mathbb{C}), \mathcal{O}^*_{X^{an}}) \xrightarrow{\delta} H^2(X(\mathbb{C}), \mathbb{Z}) \to H^2(X(\mathbb{C}), \mathcal{O}_{X^{an}}) .$$

(If X is a complete scheme, we can add zero on the left-hand side.) From the geometric point of view, the middle term $H^1(X(\mathbb{C}), \mathcal{O}^*_{X^{an}})$ is the most important. Its elements corresponds to the invertible sheaves on X^{an}. Given such a sheaf L, its image in the group $H^2(X, \mathbb{Z})$ is called the *Chern class*, and is denoted by $c_1(L)$. Given a Cartier divisor D on X, the Chern class $c_1(\mathcal{O}(D))$ is Poincaré dual to the class $\mathrm{cl}(D)$.

We now assume that X is a complete variety. Then, by GAGA, the invertible sheaves on X^{an} as well as the cohomology of $\mathcal{O}_{X^{an}}$ are algebraizable, and we can rewrite the previous sequence as

$$0 \to H^1(X, \mathbb{Z}) \to H^1(X, \mathcal{O}_X) \to \operatorname{Pic} X \xrightarrow{\delta} H^2(X, \mathbb{Z}) \to H^2(X, \mathcal{O}_X).$$

The right-hand side of that sequence gives the famous *Lefschetz-Hodge theorem* to the effect that an integral cohomology class $c \in H^2(X, \mathbb{Z})$ is a Chern class of an invertible sheaf if and only if c maps to zero in $H^2(X, \mathcal{O}_X)$. Further, let $\operatorname{Pic}^0 X$ denote the kernel of δ. It is a generalization to higher-dimensional varieties of the group of divisors of degree zero on a curve. We see that $\operatorname{Pic}^0 X$ is a quotient group of the finite-dimensional $\mathbb{C}$-space $H^1(X, \mathcal{O}_X)$ by the lattice $H^1(X, \mathbb{Z})$.

Thus, the group $\operatorname{Pic}^0 X$ is equipped with a structure of a complex-analytic variety. A deeper fact is that this variety is algebraizable; $\operatorname{Pic}^0 X$ is called the *Picard variety* of X. In particular, every line bundle with trivial Chern class can be deformed to a trivial line bundle. Finally, the quotient group

$$\mathrm{NS}(X) = \operatorname{Pic} X / \operatorname{Pic}^0 X,$$

called the *Neron-Severi group*, is embedded into $H^2(X, \mathbb{Z})$, hence it has a finite number of generators.

All the facts described above for $\mathbb{C}$-schemes can be stated algebraically and admit a generalization to algebraic varieties over arbitrary fields.

§ 3. Weights in Cohomology

3.1. Weight Filtration. From the topological point of view, algebraic varieties are, in a sense, most accessible. This manifests itself in the fact that their spectral sequences degenerate at initial terms, because of the existence of a remarkable weight filtration in the cohomology. We will describe the main facts following (Deligne (1974b)); for details, see (Deligne (1974a)).

Throughout this section, we will consider cohomology $H^k(X(\mathbb{C}), \mathbb{Q})$ with rational coefficients denoted, by abuse of notation, by $H^k(X)$. Given a scheme X, the space $H^k(X)$ admits an increasing (*weight*) filtration W:

$$\ldots \subset W_{-1} \subset W_0 \subset \ldots \subset H^k(X),$$

consisting of vector $\mathbb{Q}$-spaces. Here $W_{-1} = 0$ and $W_{2k} = H^k(X)$. The filtration depends on the scheme structure of X, and not only the topological space

$X(\mathbb{C})$. The quotients of the filtration, $\mathrm{gr}_r H^k(X) = W_r/W_{r-1}$, are said to be *components of weight r*. The *weights* of H^k are the integers r such that the corresponding components are nontrivial. If the only nontrivial component of H^k has weight r, we say that H^k is *pure of weight r*.

In fact, one can define the weight filtration on cohomology for any pair $X \subset Y$ (and even any morphism of schemes $X \to Y$). This objects, however, play an axillary role only.

3.2. Functoriality of Weights. A remarkable property of the weight filtration is its compatibility (even in a strict sense) with all natural operations on cohomology. For example, if $f\colon X \to Y$ is a morphism of schemes, then the morphism $f^*\colon H^*(Y) \to H^*(X)$ maps W_r to W_r and, in addition,

$$f^*H^*(Y) \cap W_r H^*(X) = f^*(W_r H^*(Y)).$$

Given pairs $Z \subset Y$ and $Y \subset X$, all the homomorphisms in the exact sequence

$$\ldots \to H^k(X,Z) \to H^k(Y,Z) \to H^{k+1}(X,Y) \to H^{k+1}(X,Z) \to \ldots$$

are strictly compatible with the weight filtration. Finally, the Künneth isomorphism are compatible with the weight filtration.

3.3. Assembling and Sorting out. The key fact is that for a complete smooth variety X, the space $H^k(X)$ has a pure weight k. This purity is also related to the fact that the rational homotopy type of such a variety can be recovered from the cohomology algebra $H^*(X)$ (Deligne-Griffiths-Morgan-Sullivan (1975)).

In general, the weights of $H^*(X)$ reflect how the variety is built from smooth complete pieces using additions and subtractions. We will illustrate this by examples.

Example 1. Let X be a complete variety, and $Y \subset X$ a smooth complete subvariety. It follows from the sequence

$$H^{k-1}(Y) \to H^k(X,Y) \to H^k(X)$$

that the weights in $H^k(X,Y)$ (or, equivalently, the cohomology with compact support $H^k_c(X\backslash Y)$) lie in $\{k-1, k\}$. Moreover, the weight component of $H^k_c(X\backslash Y)$ of weight k is embedded into $H^k(X)$.

Consider a concrete example: $X = \mathbb{P}^1$ and Y consists of two points, 0 and ∞, i. e., $X\backslash Y = \mathbb{C}\backslash\{0\} = \mathbb{C}^*$. We get the exact sequence

$$\begin{aligned} 0 \to H^0_c(\mathbb{C}^*) \to H^0(\mathbb{P}^1) \to \mathbb{Q}^2 \to H^1_c(\mathbb{C}^*) \to H^1(\mathbb{P}^1) \\ \to 0 \to H^2_c(\mathbb{C}^*) \to H^2(\mathbb{P}^1) \to 0\,. \end{aligned}$$

Hence $H^0_c(\mathbb{C}^*) = 0$, $H^1_c(\mathbb{C}^*) = \mathbb{Q}$ and has weight 0, and $H^2_c(\mathbb{C}^*) = \mathbb{Q}$ and has weight 2. By duality, $H^0(\mathbb{C}^*) = \mathbb{Q}$ and has weight 0, $H^1(\mathbb{C}^*) \simeq \mathbb{Q}$ and has weight 2, and $H^2(\mathbb{C}^*) = 0$.

Example 2. We assume that X consists of two smooth components, X_1 and X_2, intersecting along a smooth subvariety $X_1 \cap X_2$. It follows from the Mayer-Vietoris sequence

$$H^{k-1}(X_1 \cap X_2) \to H^k(X) \to H^k(X_1) \oplus H^k(X_2)$$

that the weights of $H^k(X)$ also lie in $\{k-1, k\}$. Thus, even so the elements of H^{k-1} are mapped to H^k, they do not mix with the remaining cohomology of "correct" weight k and remember their origin.

One can generalize Example 2. Consider a variety X covered by finitely many smooth complete varieties X_i whose various intersections are also smooth. As for any closed covering, we have a spectral sequence (Chap. 1, Sect. 4.4):

$$E_1^{p,q} = \bigoplus_{i_0,\ldots,i_p} H^q(X_{i_0} \cap \ldots \cap X_{i_p}) \Rightarrow H^{p+q}(X).$$

Notably, in our case, the sequence degenerates in E_2. This easily follows from the weight consideration, namely, for $r \geq 2$, the differentials d_r should have decreased the weights, whence they are trivial.

Example 3. We assume that a complete variety X has only one singular point denoted by P. Let $\pi\colon \widetilde{X} \to X$ be a resolution of this singularity. We get an exact sequence

$$H^{k-1}(\pi^{-1}(P)) \to H^k(X) \to H^k(\widetilde{X}) \oplus H^k(P).$$

If the variety $\pi^{-1}(P)$ is smooth, the weights of $H^k(X)$ lie in $\{k-1, k\}$.

One can show that for an arbitrary complete variety X, the weights of $H^k(X)$ lie between 0 and k. In general, for an arbitrary variety X, the weights of $H_c^k(X)$ lie between 0 and k.

3.4. Smooth Varieties. One can study the weight structure on the cohomology of smooth varieties of dimension n with the help of Poincaré-Lefschetz duality

$$H^k(X) \times H_c^{2n-k}(X) \to H_c^{2n}(X) \simeq \mathbb{Q}.$$

Note that $H_c^{2n}(X)$ has weight $2n$. In particular, the weights of $H^k(X)$ lie between k and $2n$ provided X is smooth.

Proposition. *Let $\overline{X}$ be a smooth compactification of X. Then $W_k H^k(X)$ coincides with the image of $H^k(\overline{X})$ under the restriction map $H^k(\overline{X}) \to H^k(X)$.*

By duality, it will suffice to verify that the kernel of the map $H_c^n(X) \to H_c^m(\overline{X}) = H^m(\overline{X})$ coincides with $W_{m-1}H_c^n(X)$. This, however, follows at once from the exact sequence of Example 1 of Sect. 3.

3.5. Continuity of Weights. The weight filtration depends continuously on a variety X. Precisely, consider a proper morphism $f\colon X \to S$. Given a point $t \in S$, we identify the fibers of the sheaf $R^k f_* \mathbb{Q}$ over t with the space $H^k(X_t)$, where $X_t = f^{-1}(t)$. If all the spaces $H^k(X_t)$ are isomorphic, i. e., $R^k f_* \mathbb{Q}$ is a local system on S, then the weight components W_r in the fibers $H^k(X_t)$ are glued together to form a local subsystem of the sheaf $R^k f_* \mathbb{Q}$.

Theorem (Deligne). *Let $f\colon X \to Y$ be a smooth proper morphism. Then the Leray spectral sequence*

$$E_2^{p,q} = H^p(S, R^q f_* \mathbb{Q}) \Rightarrow H^{p+q}(X, \mathbb{Q})$$

degenerates at E_2.

This confirms once again that algebraic varieties have a rather simple topological structure. In particular, we get a surjection

$$H^k(X, \mathbb{Q}) \to H^0(S, R^k f_* \mathbb{Q}) .$$

The elements of $H^0(S, R^k f_* \mathbb{Q})$ are identified with the elements of $H^k(X_t)$ that are invariant with respect to the *monodromy* (i. e., the action of the fundamental group $\pi_1(S, t)$ on $H^k(X_t)$). So, the above surjection means that the invariant class in the cohomology of X_t can be extended to a cohomology class on the whole X. More generally, such a class extends to any smooth compactification $\overline{X}$ of X (*theorem on invariant subspace*). Indeed, each element from $H^k(X_t)$ has weight k. Therefore an invariant class can be extended to a class from $W_k H^k(X)$, and we can apply Proposition of Sect. 3.4.

A local version of the preceding assertion is also true. Let $f\colon X \to S$ be a proper morphism of smooth varieties that is smooth everywhere except over a point $t_0 \in S$. If a class from $H^k(X_t)$ is invariant with respect to the local monodromy group (i. e., the group $\pi_1(S \backslash t_0)$), then it can be extended to a class on the whole X.

3.6. Existence of Weights. There are two approaches to a definition of weight filtration. The first approach uses mixed Hodge structures, and a weight filtration is an integral part of such a structure. The second approach uses l-adic cohomology and a weight filtration they equipped with (Chap. 4, Sect. 8). These cohomology are defined in a purely algebraic manner, and make sense over an arbitrary field and not necessary over $\mathbb{C}$. In the next section, we will describe a more algebraic approach to the classical topology, which will then allow us to extend this approach to abstract varieties.

§ 4. Algebraic Approach to Classical Topology

We recommend reading this section before proceeding to the next chapter, because here we explain *why* the etale approach makes sense and *how* it comes up naturally.

4.1. What the Zariski Topology Gives. In the simplicial approach to homology, the chain is a continuous image of a polyhedron. The homology of a polyhedron can be transfered to varieties by covariance via those maps. This approach makes sense if one has many maps of polyhedra to varieties. There is a dual approach based on continuous maps to polyhedra. In that approach, the cohomology of a polyhedron are transfered to a variety by contravariance. Both approaches are equivalent for triangulable spaces. In the abstract case, however, the second approach is more prospective.

The point is that the second approach is, in fact, related to a consideration of open sets. As we have seen in Chap. 1, Sect. 4.4, if each U_i and the intersection $U_{i_0} \cap \ldots \cap U_{i_p}$ have a simple structure from the cohomological point of view, then we are able to define the cohomology in a purely combinatorial fashion as the cohomology of a covering. This is indeed the case when we consider a covering by stars of a triangulation.

In the abstract case, we have the Zariski open subsets only. Can one expect that such subsets may be chosen sufficiently acyclic? In fact, the situation simplifies somewhat when we pass to "small" Zariski open subsets. For instance, we know that if U is affine, then $H^k(U)$ vanishes provided $k > \dim U$. However, further shrinking of U would not make it more acyclic. Consider the simplest case that of a curve. From the topological point of view, an affine curve is a Riemann surface with a finite number of punctures (Fig. 3).

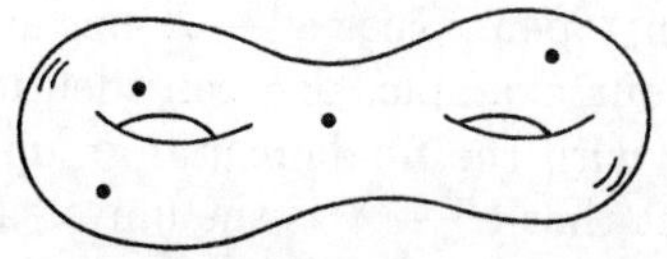

Fig. 3

From the homotopy point of view, we get a 1-dimensional polyhedron that is bouquet of several circles. Therefore such open pieces have trivial cohomology in dimension at least two. The 1-dimensional cohomology are, however, nontrivial, and they would not disappear by further shrinking of the curve, i. e., by making additional punctures.

4.2. Grothendieck's Idea. His idea was to kill 1-cycles by passing to an unramified covering $U' \to U$ instead of shrinking U. We will now discuss this idea, leaving aside, for a moment, the question of an algebraic realization of a covering. The difficulty – mostly psychological – is that an "open" piece is not a subset of an original variety X but a map to X. In any case, one should reconsider many familiar notions. For instance, what should be the intersection of two "open pieces", $U \to X$ and $U' \to X$? It is rather easy to presume that the correct object is the fiber product $U \times_X U'$. Then, however, one should not be surprized that the "self-intersection" $U \times_X U$ differ from U and carries an interesting information.

Example. Let $X = S^1$ be a circle, $U = \mathbb{R}$ a line, and $e\colon \mathbb{R} \to S^1$ the usual winding (say, $e(t) = \exp(2\pi i t)$). Then $U \times_X U'$ is the subset in $\mathbb{R} \times \mathbb{R}$ of pairs (t, t') such that $t - t'$ is an integer. Therefore $U \times_X U'$ consists of the diagonal $\Delta = \{(t,t) | t \in \mathbb{R}\}$ and its integral translations (Fig. 4).

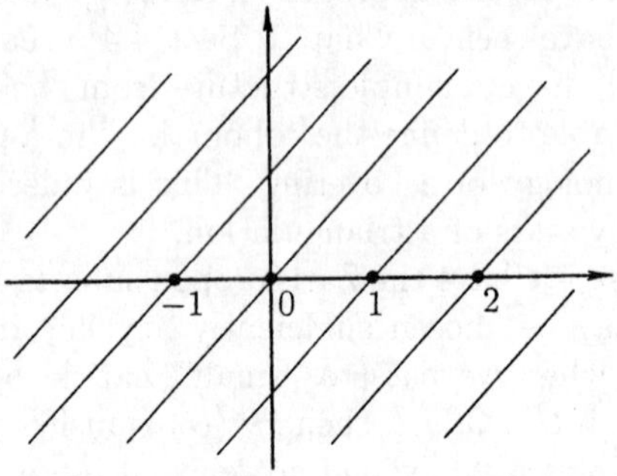

Fig. 4

In Chap. 4, we will encounter a similar splitting of the point Spec $\mathbb{F}_{q^n}$. The reason for this is that an "open" piece $U \to X$ has automorphisms.

We observe that in our example, one can identify the set of connected components of $U \times_X U$ with the fundamental group of $X = S^1$. Of course, this follows from the fact that $U \to X$ is the universal covering of X.

4.3. Nice Neighborhoods. The first crucial observation that helped to realize the program described above is the following fact discovered by Lefschetz and proved by M. Artin. We say that a scheme U is "nice" if $U(\mathbb{C})$ has the homotopy type of $K(\pi, 1)$, i. e., the universal covering of $U(\mathbb{C})$ is contractible.

Theorem. *Let X be a smooth variety. Then each point of X admits a "nice" Zariski neighborhood.*

The statement holds for curves. In general, we represent X as a fibering in curves and argue by induction on dimension. Assume X is projective, and consider a divisor D on X such that our point $x \in X$ does not belong to D. Take a general linear projection $f\colon X \to \mathbb{P}^{n-1}$, where $n = \dim X$. Then the fiber $f^{-1}(f(x))$ is smooth and transversal to D. The same is true for nearby fibers. Now, let V be a sufficiently "small" neighborhood of $f(x)$ in $\mathbb{P}^{n-1}$. We get a (topologically) locally trivial fibering $f^{-1}(V) \backslash D \to V$ whose fibers are open curves. Since the base V and the fiber have the type of $K(\pi, 1)$, the same is true for $f^{-1}(V) \backslash D$.

4.4. Idealized Reconstruction Procedure. To reconstruct the classical topology of a scheme X, we proceed as follows. First, we cover X by "nice" Zariski open subsets $U_1, \dots, U_N$, and for each U_i, take its universal covering $\widetilde{U}_i$. We get a "first level" covering $\{\widetilde{U}_i \to X\}$. The nerve of this covering, i. e., a simplicial set

$$\pi_0(\widetilde{U}) \leftleftarrows \pi_0(\widetilde{U} \underset{X}{\times} \widetilde{U}) \overset{\leftarrow}{\leftleftarrows} \pi_0(\widetilde{U} \underset{X}{\times} \widetilde{U} \underset{X}{\times} \widetilde{U}) \overset{\leftleftarrows}{\leftleftarrows} \dots$$

where $\widetilde{U} = \coprod \widetilde{U}_i$, gives correct groups H^0 (which is not surprising) and H^1 (which is essential), however, incorrect H^2 and so on, because the "intersections" $\widetilde{U}_i \times_X \widetilde{U}_j$ fail to be acyclic in general. Therefore, one should cover each such intersection by "nice" open neighborhoods, take the corresponding universal coverings, etc. Two realizations of this idea can be found in (Lubkin (1967) and Sullivan (1970); and Deligne (1974a) and Grothendieck et al. (1972–1973)). Here we will only clarify the idea by an example.

Example. Let $A = \mathbb{P}^1$ be the Riemann sphere. We cover it by two charts, $U_0 = \mathbb{P}^1 \setminus \{\infty\}$ and $U_1 = \mathbb{P}^1 \setminus \{0\}$. Since each chart is contractible, the first step is trivial. However, we cannot restrict ourselves to this covering, because a geometric realization of its nerve is a segment which has trivial 2-dimensional cohomology. So, we add to our construction the universal covering of $U_0 \cap U_1 = \mathbb{P}^1 \setminus \{0, \infty\}$. Since the intersection is homotopy equivalent to a circle, the geometric realization is homotopy equivalent to the suspension over $K(\mathbb{Z}, 1)$, which looks like S^2 and gives the correct $H^2 = \mathbb{Z}$ (see Fig. 5).

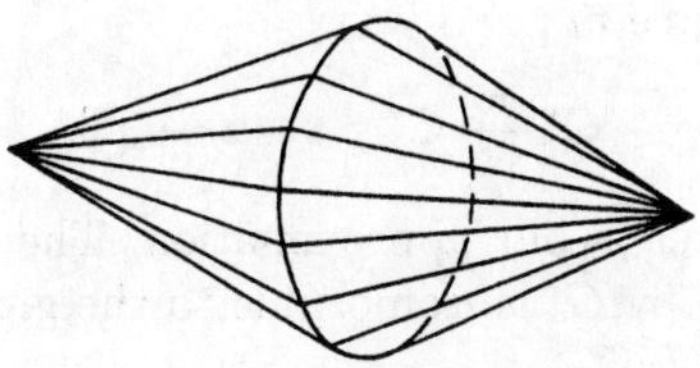

K(ℤ,1)

Fig. 5

4.5. Algebraic Coverings. Now, we go back to the question concerning an algebraic realization of universal coverings and make appropriate corrections to the procedure described in Sect. 4.4.

Let $f\colon Y \to X$ be a morphism of schemes such that the map $f(\mathbb{C})\colon Y(\mathbb{C}) \to X(\mathbb{C})$ is an unramified covering. Then f is a finite etale covering; the converse is also true. Thus, one may expect to be able to algebraically realize a universal covering for varieties with a *finite* fundamental group only. This, however, occurs very seldom, as we have already seen in the case of curves.

One is still hopeful to be able to algebraically realize any "finite" approximation to a universal covering. Here we are fortunate again. Indeed, as we have explained in Sect. 2.7, any finite unramified covering of any algebraic variety can be algebraically realized.

To obtain a more realistic picture, we make appropriate corrections to the program of Sect. 4.4. In place of the universal coverings $\widetilde{U}_i \to U_i$, one should take their finite approximations; the same for intersections, etc. Of course, one cannot expect to obtain, at once, correct cohomology with "hypercoverings" – we get approximations only. To obtain a better approximation, one should take a finer hypercovering. Thus, we are able to reconstruct not the homotopy type of $X(\mathbb{C})$ merely its "profinite completion". We are not going to explain this notion here; see (Sullivan (1970)). Note that this approach has its own virtues. For instance, if our variety is defined over $\mathbb{Q}$ – like $\mathbb{P}^n$ and Grassmannians – we can make all constructions over $\mathbb{Q}$. Then the Galois group of $\mathbb{Q}$ acts on the corresponding "profinite completion".

More significant, the preceding construction works for any variety over any field. The role of the Zariski topology is to provide "nice" open sets. The etale coverings "unwind" them. Open inclusions and etale coverings are special cases of etale morphisms. Therefore, the etale morphisms should clearly play the major role in the development of "algebraic topology" of algebraic schemes.

4.6. Instructive Example. A systematic exposition of the etale approach is given in Chap. 4. Now, we will briefly discuss a simple example, namely, the case when $X = \mathbb{C}^* = \mathbb{C}\backslash\{0\}$, which should prepare us for the etale universe. From the homotopy point of view, $\mathbb{C}^*$ is a circle, so its universal covering cannot be realized algebraically. As a good approximation, one should consider cyclic coverings of degree m

$$\mathbb{C}^* \xrightarrow{m} \mathbb{C}^*, \qquad z \mapsto z^m .$$

The bigger m the beter is our approximation. The automorphism group of this covering, denoted by G, is isomorphic to the group μ_m of m-th roots of unity, i. e. the group $\mathbb{Z}/m\mathbb{Z}$.

By constructing the nerve of the "covering" $\mathbb{C}^* \xrightarrow{m} \mathbb{C}^*$, we conclude that its cohomology (which should approximate the cohomology of $\mathbb{C}^*$) are nothing but the cohomology of the group G. One can learn about the group cohomology in (Serre (1964), Shafarevich (1986)). In our simple case, the calculations are not too complicated because $G \simeq \mathbb{Z}/m\mathbb{Z}$ is a cyclic group. We denote by γ its generator. Then the cohomology of G with coefficients in an Abelian G-group A are, in fact, the cohomology of the complex

$$A \xrightarrow{T} A \xrightarrow{N} A \xrightarrow{T} A \xrightarrow{N} \ldots ,$$

where $T(a) = a - \gamma a$ and $N(a) = \sum_{g \in G} ga = a + \gamma a + \gamma^2 a + \ldots + \gamma^{m-1} a$. We are interested in the case when G acts trivially on A. Then our complex takes even a simpler form:

$$A \xrightarrow{0} A \xrightarrow{m} A \xrightarrow{0} A \xrightarrow{m} \dots .$$

It has the following cohomology: A, $\operatorname{Ker} m$, A/mA, $\operatorname{Ker} m$, etc. So, as an approximation to $H^*(\mathbb{C}^*, \mathbb{Z})$, we get the groups $\mathbb{Z}$, 0, $\mathbb{Z}/m\mathbb{Z}$, etc. – not a very encouraging result, except for H^0!

The reason for the failure is that we have taken the group $A = \mathbb{Z}$ as coefficients. Clearly the group $H^1(X, \mathbb{Z})$ classifies unramified coverings of X with the fiber $\mathbb{Z}$, and one cannot get hold on such coverings using finite approximations. The right thing is to work with finite coefficients.

Let A be a finite group with the trivial action of G. If the order of the group A is divisible by m, the Galois cohomology, $H^*(G, A)$, are given by the following sequence of groups: A, A, 0, A, ... We get even a better result by passing to the limit over m. If $m' = km$, then the homomorphism $H^*(\mathbb{Z}/m\mathbb{Z}, A) \to H^*(\mathbb{Z}/m'\mathbb{Z}, A)$ takes the form

$$\begin{array}{ccccccc} A & A & 0 & A & 0 & A & \dots \\ \downarrow 1 & \downarrow 1 & \downarrow k & \downarrow k & \downarrow k^2 & \downarrow k^2 & \\ A & A & 0 & A & 0 & A & \dots . \end{array}$$

In the limit, as $m \to \infty$, we get quite a reasonable answer for $H^*(\mathbb{C}^*, A)$, namely: A, A, 0, 0, 0, ...

Chapter 4
Etale Cohomology

This chapter is devoted to the etale topology of schemes and the corresponding cohomology. They are of interest, because for an arbitrary algebraic variety, we get cohomology analogous to the cohomology $H^*(X, \mathbb{Z})$ of a $\mathbb{C}$-scheme X. The need for such a cohomology theory has been foreseen by A. Weil – see Sect. 1 below; Sect. 2 also contains several preliminary observations. Then we proceed to a systematic exposition of the theory, which culminates in the proof of Weil's conjectures.

§ 1. The Weil Conjectures

1.1. Finite Fields. Let $F(T_1, \dots, T_n)$ be a polynomial with integer coefficients. If we are interested in integer solutions of the equation $F(T_1, \dots, T_n) = 0$, it is natural to study the solutions in $\mathbb{R}$ as well as modulo prime numbers p. For example, the equation $Y^2 = X^3 - X - 1$ has no integer solutions since it has no solutions modulo 3. So, we are lead to study the solutions of equations over finite fields $\mathbb{F}_p = \mathbb{Z}/p\mathbb{Z}$ or, in the geometric language, the $\mathbb{F}_p$-valued points of the corresponding $\mathbb{F}_p$-schemes. If we are also interested in solutions in rings of algebraic numbers, we have to deal with points with values in finite extensions of $\mathbb{F}_p$. We have already encountered finite fields in Chap. 3, Sect. 6. We will now consider such fields in detail.

Recall first the structure of finite fields. Such a field is a finite extension of a prime field. Let K denote a finite extension of $\mathbb{F}_p$ of degree m, i. e., $\dim_{\mathbb{F}_p} K = m$. Then K has $q = p^m$ elements. The multiplicative group $K^* = K \backslash \{0\}$ has $q - 1$ elements, hence for every $x \in K^*$, we get $x^{q-1} = 1$, i. e., $x^q = x$. In fact, all the elements of K satisfy the latter equation. Therefore, K coincides with the set of solutions of the equation $X^q - X = 0$, and is unique (up to an isomorphism). Conversely, for every $q = p^m$, there exists a field with q elements, which is denoted by $\mathbb{F}_q$.

In $\mathbb{F}_q$, we have a relation $(x + y)^p = x^p + y^p$. So, the map

$$x \mapsto \Phi(x) = x^p$$

is an endomorphism of the field $\mathbb{F}_q$, called the *Frobenius endomorphism*. Since $\Phi\colon \mathbb{F}_q \mapsto \mathbb{F}_q$ is injective, it is also surjective, i. e., Φ is an automorphism of $\mathbb{F}_q$. Furthermore, the elements of the prime field $\mathbb{F}_p \subset \mathbb{F}_q$ remain fixed, and $\Phi^m = id$ where $q = p^m$. It follows that the Galois group $\mathrm{Gal}(\mathbb{F}_q/\mathbb{F}_p)$ has order m and is generated by the Frobenius automorphism Φ.

Now, we will consider this from a slightly more geometric point of view. The scheme $X = \operatorname{Spec} \mathbb{F}_q$ consists of a unique point and is a scheme over $\mathbb{F}_p$. However, as we mentioned in (Danilov (1988)), we get a correct picture of the scheme only after geometrization, i. e., after replacing $\mathbb{F}_p$ by its algebraic closure $\overline{\mathbb{F}}_p = \mathbb{F}_{p^\infty}$. By a base change

$$\begin{array}{ccc} X & \leftarrow & \operatorname{Spec}(\mathbb{F}_q \otimes \overline{\mathbb{F}}_p) = \overline{X} \\ \downarrow & & \downarrow \\ \operatorname{Spec} \mathbb{F}_p & \leftarrow & \operatorname{Spec} \overline{\mathbb{F}}_p \end{array}$$

the point X splits into m points of the scheme $\overline{X}$, each point isomorphic to $\operatorname{Spec} \overline{\mathbb{F}}_p$. Imagine we have taken a look through a telescope on a bright star and discovered that we had been dealing not with a star but a whole constellation of a regular form (Fig. 6).

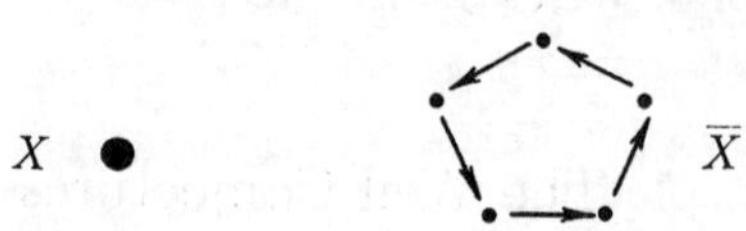

over $\mathbb{F}_p$ over $\overline{\mathbb{F}}_p$

Fig. 6

The symmetry comes from the action of the Galois group on $\overline{X}$, which cyclicaly permutes its points.

1.2. Equations over Finite Fields. Let us return to equations. Given a polynomial $f \in \mathbb{F}_q[T_1, \dots, T_n]$, we will study the solutions of the equation $f = 0$ in the field $\mathbb{F}_q$. The solution set is finite, and we denote its cardinality by $N(f)$. This number vary between 0 and q^n, depending on f. Intuitively, however, the number $N(f)$ should be close to q^{n-1} for an "ordinary" equation. How to give this statement a more precise meaning?

Let M_d denote the set of all polynomials of degree at most d in the ring $\mathbb{F}_q[T_1, \dots, T_n]$. Then the mean value of $N(f)$ is q^{n-1}, namely:

$$\frac{1}{|M_d|} \sum_{f \in M_d} N(f) = q^{n-1} .$$

Indeed, consider the "universal" polynomial F over $\mathbb{F}_q^n \times M_d$. Its zero set, $\{F = 0\}$, projects onto M_d as well as onto $\mathbb{F}_q^n$. It has exactly $\sum N(f)$ points. Since the fibers of the projection onto $\mathbb{F}_q^n$ are hyperplanes in M_d, the number of points in $\{F = 0\}$ equals $|M_d| \cdot q^{n-1}$.

One can show that the mean quadratic deviation of $N(f)$ with respect to q^{n-1} equals $q^{n-1} - q^{n-2}$ (Lidl-Neiderreiter (1983)). Therefore, one may expect that $N(f) - q^{n-1}$ is of order $\sqrt{q^{n-1}}$. The Weil conjectures give precise estimates for the individual f in cohomological terms. Before proceeding to the conjectures, we would like to briefly discuss equations of small degrees (there is an enormous amount of literature on the subject; see (Lidl-Neiderreiter (1983))).

Theorem (Warning). *We assume the degree of f is less than n. Then $N(f)$ is divisible by $p = \operatorname{char} \mathbb{F}_q$.*

To prove the statement, we form an axillary polynomial $F = 1 - f^{q-1}$. Then F equals 1 at roots of f, and 0 in the other points of $\mathbb{F}_q^n$. Hence

$$N(f) = \sum_{x \in \mathbb{F}_q^n} F(x) .$$

We will show that the sum on the right-hand side equals 0 for every polynomial of degree $< n(q-1)$, in particular, for our F. By linearity, it will suffice to verify this for monomials $T_1^{a_1} \cdot \ldots \cdot T_n^{a_n}$. For monomials the sum is equal to

$$\prod_{i=1}^{n} \Bigl(\sum_{x \in \mathbb{F}_q} x^{a_i} \Bigr) .$$

Since $\sum a_i < n(q-1)$, we get $a_i < q-1$ for some i. Now, the assertion follows from a rather simple fact:

$$\sum_{x \in \mathbb{F}_q} x^a = 0$$

for $a < q - 1$.

Corollary (Chevalley). *Let $f \in \mathbb{F}_q[T_1, \dots T_n]$ be a homogeneous polynomial whose degree satisfies the inequality $0 < \deg(f) < n$. Then the equation $f = 0$ has a nontrivial solution in $\mathbb{F}_q$.*

In general, a field K is said to be *quasi-algebraically closed* if every homogeneous equation $f(T_1, \dots, T_n) = 0$ of degree $< n$ has a nontrivial solution in K (Shafarevich (1986)). We shall encounter one important example of such a field in Sect. 5.

1.3. Zeta Function. Generalizing the preceding section, we consider an arbitrary scheme X over a finite field $\mathbb{F}_q$. For each integer $n \geq 1$, the scheme X has a finite number of points with values in $\mathbb{F}_{q^n}$ (i. e., morphism of $\operatorname{Spec} \mathbb{F}_{q^n}$ to X over $\mathbb{F}_q$). We denote this number by $N(X, n)$. The finiteness of the number of points makes finite fields an extraordinary turf for investigating algebraic varieties (Chap. 2, Sect. 6.4).

Example. There is one case where it is easy to calculate the numbers $N(X, n)$. Namely, let $X = \mathbb{A}^d$ be an affine space over $\mathbb{F}_q$. Then $X(\mathbb{F}_{q^n}) = \mathbb{F}_{q^n}^d$ consists of q^{nd} elements.

In general, assume X admits a cellular decomposition, i. e., X is decomposable into cells isomorphic to affine spaces; the number of cells isomorphic to $\mathbb{A}^i$ is denoted by b_{2i}. Then

$$N(X, n) = \sum_{i=0}^{\dim X} b_{2i} \cdot q^{ni} .$$

This is a first hint that $N(X, .)$ is related to cohomology of X, because it is reasonable to assume that the b_{2i}'s should be the Betti numbers of X (compare Chap. 3, Sect. 1.4).

The numbers $N(X, n)$ depend rather regularly on n, and it is convenient to form a formal power series (*zeta function*)

$$Z(X, t) = \exp\left(\sum_{n=1}^{\infty} \frac{N(X, n)}{n} t^n \right) \in \mathbb{Q}\,[[t]] ,$$

which contains all the information about $N(X, n)$. One can show that

$$Z(X, t) = \prod_{x \in X} (1 - t^{\deg(x)})^{-1} \in \mathbb{Z}\,[[t]] .$$

Here the product is taken over all closed points $x \in X$, and $\deg(x) = [k(x) : \mathbb{F}_q]$ is the "relative size" of x, where $k(x)$ is the residue field of x. In other words, for each point x, we consider a power series

$$Z(x, t) = 1 + t^{\deg(x)} + t^{2\deg(x)} + \dots ,$$

and then take their product. Even so we have infinitely many points, the product makes sense, because only finitely many points effect every fixed power t^k.

For example,

$$Z(\mathbb{A}^d, t) = \exp\left(\sum \frac{q^{nd}}{n} t^n\right) = \frac{1}{1-q^d t}.$$

For $\mathbb{P}^d = \mathbb{A}^d \cup \mathbb{A}^{d-1} \cup \ldots \cup \mathbb{A}^0$

$$Z(\mathbb{P}^d, t) = \frac{1}{(1-t)(1-qt)\cdot \ldots \cdot (1-q^d t)}.$$

The substitution $t = q^{-u}$ transforms the zeta function into the so-called *ζ-function*

$$\zeta(X,u) = Z(X, q^{-u}) = \prod_{x \in X} (1 - q^{-\deg(x) u})^{-1}.$$

If we denote the "absolute size" of x by $N(x) = q^{\deg(x)} = |k(x)|$, then $\zeta(X,u)$ is a product of the power series $(1 - N(x)^{-u})^{-1}$ over all closed points $x \in X$. In this form, the definition makes sense for every $\mathbb{Z}$-scheme of finite type. For example, if $X = \operatorname{Spec} \mathbb{Z}$, we get the usual Riemann ζ-function

$$\zeta(u) = \prod_p \frac{1}{1-p^{-u}} = \frac{1}{2^u} + \frac{1}{3^u} + \frac{1}{4^u} + \ldots.$$

1.4. Weil's Theorem. We consider the first nontrivial case that of a (connected, complete, and smooth) curve X of genus g over $\mathbb{F}_q$. It follows from the Riemann-Roch theorem that

$$Z(X,t) = \frac{P(t)}{(1-t)(1-qt)},$$

where $P(t)$ is a polynomial of degree $2g$ with integer coefficients, and $P(0) = 1$. Let $P(t) = \prod_j (1 - \alpha_j t)$. By analogy with the classical Riemann conjecture, E. Artin conjectured (and verified for $g = 1$) that the *absolute values of the "reciprocal roots" of the polynomial P, $\alpha_1, \ldots, \alpha_{2g}$, equal $\sqrt{q}$.* In view of the relation

$$N(X,n) = 1 + q^n - \left(\sum_{j=1}^{2g} \alpha_j^n\right)$$

we get the following estimate for the number of points on the curve X:

$$|N(X,n) - (q^n + 1)| \le 2g\sqrt{q^n}.$$

In 1940-1941, A. Weil proved this conjecture. We will sketch one of his proofs below. To begin with, consider

Example. Let X be an elliptic curve and $\alpha, \overline{\alpha}$ a pair of the "reciprocal roots". Then $\alpha\overline{\alpha} = q$ and $\alpha + \overline{\alpha} = N(X,1) - (q+1)$. Thus the number $N(X,1) = |X(\mathbb{F}_q)|$ determine all the numbers $N(X,n)$.

Now, we take a concrete curve, namely, $y^2 = x^3 - x$ over $\mathbb{F}_3$. Since $|X(\mathbb{F}_3)| = 4$, we get $\alpha + \overline{\alpha} = 0$, and $\alpha, \overline{\alpha} = \pm\sqrt{-3}$. Therefore

$$N(X,n) = |X(\mathbb{F}_{3^n})| = \begin{cases} 3^n + 1\,, & \text{if } n \text{ odd,} \\ (3^{n/2} \pm 1)^2\,, & \text{if } n \text{ even.} \end{cases}$$

Note that the numbers $N(X,n)$ are not divisible by 3, so the group of points of finite order on X has no points of order 3. This is rather unusual and implies that X is a so-called *supersingular elliptic curve* (one can read about those curves in (Hartshorne (1977), Mumford (1970))).

1.5. Proof of Weil's Theorem. The first part of the proof works for an arbitary scheme X over $\mathbb{F}_q$. Let $\overline{X} = X \otimes_{\mathbb{F}_q} \overline{\mathbb{F}}_q$ be the geometrization of X, and $\Phi\colon \overline{X} \to \overline{X}$ the Frobenius morphism over $\mathbb{F}_q$; in coordinates, the action is given by the formula

$$\Phi(x_1, \ldots, x_k) = (x_1^q, \ldots, x_k^q)\,.$$

In Sect. 1.1, we have already seen that under the projection $\overline{X} \to X$, each point $x \in X$ splits into $\deg(x)$ points of $\overline{X}$ which are permuted by Φ. Moreover, the fixed points of Φ are identified with the rational points of X, i. e. with $X(\mathbb{F}_q)$. Similarly, for each n, the set $\mathrm{Fix}(\Phi^n)$ of the fixed points of Φ^n is identified with $X(\mathbb{F}_{q^n})$. Thus $N(X,n) = |\,\mathrm{Fix}(\Phi^n)|$. It remains to find the number of fixed points of Φ^n. Of course, this is the essence of the problem.

Now, going back to curves, we consider the problem of determining the number of fixed points of the map $f\colon C \to C$, where C is a curve over an algebraically closed field K (the fact that $K = \overline{\mathbb{F}}_q$ is not essential at this point). According to the definition of Chap. 3, Sect. 1.6, this number is equal to $(\Gamma_f.\Delta)$.

We consider a more general problem concerning the intersection of divisors on a surface $C \times C'$, where C and C' are curves. We will deal with divisors up to numerical equivalence. The simplest (degenerate) divisors on $C \times C$; are the "horizontal" fibers, $C \sim C \times \{p'\}$, and "vertical" fibers, $C' \sim \{p\} \times C'$. Each divisor D admits an orthogonal decomposition

$$D = \delta(D) + \varepsilon(D)\,,$$

where $\delta(D) = (D.C')C + (D.C)C'$ is the degenerate part of D, and $\varepsilon(D)$ is the "remainder". Then, for divisors D and D' on $C \times C'$, we get

$$(D.D') = (\delta(D).\delta(D')) + (\varepsilon(D).\varepsilon(D'))\,.$$

Usually it is not difficult to calculate the first term. Weil was able to estimate the second term. Indeed, according to the Hodge theorem (Chap. 2, Sect. 5.4), the intersection form is negatively defined on the orthogonal complement of C and C'. Hence, by the Cauchy-Bunjakovski inequality,

$$(\varepsilon(D).\varepsilon(D'))^2 \leq \varepsilon(D)^2 \cdot \varepsilon(D')^2 .$$

Returning to the morphism $f\colon C \to C$, we set $D = \Gamma_f$. Since Γ_f is inverse image of the diagonal $\Delta \subset C \times C$ under the morphism $(f,1)\colon C\times C \to C\times C$, we get $(\Gamma_f.\Gamma_f) = (2-2g)\deg(f)$, where g is the genus of C. It is easy to deduce that $\varepsilon(\Gamma_f)^2 = 2g\cdot\deg(f)$ which, in turn, allows us to show that

$$((\Gamma_f.\Delta) - 1 - \deg(f))^2 \leq 4g^2 \deg(f) .$$

Now, we apply this inequality to $f = \Phi^n$, where Φ is the Frobenius morphism over $\mathbb{F}_q$. Since Φ has degree q, we deduce the desired inequality

$$\| \operatorname{Fix}\Phi^n| - (1+q^n)| \leq 2g\sqrt{q^n} .$$

1.6. The Weil Conjectures. In 1949, A. Weil stated a generalization of E. Artin's conjecture for smooth projective varieties of arbitrary dimension over $\mathbb{F}_q$. The main part of his conjectures stated that for a variety X of dimension d:

$$Z(X,t) = \frac{P_1(t)\cdot P_2(t)\cdot\ldots\cdot P_{2d-1}(t)}{P_0(t)\cdot P_2(t)\cdot\ldots\cdot P_{2d}(t)} ,$$

where $P_i(t)$ are polynomials with integer coefficients and $P_i(0) = 1$. Furthermore, if $P_i(t) = \prod_j (1-\alpha_{ij}t)$ is a decomposition in linear factors (with complex α_{ij}), then $|\alpha_{ij}| = q^{i/2}$.

In terms of $N(X,n)$, the conjecture have the following interpretation. The first part, concerning the rationality of $Z(X,t)$, gives a "harmonic" expansion

$$N(X,n) = \sum_{i=0}^{2n} (-1)^i \left(\sum_{j=1}^{b_i} \alpha_{ij}^n \right) ,$$

where $b_i = \deg P_i$. Thus, the variety is, in a sense, decomposable into elementary pieces associated to the roots α_{ij} (or their collections), and each root gives a contribution, equal to $(-1)^i\alpha_{ij}^n$, to the number of points. The second part of the conjectures – quantative or Riemann's part – gives an estimate for the collection of terms of "weight" i:

$$\left| \sum_{j=1}^{b_i} \alpha_{ij}^n \right| \leq b_i \sqrt{q^{in}} .$$

The main contribution is provided by the term of weight $2d$, for which $b_{2d} = 1$ (provided X is irreducible), so

$$N(X,n) = q^{n \dim X} + O(q^{n(\dim X - 1/2)}) .$$

1.7. Weil's Cohomology. A. Weil also proposed an almost fantastic approach to the verification of his conjectures. He suggested the existence of a hypothetical cohomology theory $X \mapsto H^*(X)$ for abstract algebraic varieties that is similar to the classical cohomology theory for complex manifolds. Precisely, $H^*(X)$ should take values in the finite-dimensional algebras over a field F, and should satisfy properties I–III of Chap. 2, Sect. 6.7. Since the Lefschetz formula (Chap. 3, Sect. 1.6) is a formal consequence of those properties, we would get the relation

$$|\operatorname{Fix} \Phi^n| = \sum_i (-1)^i \operatorname{Tr}(\Phi^n | H^i(X)),$$

where Tr denotes the trace of the corresponding homomorphism in cohomology. Then the polynomials P_i of the Weil conjectures are the characteristic polynomials of the action of Φ in the cohomology H^i, $P_i(t) = \det(1 - t\Phi | H^i(\overline{X}))$, and α_{ij} are the eigenvalues of Φ on the space $H^i(\overline{X})$.

The above mentioned facts about varieties with cellular decompositions and curves fit very well into this picture.

Weil's idea had a great impact on the development of algebraic geometry. It inspired not only the development of etale cohomology, which is described in this chapter, but also initiated three successive revisions of the "foundations" of algebraic geometry by Weil himself, Serre, and finally Grothendieck.

We should point out that when Weil formulated his idea there were no cohomology of abstract algebraic varieties – even the notion of an abstract variety was yet to be conceived. We have already encountered one such cohomology theory, namely, the de Rham cohomology. This theory has the required formal properties and gives a formula for the number of fixed points, unfortunately, only in $K = \overline{\mathbb{F}}_p$, i. e. modulo p. Of course, even this gives some assertions similar to the Warning theorem of Sect. 1.2. Their generalizations that utilize the crystalline cohomology are discussed in (Mazur (1975)). However, in order to obtain a genuine formula for the number of fixed points, the coefficient field F of a cohomology theory should have characteristic zero.

We would also like to point out yet another restriction on the coefficient field F. It was pointed out by Serre that $\mathbb{R}$ (and any subfield of $\mathbb{R}$) cannot be such a field. Indeed, for every variety X, the ring $\operatorname{End}(X)$ (and consequently $\operatorname{End}(X) \otimes \mathbb{R}$) must act, by funtoriality, on the space $H^1(X)$. In particular, this is true for a curve of Sect. 1.4. For that curve, however, $\operatorname{End}(X) \otimes \mathbb{R}$ is the quaternion field (Mumford (1970)), and cannot act on the 2-dimensional space $H^1(X)$. Similarly, one cannot take the field $\mathbb{Q}_p$ of p-adic numbers as the coefficient field F, where p is the characteristic of the ground field.

On the other hand, we shall see in Sect. 7 that for each prime l, $l \neq p$, there exists an *l-adic cohomology theory with coefficients in* $\mathbb{Q}_l$ (as well as the *crystalline cohomology theory*). Such a variety of theories suggests that all these theories are manifestations of some universal cohomology theory. For instance, one may ask whether there is a cohomology theory with the coefficient field $\overline{\mathbb{Q}}^{ab}$ or $\overline{\mathbb{Q}}$, where $\overline{\mathbb{Q}}^{ab}$ is the maximal Abelian extension of

$\mathbb{Q}$, and $\overline{\mathbb{Q}}$ is the field of algebraic numbers. Presumably, those questions are intimately related to the theory of motives (Manin (1968)).

§ 2. Algebraic Fundamental Group

Before proceeding to etale topology, we would like to describe the notion of the fundamental group. This important notion is also closely related to the 1-dimensional cohomology (later we shall see that $H^1(X, A) = \mathrm{Hom}(\pi_1(X), A)$ for a finite Abelian group A). This is, in fact, an important step because all the remaining cohomology are more or less reduced to the 1-dimensional cohomology.

Henceforth, all schemes are separated and Noetherian; for simplicity, one may even restrict himself to algebraic schemes (over fields, or $\mathbb{Z}$ in order not to exclude arithmetic applications).

2.1. Etale Morphisms. The etale morphisms play an exceptional role in etale topology. We will recall the definition and main properties; see details in (Danilov (1988), Grothendieck (1971), Milne (1980), Raynaud (1970)).

A morphism of finite type $f\colon X \to Y$ is said to be *etale* if it is flat and the diagonal in $X \times_Y X$ is open and closed. Equivalently, it is a smooth morphism of relative dimension 0. For schemes over an algebraically closed field, we have a less sophisticated definition: for every point $x \in X$, the induced map of tangent cones, $d_x f\colon C_x X \to C_{f(x)} Y$ is an isomorphism.

We will give several well-known examples of etale morphisms. An open inclusion is obviously etale. Let $Y = \mathrm{Spec}\, K$, where K is an algebraically closed field. Then X is etale over Y if and only if X is a finite union of copies of Y, i.e., X is a finite reduced scheme. Now, let $Y = \mathrm{Spec}\, K$, where K is an arbitrary field. Then X is etale over Y if and only if $X \times_Y \mathrm{Spec}\, \overline{K}$ is etale over $\mathrm{Spec}\, \overline{K}$, where $\overline{K}$ is the algebraic closure of K.

For schemes over $\mathbb{C}$, a morphism $f\colon X \to Y$ is etale if and only if the map $f(\mathbb{C})\colon X(\mathbb{C}) \to Y(\mathbb{C})$ is a local isomorphism in the classical topology. This suggests that in the abstract case, etale morphism are local isomorphism in some topology, which is stronger than the Zariski topology.

Here we list several additional properties of etale morphism. A composition of etale morphisms is an etale morphism. A morphism remains etale under a base change. Finally, if X and Y are etale S-schemes, then any S-morphism $X \to Y$ is etale.

2.2. Etale Coverings. Let T be a "nice" topological space (say, a polyhedron). There are two definitions of the fundamental group $\pi_1(T)$, namely, utilizing closed loops or utilizing unramified coverings. Only the second approach makes sense in the abstract case. First, we recall the relationship between those two definitions of π_1.

Let $f\colon T' \to T$ be an unramified covering of topological spaces, and $t \in T$. Then the fundamental group $\pi_1(T, t)$ acts, via monodromy, on the fiber $F = f^{-1}(t)$, and this action determines uniquely the covering. Thus, to give an unramified covering of the space T is the same as to give an action of the group π_1 on the set F. The connected components of T', then, correspond to the orbits of $\pi_1(T)$ on F. The action of π_1 on itself by left translations determines the universal (connected) covering $\widetilde{T} \to T$. So, one can define $\pi_1(T)$ as the automorphism group of the universal covering $\widetilde{T} \to T$.

Now, we turn to an algebrization of those notions. The algebraic analogue of an unramified covering is a finite etale morphism (in short, an *etale covering*). A connected scheme X is said to be *simply connected* if every etale covering of X is trivial, i. e., a direct sum of several copies of X.

Example 1. We will show that $\mathbb{P}^1$ over an algebraically closed field is simply connected, as one should expect by analogy with the complex case. Indeed, let $f\colon X \to \mathbb{P}^1$ be an etale covering of degree d. Then we have a relation between the corresponding canonical sheaves: $\omega_X = f^*\omega_{\mathbb{P}^1}$. The degree of ω_X is d times the degree of $\omega_{\mathbb{P}^1}$. Sine $\deg \omega_{\mathbb{P}^1} = -2$, we get $\deg \omega_X = -2d$. On the other hand, by the Riemann-Roch theorem, $\deg \omega_X = 2g - 2 \geq -2$, where g is the genus of the curve X. So $-2d \geq -2$ whence $d \leq 1$.

Example 2. We will show that $\operatorname{Spec} \mathbb{Z}$ is simply connected. This is one the fundamental facts in number theory – Minkowski's theorem. Let $\operatorname{Spec} A \to \operatorname{Spec} \mathbb{Z}$ be a finite covering of degree n, where A is a ring of algebraic numbers. The ramification of the morphism is given by the discriminant D. For example, for the Gauss ring $A = \mathbb{Z} + \mathbb{Z}\sqrt{-1}$, the discriminant equals -4, so the corresponding covering is ramified over the point $(2) \in \operatorname{Spec} \mathbb{Z}$. There are two more double coverings of $\operatorname{Spec} \mathbb{Z}$ ramified at (2), namely, $\operatorname{Spec} \mathbb{Z}[\sqrt{\pm 2}]$.

A covering is unramified if its discriminant D is invertible in $\mathbb{Z}$, i. e., $D = \pm 1$. In number theory, we have a rather nontrivial estimate on the discriminant (Borevich-Shafarevich (1985), p. 174):

$$|D| \geq \left(\frac{\pi}{4}\right)^n \frac{n^{2n}}{(n!)^2}.$$

It follows that $|D| = 1$ only for $n = 1$.

In a sense, $\operatorname{Spec} \mathbb{Z}$ looks topologically like a 3-dimensional sphere viewed as the Hopf fibration over S^2.

2.3. Algebraic Fundamental Group. Let X be a connected scheme. An etale covering $\widetilde{X} \to X$ is said to be *universal* if the scheme $\widetilde{X}$ is simply connected. This definition comes from the fact that every etale covering $X' \to X$ is dominated by a universal one. Indeed, consider the fiber product

$$\begin{array}{ccc} X' & \leftarrow & X' \underset{X}{\otimes} \widetilde{X} \\ \downarrow & & \downarrow \\ X & \leftarrow & \widetilde{X}\,. \end{array}$$

Let Z denote a connected component of $X' \times_X \widetilde{X}$. Then the projection $Z \to \widetilde{X}$ is an etale covering hence an isomorphism, because $\widetilde{X}$ is simply connected. Then the second projection gives the morphism $\widetilde{X} \overset{\sim}{\leftarrow} Z \to X'$.

It follows that a universal covering is determined uniquely up to an isomorphism. Moreover, there are exactly $\deg(\widetilde{X}|X)$ automorphisms of the scheme $\widetilde{X}$ over X. Finally, one can represent X' as a quotient space of $\widetilde{X}$ by an action of a finite subgroup $\mathrm{Aut}(\widetilde{X}|X') \subset \mathrm{Aut}(\widetilde{X}|X)$. All this justifies to call $\mathrm{Aut}(\widetilde{X}|X)$ the algebraic *fundamental group* of the scheme X, and denote it by $\pi_1(X)$.

Unfortunately, the universal coverings seldom exist. We have already observed this for $\mathbb{A}^1 \backslash \{0\}$ (in the complex case). We get another example by taking $\operatorname{Spec} \mathbb{F}_p$, which has etale covering – essentially, finite extensions of $\mathbb{F}_p$ – of arbitrary large degree. The example of the field $\mathbb{R}$ – which has a universal covering $\operatorname{Spec} \mathbb{C} \to \operatorname{Spec} \mathbb{R}$ – is almost extraordinary. The solution is to replace a universal covering by its finite approximations, as in Chap. 3, Sect. 4.

Definition. The etale covering $\overline{X} \to X$ is said to be a *Galois covering* if $\overline{X}$ is connected and $\mathrm{Aut}(\overline{X}|X)$ has $\deg(\overline{X}|X)$ elements.

Thus, the Galois coverings are maximally symmetric. In topology, such coverings are called regular; they are in a one-to-one correspondence with normal subgroups of π_1. We have sufficiently many Galois coverings, namely, every etale covering (and even a finite collection of such coverings) is dominated by a Galois covering.

Thus, the abstract case is completely analogous to the topological one, only the coverings of the scheme X are controlled not by $\pi_1(X)$ but by the inverse system of finite groups $\mathrm{Aut}(\overline{X}|X)$, where $\overline{X}$ runs over all Galois coverings of X. We usually pass to the limit, and obtain the (profinite) group $\varprojlim \mathrm{Aut}(\overline{X}|X)$, which we call the *algebraic fundamental group* of X and denote by $\pi_1(X)$ or $\pi_1^{\alpha}(X)$. We can summarize the above discussion in the following theorem.

Theorem. *The category* $\mathrm{Fet}(X)$ *of finite etale coverings of a scheme* X *is equivalent to the category of finite* $\pi_1(X)$*-sets, i. e., sets equipped with a continuous action of* $\pi_1(X)$.

Example 1. Let $X = \operatorname{Spec} K$ be the spectrum of a field K. The connected etale coverings $X' \to X$ are in a one-to-one correspondence with the finite separable extensions $K \subset K'$. Moreover, the Galois coverings correspond to the Galois extensions. So, the fundamental group $\pi_1(\operatorname{Spec} K)$ coincides with the Galois group $\mathrm{Gal}(\overline{K}/K)$ of K. Thus, from the "topological" point of view, the one element scheme $\operatorname{Spec} K$ looks like the space $K(\pi, 1)$, where $\pi = \mathrm{Gal}(\overline{K}/K)$. Only for an algebraically closed field K, $\operatorname{Spec} K$ is a topologically trivial object.

We know the structure of extensions of $\mathbb{F}_p$ (Sect. 1.1). The fundamental group $\pi_1(\operatorname{Spec}\mathbb{F}_p)$ is isomorphic to the profinite completion of $\mathbb{Z}$. Therefore, $\operatorname{Spec}\mathbb{F}_p$ topologically looks more or less like a circle S^1.

Example 2. Let X be an algebraic scheme over $\mathbb{C}$. Then $\pi_1^{\mathrm{alg}}(X)$ coincides with the profinite completion of $\pi_1(X(\mathbb{C}))$. This means, in effect, that the scheme X and the topological space $X(\mathbb{C})$ have the same finite unramified coverings, which we have observed in Chap. 3, Sect. 2.

2.4. Functorial Properties of the Fundamental Group. The fundamental group is just the way to store the information about etale coverings in a compact and natural form. For instance, a morphism of connected schemes $f\colon X \to Y$ induces a base change functor $f^*\colon \operatorname{Fet}(Y) \to \operatorname{Fet}(X)$. In terms of the fundamental group this means the existence of a homomorphism

$$f_* : \pi_1(X) \to \pi_1(Y)\,.$$

We will present a few examples of translations from the language of coverings to the language of fundamental groups:

a) $f_*\colon \pi_1(X) \to \pi_1(Y)$ is surjective if and only if for every connected etale covering $Y' \to Y$, the variety $X \times_Y Y'$ is connected;

b) $f_*\colon \pi_1(X) \to \pi_1(Y)$ is injective if and only if every covering $X' \to X$ is a direct summand of a suitable induced covering $X \times_Y Y' \to X$;

c) $f_*\colon \pi_1(X) \to \pi_1(Y)$ equals zero if and only if every etale covering $Y' \to Y$ gives rise to a trivial covering over X.

Example. We consider a homomorphism $\pi_1(\mathbb{P}^1) \to \pi_1(\mathbb{P}^n)$ induced by an embedding $\mathbb{P}^1 \subset \mathbb{P}^n$. We claim it is surjective; this together with Example 1 of Sect. 2.2 implies that $\mathbb{P}^n$ is simply connected. According to (a), one has to verify that given a connected etale covering $f\colon X \to \mathbb{P}^n$, the scheme $f^{-1}(\mathbb{P}^1)$ is connected. This follows from the connectedness theorem; see Chap. 2, Sect. 5.6, or (Danilov (1988)).

In general, if X is an irreducible projective variety of dimension ≥ 2 and $H \subset X$ is a hyperplane, then the homomorphism $\pi_1(H) \to \pi_1(X)$ is surjective. If, in addition, X is smooth and $\dim X \geq 3$, then it is an isomorphism (Grothendieck (1968b)). In particular, every hypersurface in $\mathbb{P}^n$, $n \geq 3$, is simply connected. (Details on the technique for calculating π_1 can be found in (Grothendieck (1971))).

2.5. Construction of Coverings. In many cases one has to explicitly present etale morphisms. We have two methods at hand. First method is based on the comparison with characteristic zero; see Example 2 of Sect. 2.3. The second method utilizes algebraic groups. Let G be an algebraic group, and N a finite subgroup of G. Then the quotient morphism $G \to G/N$ gives an etale covering whose fibers are isomorphic to N. Therefore, a morphism $X \to G/N$ induces a covering of X, perhaps a trivial one.

Most often, this method is applied to the case when $G = \mathbb{G}_m$, the multiplicative group. If n is not divisible by the characteristic of the ground field, the n-th power homomorphism

$$\mathbb{G}_m \xrightarrow{n} \mathbb{G}_m , \qquad Z \mapsto Z^n ,$$

is an etale morphism.

The following version of this construction is also useful. Let L be a line bundle on X. Then there is a canonical morphism of X-schemes

$$\eta : L \to L^{\otimes n}$$

that maps a point with coordinates (x, t) to a point with coordinates (x, t^n). Now, consider a section f of the bundle $L^{\otimes n}$. The subscheme $X' = \eta^{-1}(f)$ is finite over X and etale provided n and f are invertible. Note that f is invertible if and only if $L^{\otimes n}$ is a trivial bundle. This shows the existence of a close relationship between etale coverings of X and elements of finite order in the Picard group $\operatorname{Pic} X$ (Kummer theory; see Sect. 4.6).

§ 3. Etale Topology

As we explained in Chap. 3, Sect. 4, the main idea of the etale topology is to declare etale morphism to be "locally trivial" maps. We would like to understand how to generalize the notions of a sheaf, a stalk at a point, etc. As we shall see below, no changes are, in fact, required.

3.1. Etale Presheaves. By definition, a presheaf on a topological space T is a contravariant functor on the category $O(T)$ of open subsets of T. In the etale topology, the role of $O(T)$ is played by the category $\operatorname{Et}(X)$ of all schemes etale over X.

Definition. A contravariant functor from $\operatorname{Et}(X)$ to the category of sets (groups, Abelian groups, etc.) is said to be an *etale presheaf* of sets (groups, Abelian groups, etc.).

The point is that various objects can be associated to any (or etale only) X-scheme and not just the open subscheme of X.

Example 1. Take a fixed scheme Y over X. For each X-scheme U, we consider the set

$$h_Y(U) = \operatorname{Hom}_X(U, Y) .$$

Clearly h_Y is a contravariant functor with respect to U. It restriction to $\operatorname{Et}(U)$ gives an etale presheaf on X. Note that such a functor is said to be *representable* (by the scheme Y).

Example 2. Let F be a sheaf of O_X-modules. Given an X-scheme $f \colon U \to X$, we consider the group

$$F(U) = (f^{-1}F \underset{f^{-1}\mathcal{O}_X}{\otimes} \mathcal{O}_U)(U).$$

We get a functor hence an etale presheaf, which we denote by F_{et} (compare Chap. 3, Sect. 2.1).

Example 3. For an X-scheme U, we consider its Picard group $\operatorname{Pic} U$. We get a presheaf.

3.2. Etale Sheaves. In Sect. 3 of Chap. 1, we have already explained that the sheaf axiom is, in effect, a condition of commutativity with limits. This condition can be translated directly to the category $\operatorname{Et}(X)$, and it looks even more natural there. In fact, $\operatorname{Et}(X)$ contains natural finite direct limits (if we admit non-separable schemes).

Definition. An etale presheaf F is said to be an *etale sheaf* (or just a *sheaf*) if it transforms direct images into inverse images, i. e.,

$$F(\varinjlim U_\alpha) = \varprojlim F(U_\alpha).$$

We will explain the meaning of this condition. It can be reduced essentially to the following two conditions:

a) $F(U \coprod U') = F(U) \times F(U')$;

b) if $U' \to U$ is a surjective morphism in the category $\operatorname{Et}(X)$, then we get an exact sequence

$$F(U) \to F(U') \rightrightarrows F(U' \times_U U').$$

Indeed, if $U' \to U$ is a surjective etale morphism, we can identify U with the coequalizer of the pair of morphisms $U' \times_U U' \rightrightarrows U'$, i. e., the quotient scheme of U' by the equivalence relation $U' \times_U U' \subset U' \times U'$.

One can show that each presheaf of Example 1 in Sect. 3.1 is a sheaf in the etale topology. This important observation can be stated as follows: If a contravariant functor is representable, then it is a sheaf in the etale topology. Conversely, every etale sheaf is representable, in this sense, by an "etale" X-scheme Y; we use quotation marks because the scheme Y may be nonseparable and of infinite type over X. One may also say that the category of etale sheaves over X is obtained from $\operatorname{Et}(X)$ by adding various inductive limits.

The presheaves of Example 2 in Sect. 3.1 are also etale sheaves. On the other hand, the presheaf Pic of Example 3 in Sect. 3.1 is not a sheaf – nether in the etale topology, no in the Zariski topology (take the standard covering of $\mathbb{P}^1$ and the sheaf $\mathcal{O}(1)$). To get a better handle on etale sheaves, we will treat in detail the schemes consisting of a single point.

Example. Let $X = \operatorname{Spec} K$, where K is a field. First, we assume that K is algebraically closed. As we explained in Sect. 2.1, every etale scheme over X is isomorphic to a direct sum of several copies of X (and the category $\operatorname{Et}(X)$ is equivalent to the category of finite sets). According to condition (a), a presheaf on X is determined by a set $F(X)$, and we do not make a distinction between the sheaf F and the set $F(X)$.

Now, let K be an arbitrary field. Then the category $\mathrm{Et}(X)$ is equivalent to the category of finite G-sets, where $G = \mathrm{Gal}(\overline{K}/K)$ is the Galois group of K. An etale sheaf F associates a set $F(X')$ to an etale X-scheme X', and those sets are related by various diagrams. In particular, for each X', we get a canonical map $F(X) \to F(X')$. Moreover, if X' is a Galois object, then the group $\mathrm{Aut}(X'|X)$ acts on $F(X')$ leaving $F(X)$ fixed.

What are the restrictions imposed on this collection of data by the sheaf axiom? Condition (a) allows us to restrict ourselves to the connected X''s. Let X' be a Galois covering with the group G. Then the axiom (b) applied to the covering $X' \to X$ means that $F(X)$ coincides with the set of elements of $F(X')$ fixed under the action of G. Now, it is clear that to give an etale sheaf F on X is the same as to give a set $\varinjlim F(X')$ (where X' runs over the Galois coverings of X) equipped with a continuous action of the profinite group $\pi_1(X) = \mathrm{Gal}(\overline{K}/K)$.

3.3. Category of Sheaves. As in case of topological spaces, to an etale presheaf F on X one may associate an etale sheaf $\widetilde{F}$. As before, this is a formal exercise on inductive limits. For example, given X as in Sect. 3.2 and an etale sheaf F on X, then $\widetilde{F}$ corresponds to a G-set $\varinjlim F(X')$.

A morphism of sheaves (or presheaves) is defined as a morphism of functors. Thus etale sheaves form a category (which is equivalent to the category of "etale" X-schemes by the previous section). In particular, a morphism of X-schemes $Y \to Z$ induces a morphism of sheaves $h_Y \to h_Z$. Recall also that an epimorphism of sheaves $u: F \to G$ does not necessarily mean that u is an epimorphism of the corresponding presheaves, and a section $t \in G(X)$ can be lifted only locally to a section of F: there is an etale covering $U \to X$ and $s \in F(U)$ such that $u(s) = t|U$.

Example. Let $u: Y \to Z$ be a smooth surjective morphism of schemes over X. Then the corresponding morphism of sheaves $h_Y \to h_Z$ is an epimorphism. In fact, a smooth morphism always has an etale quasi-section, while global (or even local in the Zariski topology) sections very seldom exist.

The notions of direct and inverse image makes sense for etale sheaves. Everything is very simple for direct images. Let $f: X \to Y$ be a morphism of schemes. Given an "open" $V \in \mathrm{Et}(Y)$, its "inverse" image $f^{-1}(V) = X \times_Y V$ is also "open", i. e. belongs to $\mathrm{Et}(X)$. Therefore, f is "continuous" in the etale sense, and for a sheaf F on X, we can define a sheaf f_*F on Y as follows:

$$(f_*F)(V) = F(f^{-1}(V)).$$

It is indeed a sheaf, and it is called the *direct image* of F under the morphism f. By formal considerations, the functor f_* has the left adjoint functor f^{-1}, which is called the *inverse image*. It transforms the sheaves on Y into the sheaves on X. If a sheaf G on Y is representable by an "etale" Y-scheme, which is also denoted by G, then $f^{-1}G$ is representable by the "etale" X-scheme $G \times_Y X$.

3.4. Stalk of Sheaf at a Point. There is one especially important case of the inverse image operation. A *geometric point* of a scheme X is a morphism $u\colon \xi \to X$, where ξ is the spectrum of an algebraically closed field. Let F be a sheaf on X. Then the sheaf $F_\xi = u^{-1}(F)$ on ξ is said to be *the stalk of the sheaf* F at ξ. As we explained in Sect. 3.2, this sheaf can be identified with a set which admits an explicit description.

Every commutative diagram

$$\begin{array}{ccc} & \nearrow & U \\ & & \downarrow \\ \xi & \longrightarrow & X \end{array}$$

with an etale morphism $U \to X$ is said to be an *etale neighborhood* of the geometric point $\xi \to X$. The etale neighborhoods form a directed category, and F_ξ is nothing but $\varinjlim F(U)$, where the limit is taken over this category. The role of the stalk functor is illustrated by the following

Proposition. *A sheaf morphism $F \to G$ is an isomorphism (monomorphism, epimorphism) if and only if the corresending morphism of fibers $F_\xi \to G_\xi$ is an isomorphism (monomorphism, epimorphism) for every geometric point ξ.*

3.5. Etale Localization. Let $\xi \to X$ be a geometric point of a scheme X. It is often convenient to pass to the projective limit over all etale neighborhoods of ξ. In the category of schemes, the limit

$$\widetilde{X}_\xi = \varprojlim U$$

exists, and is called the *strict localization* of X at geometric point ξ. This is just the spectrum of the ring $\varinjlim H^0(U, \mathcal{O}_U)$. That is, in a sense, the "smallest etale neighborhood" of ξ, though, strictly speaking, the morphism $\widetilde{X}_\xi \to X$ is not etale but pro-etale. Intuitively, $\widetilde{X}_\xi$ corresponds to the notion of a small ε-neighborhood for complex schemes. Of course, it is only relatively small: any two such neighborhoods, $\widetilde{X}_\xi$ and $\widetilde{X}_{\xi'}$, always "intersect" over a generic point of X. The main point is that the scheme $\widetilde{X}_\xi$ represents a homotopy trivial object, as a small ball. This follows from the following, essentially tautological, property: every etale covering $U \to \widetilde{X}_\xi$ has a section.

The scheme $\widetilde{X}_\xi$ is local, and has yet another very important property. If $Y \to \widetilde{X}_\xi$ is a finite morphism of schemes, then Y is a sum of local schemes. Such schemes and the corresponding rings are said to be *Henselian* (because Hensel established this property for the ring $\mathbb{Z}_p$ of p-adic numbers). For more details on Henselian rings and schemes, see (Grothendieck-Dieudonné (1967), Milne (1980), Raynaud (1970)).

§ 4. Cohomology of Etale Sheaves

4.1. Abelian Sheaves. The notion of cohomology makes sense for sheaves of Abelian groups, so we begin with them. Note that H^0 is defined for any sheaf of sets, and H^1 for any sheaf of groups (there are generalizations to higher H^q as well), however, we will consider Abelian sheaves only.

A sheaf on a scheme X with values in the category of Abelian groups is said to be an *Abelian* (etale) *sheaf*. One can also say that it is a sheaf of sets with an Abelian group structure. Since the category of Abelian sheaves is Abelian, it makes sense to speak about exact sequences of Abelian sheaves (we will often omit the word "Abelian"). A sequence of sheaves $F \to G \to H$ is exact if and only if for every geometric point ξ, the corresponding sequence of Abelian groups $F_\xi \to G_\xi \to H_\xi$ is exact (see Sect. 3.4). We will present two main examples of Abelian sheaves:

Example 1. Let A be an Abelian group. For every scheme X, one can define a constant sheaf A_X by setting $A_X(U) = A^{\pi_0(U)}$ for $U \in \mathrm{Et}(U)$. This sheaf is representable by an X-scheme $A \times X$. Its fiber equals A at every point. Most often we taken $\mathbb{Z}/n\mathbb{Z}$ as A; the case $A = \mathbb{Z}$ is less interesting.

Example 2. Let A denote a commutative group scheme. Then the sheaf representable by the scheme $A \times X$ is obviously Abelian. The main example is the following scheme:

$$A = \mathbb{G}_m = \operatorname{Spec} \mathbb{Z}\,[T, T^{-1}]$$

(the multiplicative group). The corresponding sheaf associates to any $U \in \mathrm{Et}(X)$ the group $H^0(U, \mathcal{O}_U^*)$ of invertible functions on U, and is denoted by $\mathcal{O}^*$. Let

$$\mathcal{O}^* \xrightarrow{n} \mathcal{O}^*$$

be the multiplication by n in that sheaf (in the multiplicative notation, the n-th power map). The kernel of this homomorphism is denoted by μ_n. This Abelian sheaf is representable by the group scheme $\operatorname{Spec} \mathbb{Z}\,[T]/(T^{n-1})$. If n is invertible on the scheme X, then the sheaf μ_n is locally isomorphic to the constant sheaf $\mathbb{Z}/n\mathbb{Z}$.

Lemma. *If n is invertible on X, the sequence*

$$0 \to \mu_n \to \mathcal{O}^* \xrightarrow{n} \mathcal{O}^* \to 1$$

of Abelian groups on X is exact.

The lemma follows from the observation that in our case, $\mathbb{G}_m \xrightarrow{n} \mathbb{G}_m$ is an etale covering (see Sect. 2.5 or 3.3). The sequence, however, is not exact in the

Zariski topology. It is called the *Kummer sequence.* It partially replaces the exponential sequence on complex varieties.

4.2. Cohomology. As with any scheme, the most interesting objects are the global sections of etale sheaves. The group $F(X)$ of global sections of a sheaf F on X is often denoted by $H^0(X, F)$. The functor H^0 is left exact but not right exact. To control this phenomenon, we introduce the cohomology functors H^q as the derived functors of H^0 (see Chap. 1, Sect. 1 and 4). This is possible because the category of etale Abelian sheaves has enough injective objects. One can utilize the etale version of the flabby Godement resolution by associating to F the flabby sheaf $\prod F_\xi$, where ξ runs over the points of X.

The main point is that any short exact sequence of Abelian sheaves

$$0 \to A \to B \to C \to 0$$

yields a long exact cohomology sequence

$$0 \to H^0(X, A) \to H^0(X, B) \to H^0(X, C) \to H^1(X, A) \to \dots$$
$$\dots \to H^q(X, A) \to H^q(X, B) \to H^q(X, C) \to H^{q+1}(X, A) \to \dots .$$

Similarly, one can define the direct image functors $R^q f_*$ for any morphism of schemes $f\colon X \to Y$. As in Sect. 4 of Chap. 1, there is a *Leray spectral sequence*

$$E_2^{p,q} = H^p(Y, R^q f_* F) \Rightarrow H^{p+q}(X, F) .$$

We seldom compute cohomology utilizing the definitions, although, in the final analysis, everything may be reduced to the definitions. We will describe below two important cases when it is possible to reduce the etale cohomology to familiar objects.

4.3. Galois Cohomology. Let X be the simplest scheme, namely, the spectrum of the field K. As we explained in Sect. 2.2, an Abelian sheaf F on X is in fact an Abelian group A equipped with an action of the Galois group $G = \mathrm{Gal}(\overline{K}|K)$ of K. Further, $H^0(X, F)$ coincides with the subgroup of invariant elements, A^G or $\mathrm{Hom}_G(\mathbb{Z}, A)$, where G acts trivially on $\mathbb{Z}$. Clearly $H^q(X, F)$ coincide with the cohomology of the G-module A. The latter is denoted by $H^q(G, A)$; see (Serre (1964)), as well as Chap. 3, Sect. 4. We observe that one often takes a projective resolution of a trivial G-module $\mathbb{Z}$ instead of an injective resolution of a G-module A. A standard form of such a resolution – a free Abelian group over the simplicial G-set

$$\{*\} \leftarrow G \leftleftarrows G \times G \leftleftarrows\!\!\!\!\leftarrow \dots$$

– arises allusions to the complex of the "universal covering" $\widetilde{X} \to X$ (compare Chap. 3, Sect. 4.4). In particular, $H^1(G, A)$ consists of 1-cocycles modulo 1-coboundaries. A 1-cocycle is a map $\varphi\colon G \to A$ such that $\varphi(gg') = g\varphi(g') + \varphi(g)$. If $\varphi(g) = ga - a$ for $a \in A$, then φ is said to be a *coboundary.* For instance, if

G acts trivially on A, then $H^1(G, A)$ coincides with the group $\mathrm{Hom}(G, A)$ of group homomorphisms of G to A.

Since the Galois group G is profinite, $H^1(G, \mathbb{Z}) = \mathrm{Hom}(G, \mathbb{Z}) = 0$ provided $\mathbb{Z}$ is a trivial G-module.

4.4. Cohomology of Coherent Sheaves. By analogy with GAGA, one may expect that the cohomology of the coherent etale sheaves F_{et}, defined in Example 2 of Sect. 3.1, are intimately related to the cohomology of the coherent sheaves F in the Zariski topology. They, in fact, coincide for arbitrary schemes not necessarily complete, in contrast to GAGA.

Theorem. *Let F be a quasi-coherent sheaf in the Zariski topology on X. Then we have a canonical isomorphism*

$$H^q(X_{\mathrm{zar}}, F) \xrightarrow{\sim} H^q(X_{\mathrm{et}}, F_{\mathrm{et}}).$$

It will suffice to verify the theorem for an affine scheme X, and then for its arbitrary etale covering $U \to X$. Let $X = \operatorname{Spec} A$, $U = \operatorname{Spec} B$, and F corresponds to an A-module M. Then the cohomology of this covering coincide with the cohomology of the complex

$$M \to M \underset{A}{\otimes} B \rightrightarrows M \underset{A}{\otimes} B \underset{A}{\otimes} B \substack{\rightarrow\\ \rightarrow\\ \rightarrow} \dots,$$

which is acyclic, as we have already mentioned in Chap. 2, Sect. 1.2.

So, the etale cohomology of coherent sheaves provide no new information compared with the Zariski topology, and we can turn all the attention to "discrete" sheaves.

Corollary. *Let K' be a Galois extension of a field K with the Galois group G. Then $H^q(G, K') = 0$ for $q > 0$.*

By the way, this also follows from the observation that K' as a G-module is isomorphic to the group algebra $K[G']$ (Serre (1964)).

4.5. Torsors. The elements of H^1 admit a geometric interpretation as torsors, which is useful in calculations as well as applications. We will explain this notion. For simplicity consider a finite group A. A twisted form of the constant sheaf A_X is said to be a *principle A-covering* or *A-torsor.* In other words, this is an X-scheme Z with an action of the group A that is locally (in etale topology) isomorphic to a trivial covering with fiber A, i.e., there is an etale covering $U \to X$ such that $Z \times_X U$ is isomorphic to $A \times U$.

One can show that Z is then also an etale covering of X, and one can just take Z as U. Utilizing Sect. 2, one can show that for a connected scheme X, the set of A-torsors (up to an isomorphism) can be identified with the set of group homomorphisms $\mathrm{Hom}(\pi_1(X), A)$ (compare with Sect. 4.3). If A is Abelian, then the latter set is also a group, and is isomorphic to $H^1(X_{\mathrm{et}}, A_X)$.

Similar results hold if we replace A by either: (a) any affine group scheme, or (b) any Abelian sheaf (Milne (1980)).

One particular case is especially important, namely, the torsors of the sheaf $\mathcal{O}^*$, or the group scheme $\mathbb{G}_m$. As we know, in the Zariski topology, the Picard group $\operatorname{Pic} X$ has an interpretation as $H^1(X_{\text{zar}}, \mathcal{O}_X^*)$. It turns out that the same is true in the etale topology:

$$H^1(X_{\text{et}}, \mathcal{O}^*) = \operatorname{Pic} X .$$

Again this is a formal exercise in the descent theory, and we will describe one important case only, when X is the spectrum of a field K. Then the previous statement is equivalent to the triviality of the group $H^1(G, \overline{K}^*)$, where $\overline{K}$ is a finite Galois extension with the Galois group G (the latter is known as the *Hilbert theorem 90*).

Indeed, let $\varphi\colon G \to \overline{K}^*$ be a 1-cocycle. For $\theta \in \overline{K}$, we form a "resolution" $\beta = \sum_{g \in G} \varphi(g) \cdot g(\theta)$. Then for a "general" θ, the element β is nontrivial. It is easy to verify that $g(\beta) = \varphi(g^{-1})\beta$, i.e., $\varphi(g) = \beta/\varphi(\beta)$ and φ is a coboundary.

4.6. The Kummer Theory. Applying the cohomology to the Kummer exact sequence of Example 2 in Sect. 4.1 and utilizing the interpretation of $H^1(\mathcal{O}^*)$ given in Sect. 4.4, we obtain an exact sequence

$$\begin{aligned} 0 \to \mu_n(X) \to H^0(X, \mathcal{O}^*) \xrightarrow{n} H^0(X, \mathcal{O}^*) \to H^1(X, \mu_n) \to \\ \to \operatorname{Pic} X \xrightarrow{n} \operatorname{Pic} X \to H^2(X, \mu_n) \to H^2(X, \mathcal{O}^*) . \end{aligned}$$

Now, assuming that X is a scheme over an algebraically closed field K (as a matter fact, it will suffice to assume that K contains all n-th root of 1), and n is invertible in K. Then the sheaf μ_n is isomorphic to $\mathbb{Z}/n\mathbb{Z}$, and we get an exact sequence

$$0 \to H^0(X, \mathcal{O}^*)/H^0(X, \mathcal{O}^*)^n \to H^1(X, \mu_n) \to \operatorname{Pic}(X)_n \to 0 .$$

In other words, any etale covering of X with the fiber $\mathbb{Z}/n\mathbb{Z}$ comes from the construction described in Sect. 2.5. If, in addition, X is a complete variety then $H^0(X, \mathcal{O}^*) = K^*$, and we get

$$H^1(X, \mu_n) \xrightarrow{\sim} \operatorname{Pic}(X)_n = \operatorname{Ker}(\operatorname{Pic} X \xrightarrow{n} \operatorname{Pic} X) .$$

4.7. Acyclicity of Finite Morphisms. Let $f\colon X \to Y$ be a morphism of schemes, F an Abelian sheaf on X, and $\xi \to Y$ a geometric point. Utilizing the notion of etale localization, one can describe the fibers of $R^q f_* F$ at the point ξ. Let $\widetilde{Y} = \widetilde{Y}_\xi$ denote the strict localization of Y at ξ (see Sect. 3.5). Then

$$(R^q f_* F)_\xi = H^q(X \underset{Y}{\otimes} \widetilde{Y}, \widetilde{F}) , \qquad (*)$$

where $\widetilde{F}$ denotes the "restriction" of F to $X \times_y \widetilde{Y}$. Indeed, the left-hand side is the inductive limit of $H^q(X \times_Y V, F)$ as V runs over etale neighborhoods of the point $\xi \to Y$.

Theorem. *If $f: X \to Y$ is a finite morphism, then $R^q f_* F = 0$ for $q > 0$.*

We have to show that $(R^q f_* F)_\xi$ is trivial for every geometric point $\xi \to Y$. According to $(*)$, we may assume that Y is a strictly local scheme. Then, by the Hensel condition (Sect. 3.5), the scheme X is a direct sum of finitely many strictly local schemes, acyclic in the etale sense.

One can also easily describe the fibers of the sheaf $f_* F$. This may be used to deduce that nilpotents (and finite radical extensions) do not affect the etale cohomology. So, in the sequel, we may assume that all schemes are reduced.

§ 5. Cohomology of Algebraic Curves

In this section we will consider cohomology of the simplest algebra-geometric objects, namely, curves. This case is of primary importance, because utilizing various tricks – like fibering in subvarieties of lower dimension – one can reduce many problems to curves. Throughout the section, k is an algebraically closed field, n an integer invertible in k, and X a smooth irreducible curve over k.

5.1. Outline of Strategy. We are interested in the cohomology of X with coefficients in the sheaf $\mathbb{Z}/n\mathbb{Z}$. However, our approach to the problem is not straightforward. Let $\eta = \operatorname{Spec} k(X)$ be the generic (not a geometric) point of X. We first study the cohomology of the point η with coefficients in the sheaf $\mathcal{O}^*$. Here we apply the Galois cohomology methods; the main result is the following theorem.

Theorem 1. *$H^q(\eta, \mathcal{O}^*) = 0$ for $q > 0$.*

Then we proceed to the curve X but still with the sheaf $\mathcal{O}^*$.

Theorem 2. *$H^q(X, \mathcal{O}^*) = 0$ for $q > 1$.*

We are then able to conclude our task by utilizing of the Kummer theory.

Now, we will explain the reason for our indirect approach. Our aim is to obtain an etale analogue of the topological fact that an open curve is essentially 1-dimensional (passing from X to its generic point η does not affect the problem). One might even expect that $H^q(\eta) = 0$ for $q > 1$ and every sheaf on η, not only the sheaf $\mathbb{Z}/n\mathbb{Z}$. However, this is false. For example, it follows from the short exact sequence $0 \to \mathbb{Z} \to \mathbb{Q} \to \mathbb{Q}/\mathbb{Z} \to 0$ that $H^2(\eta, \mathbb{Z}) = H^q(\eta, \mathbb{Q}/\mathbb{Z})$ is a very large group – not surprising because we already realize that one may expect reasonable answers only when the coefficients are finite.

To calculate $H^q(\eta, \mathbb{Z}/n\mathbb{Z})$, it is natural to utilize the Kummer theory, which leads us to Theorem 1. This theorem granted, it is convenient to go back via $H^q(X, \mathcal{O}^*)$ instead of $H^q(\eta, \mathbb{Z}/n\mathbb{Z})$.

5.2. Tsen's Theorem. Let $K = k(X)$ denote the field of rational functions on X, and L a finite Galois extension of K. Theorem 1 of Sect. 5.1 will follow at once from the following assertion. For every L as above and $q > 0$

$$H^q(\mathrm{Gal}(L/K), L^*) = 0\,.$$

The latter is true for $q = 1$ by Sect. 4.5. Rather formal reductions (Cassels-Fröhlich (1967), Serre (1964)) imply that it will suffice to verify the vanishing of the Tate 0-dimensional cohomology

$$\widehat{H}^0(\mathrm{Gal}(L/K), L^*) = K^*/NL^*\,,$$

where $N\colon L \to K$ is the norm map. In our case, the surjectivity of N follows from a specific property of function fields on curves, namely, they are quasi-algebraic; see Sect. 1, as well as (Shafarevich (1986), § 11). Indeed, set $d = [L : K]$. The equation $N(x) = ax_0^d$ is homogeneous of degree d in $d+1$ variables $x_0, x_1, \ldots, x_d$, so it has a nontrivial solution because K is quasi-algebraic.

Thus Theorem 1 of Sect. 5.1 is reduced to the following *theorem of Tsen.*

Theorem. *A field K of transcendence degree 1 over an algebraically closed field is quasi-algebraically closed.*

We will explain a proof of this result. First, we consider the case when $K = k(T)$ is a purely transcendental extension. Let $F = \sum a_m(T)X^m$ be a homogeneous polynomial of degree $|m| \le d$, where $m = (m_0, \ldots, m_d)$ is a multiindex. Multiplying F by the denominators of the a_m's, we can assume that each $a_m(T)$ is a polynomial in T of degree $\le A$. Employing the method of indeterminate coefficients, we will look for a solution in the form of polynomials of degree N in T: $X_i = X_i(T)$ where $i = 0, \ldots, d$. Then $F(X(T))$ is polynomial of degree $A + dN$, whose coefficients are polynomials of $(d+1)(N+1)$ indeterminate coefficients of the polynomials $X_0, \ldots, X_d$. The condition that the coefficients of $F(X(T))$ are trivial gives a system of $A + dN + 1$ equations in $(d+1)(N+1)$ variables. This system obviously has a trivial solution. Further, since for $N \ge A$, we get

$$(d+1)(N+1) > A + dN + 1\,,$$

and the system has a nontrivial solution (Danilov (1988), Chap. 2, Sect. 3).

Now, let K be an extension of degree r of the field $k(T)$. Let $e_1, \ldots, e_r$ be a basis of K over $k(T)$. We write each X_i as $\sum_{j=1}^r Y_{ij}e_j$, where Y_{ij} are unknown elements of $k(T)$. The equation $F(X) = 0$ takes the form of a homogeneous equation $N_{K/k(T)}F(\sum Y_{ij}e_j) = 0$ over $k(T)$. Since the number of variables $\ge r(\deg F + 1)$ and its degree equals $r \cdot \deg F$, the equation has a solution, as above.

5.3. Cohomology of $\mathcal{O}^*$. Now, let x be a closed (geometric) point of X. We will compare the cohomology of X and $X \backslash \{x\}$ with coefficients in $\mathcal{O}^*$. We denote by $\widetilde{X}$ the strict localization of X at x. Consider the Mayer-Vietoris sequence for a covering of X by two "open" sets, $X\backslash\{x\}$ and $\widetilde{X}$,

$$H^{q-1}(\widetilde{X}\backslash\{x\}) \to H^q(X) \to H^q(X\backslash\{x\}) \oplus H^q(\widetilde{X}) \to H^q(\widetilde{X}\backslash\{x\})$$

(the coefficients are $\mathcal{O}^*$). Since $\widetilde{X}\backslash\{x\} = \operatorname{Spec} \widetilde{K}$, where $\widetilde{K}$ is the quotient field of the scheme $\widetilde{X}$, by Tsen's theorem $H^q(\widetilde{X}\backslash\{x\}, \mathcal{O}^*) = 0$ for $q > 0$. The scheme $\widetilde{X}$ is acyclic, hence for $q > 1$, we obtain an isomorphism

$$H^q(X, \mathcal{O}^*) \tilde{\to} H^q(X\backslash\{x\}, \mathcal{O}^*) .$$

So, removing one by one points we obtain the following isomorphism in the limit:

$$H^q(X, \mathcal{O}^*) \tilde{\to} H^q(\eta, \mathcal{O}^*)$$

for $q > 1$. Now Theorem 2 of Sect. 5.1 follows from Theorem 1.

5.4. Cohomology of Complete Curves. Now let X be a complete curve. By the Kummer theory (Sect. 4.6), we obtain the exact sequence

$$0 \to H^1(X, \mu_n) \to \operatorname{Pic} X \xrightarrow{n} \operatorname{Pic} X \to H^2(X, \mu_n) \to 0 ,$$

as well as the vanishing of $H^q(X, \mu_n)$ for $q > 2$. To describe H^1 and H^2, we recall several facts about the Picard group of a curve.

First, we have an exact sequence

$$0 \to \operatorname{Pic}^0 X \to \operatorname{Pic} \xrightarrow{\deg} \mathbb{Z} \to 0 .$$

Second, one can identify the group $\operatorname{Pic}^0 X$ of divisor classes of degree 0 with the group of rational points $J(k)$ of an Abelian variety J, the *Jacobian* of X. The dimension of J equals the genus of X, i. e. $\dim H^1(X, \mathcal{O}_X)$. To somewhat clarify the picture, we observe that any divisor of degree g is equivalent to an effective divisor, i. e. a sum $P_1 + \ldots + P_g$. Fixing a point $P_0 \in X$, we get a surjection

$$X(k)^g \to \operatorname{Pic}^0 X , \qquad (P_1, \ldots, P_g) \mapsto \sum_{i=1}^{g} (P_i - P_0) .$$

Now, one can easily verify that a suitable quotient space of the variety X^g has $\operatorname{Pic}^0 X$ as its points, and since X^g is complete, the variety J is complete too (Serre (1959)).

Now, consider the homomorphism multiplication by n

$$J \xrightarrow{n} J . \qquad (*)$$

Its differential is also the morphism multiplication by n, which is bijective. It follows that $(*)$ is etale, hence surjective because J is a complete variety. We get

$$H^2(X, \mu_n) = \operatorname{Coker}(\mathbb{Z} \xrightarrow{n} \mathbb{Z}) = \mathbb{Z}/n\mathbb{Z} .$$

The kernel of $(*)$ is a finite group isomorphic to $(\mathbb{Z}/n\mathbb{Z})^{2g}$ by the theory of Abelian varieties (Mumford (1970)). We now summarize our calculations:

Theorem. *Let X be a smooth complete curve. Then*

$$H^q(X,\mu_n) = \begin{cases} \mu_n \simeq \mathbb{Z}/n\mathbb{Z} & \text{for } q=0\,, \\ \mathrm{Pic}^0(X)_n \simeq (\mathbb{Z}/n\mathbb{Z})^{2g} & \text{for } q=1\,, \\ \mathbb{Z}/n\mathbb{Z} & \text{for } q=2\,, \\ 0 & \text{for } q>2\,. \end{cases}$$

5.5. Duality on Complete Curves. As before, let X be a smooth complete curve. The cup product in cohomology defines a pairing

$$H^q(X,\mu_n)\otimes H^{2-q}(X,\mathbb{Z}/n\mathbb{Z}) \to H^2(X,\mu_n) = \mathbb{Z}/n\mathbb{Z}\,.$$

Theorem. *For a smooth complete curve X, this pairing is perfect and identifies $H^{2-q}(X,\mathbb{Z}/n\mathbb{Z})$ with the group* $\mathrm{Hom}(H^q(X,\mu_n),\mathbb{Z}/n\mathbb{Z})$.

The principal nontrivial case is $q=1$. We have to verify that every homomorphism $\alpha\colon H^1(X,\mu_n)\to\mathbb{Z}/n\mathbb{Z}$ determines an element of the group $H^1(X,\mathbb{Z}/n\mathbb{Z})$, i.e. a $(\mathbb{Z}/n\mathbb{Z})$-torsor over X. We will describe the corresponding geometric construction. We can view the morphism $(*)$ as a torsor J with the structure group $J(k)_n\simeq H^1(X,\mu_n)$. Applying the homomorphism α, we obtain a $(\mathbb{Z}/n\mathbb{Z})$-torsor over J, so it remains to make a base change

$$\varphi\colon X\to J\,,\qquad \varphi(P)=\mathrm{cl}(P-P_0)\,.$$

By the third fact about the Picard group of X, φ induces an isomorphism

$$\varphi^*\colon \mathrm{Pic}^0 J \to \mathrm{Pic}^0 X = J$$

(*autoduality* of Jacobian). This implies that the pairing is perfect.

5.6. Open Curves. Utilizing Sect. 5.3, one can easily obtain the cohomology of open curves. Let X be a non-complete curve. Then $\mathrm{Pic}\,X$ is divisible, hence we get $H^2(X,\mu_n)=0$. The group $H^1(X,\mu_n)$ is a free $(\mathbb{Z}/n\mathbb{Z})$-module with $2g+s-1$ generators, where s is the number punctures $(s\geq 1)$.

In studying open varieties, we often employ the so-called *cohomology with compact support*, denoted by $H^q_c(X,\mu_n)$. We will treat them in detail in the next section. For now, we let $H^q_c(X,\mu_n)$ to be $H^q(\overline{X},j_!\mu_n)$, where $\overline{X}$ is a smooth compactification of the curve X, and $j_!\mu_n$ is the sheaf obtained by extending $\mu_{n,X}$ by zero outside X. Set $S=\overline{X}\backslash X$. We get an exact sequence of sheaves on $\overline{X}$:

$$0\to j_!\mu_n\to\mu_{n,\overline{X}}\to\mu_{n,S}\to 0\,.$$

It follows that

a) $H^2_c(X,\mu_n)=H^2(\overline{X},\mu_n)=\mathbb{Z}/n\mathbb{Z}$,

b) $H^0_c(X,\mu_n)=0$ (if $S\neq\emptyset$),

c) the following sequence is exact:

$$0 \to \mu_n \to \mu_n^S \to H_c^1(X, \mu_n) \to H^1(\overline{X}, \mu_n) \to 0 .$$

Using Sect. 5.5, one can show that the cup product establishes the perfect pairing

$$H_c^q(X, \mu_n) \otimes H^{2-q}(X, \mathbb{Z}/n\mathbb{Z}) \to H_c^2(X, \mu_n) = \mathbb{Z}/n\mathbb{Z} .$$

In short, everything is similar to the complex case. (For a detailed treatment, see (Grothendieck et al. (1977a), Milne (1980))).

§ 6. Fundamental Theorems

In the sequel, by a *scheme* we mean an algebraic scheme over a fixed field k.

6.1. Constructible Sheaves. In etale topology, we obtain reasonable answers only for "finite" sheaves like $\mathbb{Z}/n\mathbb{Z}$. We will now elaborate on this condition.

A sheaf of sets F on a scheme X is said to be *locally constant* if, after replacing X by a suitable etale converning $U \to X$, F becomes a constant sheaf with finite fibers. Then F is representable by an etale covering $Z \to X$, or is given by a representation of $\pi_1(X)$ on a finite set.

We define constructible sheaves by induction. A sheaf F on a scheme X is said to be *constructible*, if it is locally constant on a nonempty Zariski open subset $U \subset X$ and constructible on $X \backslash U$. It follows at once from the definition that the fibers of constructible sheaves are finite; moreover, the number of points in the fiber F_ξ depends constructively on ξ. Finally, we may say that the constructible sheaves are representable by quasi-finite (non-separable) schemes over X.

The class of constructible sheaves is rather versatile to work with. It is closed under extensions and direct and inverse images.

6.2. The Base Change Theorem. Let $f \colon X \to Y$ be a morphism of schemes, F an abelian sheaf on X, and ξ a geometric point of Y. The base change theorem compares the fibers of the sheaf $(R^q f_* F)_\xi$ with the cohomology $H^q(X_\xi, F_\xi)$, where $X_\xi = X \times_Y \xi$ is the fiber of f over ξ, and F_ξ the induced sheaf on X_ξ.

Theorem. *Let f be a proper morphism, and F a constructible sheaf. Then the canonical homomorphism*

$$(R^q f_* F)_\xi \to H^q(X_\xi, F_\xi)$$

is an isomorphism for all $q \geq 0$.

Note that for the usual topological spaces, a similar statement is true for any sheaf (Godement (1958)). In the etale case, the same is true for finite morphisms (Sect. 4.7), while for arbitrary proper morphisms, we must require F to be constructible.

We will only sketch the proof; for details, see (Grothendieck et al. (1972–1973), Grothendieck et al. (1977a)). Using Chow's lemma and fibering X by hyperplane sections, we are reduced to the case when the fibers of f have dimension ≤ 1. By acyclicity of finite morphisms, we may assume that $F = \mathbb{Z}/n\mathbb{Z}$, perhaps after replacing X by its finite covering. Replacing Y by its strict localization at ξ, we may assume that Y is strictly local scheme. So, it will suffice to prove that the homomorphism

$$H^q(X, \mathbb{Z}/n\mathbb{Z}) \to H^q(X_\xi, \mathbb{Z}/n\mathbb{Z})$$

is bijective. Further, by formal homological reductions, it will suffice to prove bijectivity for $q = 0$ only, and surjectivity for $q > 0$. According to Sect. 5.4, it will suffice to consider the cases $q = 0, 1, 2$, which are treated seperatly utilizing geometry.

Case $q = 0$. For any scheme X, $H^0(X, \mathbb{Z}/n\mathbb{Z}) = (\mathbb{Z}/n\mathbb{Z})^{\pi_0(X)}$ where $\pi_0(X)$ is the number of connected components. Therefore we have to establish that the map

$$\pi_0(X_\xi) \to \pi_0(X)$$

is bijective. Since f is proper, every component of X intersects X_ξ. So, it will suffice to show that X_ξ is connected provided X is connected. Using the Stein factorization theorem (Danilov (1988), Chap. 2, Sect. 3), we may assume that f is finite. Now, everything follows from the main property of Hensel rings (Sect. 3.5).

Case $q = 1$. By the above discussion, we may assume that X and X_ξ are connected. The elements of $H^1(., \mathbb{Z}/n\mathbb{Z})$ classify $(\mathbb{Z}/n\mathbb{Z})$-torsors, so it will suffice to verify that every etale covering $X'_\xi \to X_\xi$ can be extendet to an etale covering $X' \to X$.

We proceed as follows. First, we can extend (without obstructions) the covering in question to any infinitesimal neighborhood of X_ξ in X. By Grothendieck's theorem on algebraizations of formal schemes (Grothendieck-Dieudonné (1962–1963)), we then obtain an etale covering, however, not of X but of the scheme $X \otimes_A \widehat{A}$, where $\widehat{A}$ is the completion of the local ring $A = \mathcal{O}_{Y,\xi}$. Finally, we apply the M. Artin approximation theorem (Artin (1969)).

Case $q = 2$. It will suffice to show that the map

$$H^2(X, \mathbb{Z}/n\mathbb{Z}) \to H^2(X_\xi, \mathbb{Z}/n\mathbb{Z})$$

is surjective. For an arbitrary scheme X, the Kummer theory gives the morphism $\operatorname{Pic} X \to H^2(X, \mathbb{Z}/n\mathbb{Z})$. It follows from Sect. 5.4 that it is surjective for the complete curve X_ξ. So, it remains to verify that the restriction map

$$\operatorname{Pic} X \to \operatorname{Pic} X_\xi$$

is surjective, i.e., a Cartier divisor D_ξ on X_ξ can be extended to a Cartier divisor D on X. We can assume that the support of D_ξ is a single point on the curve X_ξ. By extending the local equation of D_ξ to X, we obtain a divisor D on X that cuts on X_ξ the divisor D_ξ plus something else away from the support of D_ξ. Now, we utilize the fact that Y is Henselian and take the component of D passing through D_ξ.

6.3. Cohomology with Compact Support. Let F be a constructible sheaf on a scheme X. Let $j\colon X \to \overline{X}$ be an open inclusion in a complete scheme $\overline{X}$. We denote by $j_!F$ the extension of F by zero on $\overline{X}\setminus X$. In other words, the restriction of $j_!F$ to X is F, amd $j_!F$ is zero on $\overline{X}\setminus X$. Clearly $j_!F$ is a constructible sheaf.

Definition. The groups $H^q_c(X,F) = H^q(\overline{X}, j_!F)$ are said to be the *cohomology groups with compact support.*

Obviously, $H^q_c(X,F) = H^q(X,F)$ for a complete scheme X. In general, however, it is not obvious that the definition is independent on the choice of the compactification $X \subset \overline{X}$. We will prove that fact.

Let $j'\colon X \to \overline{X}'$ be another compactification. As usual, we can assume that $j' = f \circ j$, where $f\colon \overline{X} \to \overline{X}'$ is a (proper) morphism. According to the Leray spectral sequence, it will suffice to verify that $f_*(j_!F) = j'_!$ and $R^q f_*(j_!F) = 0$ for $q > 0$. Both conditions may be verified pointwise for each point $\xi \to \overline{X}'$, and in view of Theorem of Sect. 6.2, the conditions are trivial. This proves the independence of the definition of $H^q_c(X,F)$ on the compactification.

It is often more convenient to deal with the compact cohomology than with the usual one because of their "additivity" (compare Chap. 3, Sect. 1.4): For a closed embedding $i : Y \to X$, we have an exact sesquence

$$\ldots \to H^q_c(X\setminus Y, F) \to H^q_c(X,F) \to H^q_c(Y, i^*F) \to H^{q+1}_c(X\setminus Y, F) \to \ldots .$$

Similary, if $f\colon X \to S$ is a morphism, and

$$X \xrightarrow{j} \overline{X} \xrightarrow{\overline{f}} S$$

a decomposition of f in an open inclusion j and a proper morphism $\overline{f}$, then one may define the sheaves

$$R^q f_!(F) = R^q f_*(j_!F)\,.$$

Again, this definition is independent on the choice of the decomposition, provided the sheaf is constructible. Now given an arbitrary morphism f, we get the base change theorem

$$(R^q f_! F)_\xi = H^q_c(X_\xi, F_\xi)$$

and the spectral sequence

$$E_2^{p,q} = H^p_c(S, R^q f_! F) \Rightarrow H^{p+q}_c(X,F)\,.$$

6.4 Finiteness Theorem. *Let $f\colon X \to S$ be a morphism of finite type, and F a constructible sheaf on X. Then the sheaves $R^q f_! F$ are constructible.*

By standard reductions, as in Sect. 6.2, we can assume that f is a fibering in complete smooth curves, and $F = \mathbb{Z}/n\mathbb{Z}$. Replacing S by an open subset, we can, in addition, assume that f is smooth. Then however, all fibers have the same cohomology (see Theorem of Sect. 5.4).

A similar argument proves that $R^q f_! F = 0$ for $q > 2d$, provided the dimensions of fibers are at most d.

Corollary. *Let X be an algebraic scheme over an algebraically closed field. Then $H^q_c(X, F)$ are finite groups for all q, and equal zero for $q > 2 \dim X$.*

Remark. Similar assertions hold for usual (non-compact) cohomology as well, though the corresponding arguments are more subtle (Grothendieck et al. (1977a), Grothendieck et al. (1972–1973)). Furthermore, if X is an affine scheme, then $H^q(X, F) = 0$ even for $q > \dim X$. The latter is an etale analog of Corollary of Chap. 3, Sect. 2.3; as before, this fact, together with duality, implies the weak Lefschetz theorem on hyperplan sections (see Sect. 7.7 below).

6.5. Comparison with the Classical Cohomology. The preceding results show that the etale cohomology of finite sheaves are "similar" to the classical cohomology. We can make a precise statement for complex schemes. Let F be a constructible sheaf on a $\mathbb{C}$-scheme X, and $F(\mathbb{C})$ the corresponding sheaf on $X(\mathbb{C})$. Then

$$H^q_c(X, F) = H^q_c(X(\mathbb{C}), F(\mathbb{C})) .$$

A similar equality holds for the usual cohomology H^q as well. The proof utilizes, as before, a reduction to smooth complete curves and the sheaf $\mathbb{Z}/n\mathbb{Z}$, in which case Theorem of Sect. 5.4 holds.

6.6 Specialization and Vanishing Cycles. The monodromy plays an important role in the study of complex varieties (Deligne-Katz (1973)). Its definition is based on the observation that a smooth morphism $f\colon X \to S$ is locally trivial from the topological point of view. In the etale topology we do not have the "equality" of fibers, and one may expect only the "equality" of the cohomology of fibers. We would like to compare the cohomology of nearby fibers.

Recall first how this is done in the classical case. Let $X \to S$ be a morphism of $\mathbb{C}$-schemes that is proper and smooth everywhere except for the fiber over a point $s_0 \in S$. By choosing a path Γ from $s \in S$ to s_0, we can construct a specialization map $\Gamma\colon X_s(\mathbb{C}) \to X_{s_0}(\mathbb{C})$ determind up to a homotopy. The corresponding map in cohomology, $\Gamma_*\colon H^*(X_{s_0}) \to H^*(X_s)$, is called a *specialization* too. In case of a smooth family, the specialization is an isomorphism. In general, however, several cycles may vanish, and to describe them one has to take into account higher direct images of the map Γ. One proceeds as follows. Let x be a point of X_s and B a small ball around x in $X(\mathbb{C})$. For any point $s \in S$ that is close to s_0, the intersection $B \cap X_s(\mathbb{C})$ is independent of the choice of B and s, and is called the *variety of vanishing cycles.* Its cohomology are precisely the fibers of $R^q\Gamma_*$.

The same can be done in the abstract case, and even in a somewhat more natural way. Recall that the analog of a small ball is a strict localization (Sect. 3.5). Henceforth, we restrict ourselves to the geometric case, thus assuming k is algebraically closed and the points $s_0 \in S$ and $x \in X$ are closed, i. e. geometric. We denote the corresponding localizations by $\widetilde{S}$ and $\widetilde{X}$

$$\begin{array}{ccccc} x \in X & \leftarrow & \widetilde{X} & \leftarrow & \widetilde{X} \times_{\widetilde{S}} \overline{\eta} \\ {\scriptstyle f}\downarrow & & \downarrow & & \downarrow \\ s_0 \in S & \leftarrow & \widetilde{S} & \leftarrow & \overline{\eta} \end{array}$$

It is most convenient to take a generic point η of S in place of a "nearby" point $s \in S$. Let $\overline{\eta}$ be the geometric point over η.

Definition. An S-morphism $\overline{\eta} \to \widetilde{S}$ is said to be a *specialization* (or *path*) from $\overline{\eta}$ to s_0. The scheme

$$\widetilde{X}_{\overline{\eta}} = \widetilde{X} \times_{\widetilde{S}} \overline{\eta}$$

is said to be a *variety of vanishing cycles* (at x).

We observe that the scheme $\widetilde{X}_{\overline{\eta}}$ over the field $k(\overline{\eta})$ is not algebraic, only a limit of algebraic schemes; however, this is not very essential. The cohomology of $\widetilde{X}_{\overline{\eta}}$ are said to be *vanishing cohomology* (at x). If they are trivial, i. e., $H^0(\widetilde{X}_{\overline{\eta}}, \mathbb{Z}/n\mathbb{Z}) = \mathbb{Z}/n\mathbb{Z}$ and $H^q(\widetilde{X}_{\overline{\eta}}, \mathbb{Z}/n\mathbb{Z}) = 0$ for $q > 0$, then the morphism f is said to be *locally acyclic* at x.

6.7 Acyclicity of Smooth Morphims. The local triviality of a smooth morphism is replacad by the following

Theorem. *Any smooth morphism $f : X \to S$ is locally acyclic.*

The assertion is local in the etale topology, so we may assume that the base S is a strictly local scheme, and $X = \mathbb{A}^N_S$. By induction, we may assume that $N = 1$. Thus we are reduced to the following case: Let A be a strict Henselian local ring, $A\{T\}$ the (strict) Henselization of the polynomial ring $A[T]$, $S = \operatorname{Spec} A$, and $X = \operatorname{Spec} A\{T\}$. We have to verify that the fiber of $X \to S$ over the generic point $\overline{\eta}$ is acyclic.

The fiber $X_{\overline{\eta}}$ is a pro-algebraic curve hence $H^q(X_{\overline{\eta}}) = 0$ for $q > 1$. It remains to verify that $H^0(X_{\overline{\eta}}, \mathbb{Z}/n\mathbb{Z}) = \mathbb{Z}/n\mathbb{Z}$ and $H^1(X_{\overline{\eta}}, \mathbb{Z}/n\mathbb{Z}) = 0$ (we assume that n is invertible in A).

Case H^0. We have to show that $X_{\overline{\eta}}$ is connected. This fiber is equal to $\operatorname{Spec} A\{T\} \otimes \overline{K}$, where $\overline{K}$ is the algebraic closure of the quotient field K of A. Since $\overline{K}$ is a limit of finite extensions K', it will suffice to verify that $A\{T\} \otimes_A K'$ is a domain. Let A' denote the integral closure of A in K'. This is a Hensel ring too. Since $A\{T\} \otimes_A A' = A'\{T\}$, we can replace A by A', and assume that A is normal. Now everything is clear: the ring $A[T]$ and its etale coverings are normal domains, hence its localization $A\{T\} \otimes_A K$ is also a normal domain.

Case H^1. We have to show that $X_{\overline{\eta}}$ has no nontrivial $(\mathbb{Z}/n\mathbb{Z})$-torsors. Suppose there is such a torsor. Replacing $\overline{K}$ by K' and A by A', as above, we can assume that already $X_\eta = \operatorname{Spec}(A\{T\} \otimes_A K)$ has such a torsor. By the Kummer theory, such a covering is given by an equation $Z^n = g$, where $g \in A\{T\}$. Since the covering is unramified over X_η, we get $g = ua$, where u is invertible and $a \in A$. Replacing A by $A' = A[\sqrt[n]{a}\,]$, we obtain the required trivialization of the covering.

(For details, again see (Grothendieck et al. (1972–1973), Grothendieck et al. (1977a))).

6.8. Etale Monodromy. By abuse of notation, we denote by Λ the sheaf or group $\mathbb{Z}/n\mathbb{Z}$, where n is invertible in k. Let $f\colon X \to S$ be a smooth proper morphism; this is the situation where the classical monodromy was defined. Its etale analogue is the assertion that the sheaf $R^q f_*(\Lambda_X)$ is locally constant on S.

Theorem. *In the above situation, every specialization* $\overline{\eta} \to s$ *on* S *yields an isomorphism*

$$H^q(X_s, \Lambda) \xrightarrow{\sim} H^q(X_{\overline{\eta}}, \Lambda)\,.$$

Again, we may assume that S is a strictly local scheme. Consider a cartesian diagram

$$\begin{array}{ccccc} X_{\overline{\eta}} & \xrightarrow{\varepsilon'} & X & \leftarrow & X_s \\ \downarrow & & \downarrow f & & \downarrow \\ \overline{\eta} & \xrightarrow{\varepsilon} & S & \leftarrow & \{s\}\,. \end{array}$$

By the base change theorem of Sect. 6.2, $H^*(X_s, \Lambda) = H^*(X, \Lambda)$, so it remains to verify that $\varepsilon'^*\colon H^*(X, \Lambda) \to H^*(X_{\overline{\eta}}, \Lambda)$ is a bijection. This will follow from the Leray spectral sequence, provided $\varepsilon'^*\Lambda = \Lambda$ and $R^q\varepsilon'_*\Lambda = 0$ for $q > 0$. The fibers of the sheaves in question are, in fact, the vanishing cohomology, which are trivial by Theorem of Sect. 6.7.

A similar approach is used to study the behavior of cohomology of fibers under degenerations. A unit disk Δ, $0 \in \Delta \subset \mathbb{C}$, of the classical theory is replaced by the Hensel "arrow" $S = \operatorname{Spec} k\{T\}$, and the punctered disk $\Delta^* = \Delta\backslash\{0\}$ is replaced by $S^* = S\backslash\{0\} = \operatorname{Spec} k\{T\}\,[T^{-1}]$. The fundamental group $\pi_1(\Delta^*) = \mathbb{Z}$ is replaced by $\pi_1(S^*)$, which is isomorphic to $\widehat{\mathbb{Z}}$ up to a p-torsion. For an exposition of the Picard-Lefschetz theory of the simplest quadratic degenerations that is used in the cohomological study of Lefschetz's pencils, we refer to (Deligne-Katz (1973)).

§ 7. l-Adic Cohomology

In this section, the field k is algebraically closed, and a scheme is an algebraic scheme over k. The cohomology with coefficients in the sheaves $\mathbb{Z}/n\mathbb{Z}$, constructed in the preceeding sections, provide a finite approximation to the classical integral cohomology. To make the approximation more complete, we pass to the limit as $n \to \infty$.

7.1. l-Adic Sheaves. We fix a prime number l invertible in k, i. e. $l \neq$ char k. The projective system of etale sheaves on X

$$\mathbb{Z}/l\mathbb{Z} \leftarrow \mathbb{Z}/l^2\mathbb{Z} \leftarrow \ldots \leftarrow \mathbb{Z}/l^n\mathbb{Z} \leftarrow \ldots$$

yields a projections system of the cohomology groups

$$H^q(X, \mathbb{Z}/l\mathbb{Z}) \leftarrow \ldots \leftarrow H^q(X, \mathbb{Z}/l^n\mathbb{Z}) \leftarrow \ldots .$$

The limit $\varprojlim H^q(X, \mathbb{Z}/l^n\mathbb{Z})$ is symbolically denoted by $H^q(X, \mathbb{Z}_l)$, and is called the *l-adic cohomology* group of X. Note that this is not the etale cohomology of the sheaf $\mathbb{Z}_l$ on X! The l-adic cohomology are modules over the ring $\mathbb{Z}_l = \varprojlim \mathbb{Z}/l^n\mathbb{Z}$ of l-adic integers.

The rings of this type, constructed by Hensel, play an important role in algebraic number theory. They relate zero and positive characteristics. In fact, the residue field $\mathbb{Z}_l/l\mathbb{Z}_l = \mathbb{Z}/l\mathbb{Z}$ is a finite field with l elements, while the quotient field $\mathbb{Q}_l = \mathbb{Z}_l[l^{-1}]$, an l-adic number field, has characteristic zero. We have already utilized this observation in Chap. 2, Sect. 6.3.

We now return to cohomology. In addition to constant sheaves, as $\mathbb{Z}_l$, it is convenient to employ their twisted forms. A projective system of constructible sheaves

$$F = (F_1 \leftarrow F_2 \leftarrow \ldots \leftarrow F_n \leftarrow \ldots)$$

is said to be an *l-adic sheaf* (or *sheaf of $\mathbb{Z}_l$-modules*). Each F_n is a sheaf of modules over $\mathbb{Z}/l^n\mathbb{Z}$, and F_n coincides with $F_{n+1}/l^n F_{n+1} = F_{n+1} \otimes (\mathbb{Z}/l^n\mathbb{Z})$. The $\mathbb{Z}_l$-modules

$$H^q(X, F) = \varprojlim_n H^q(X, F_n)$$

are said to be the cohomology of the sheaf F. If each F_n is locally constant in the sense of Sect. 6.1, the l-adic sheaf F is said to be *locally constant* or *lisse*. Such a sheaf is given by a representation of the fundamental group $\pi_1(X, x)$ in the fiber F_x viewed as a $\mathbb{Z}_l$-module.

In addition to the constant sheaf $\mathbb{Z}_l$, the sheaf

$$\mathbb{Z}_l(1) = (\mu_l \leftarrow \mu_{l^2} \leftarrow \ldots \leftarrow \mu_{l^n} \leftarrow \ldots)$$

plays an important role. It is (non-canonically) isomorphic to the sheaf $\mathbb{Z}_l$. Such a punctiliousness is useful if one would like to control the action of the

Galois group (see Sect. 7.9). We denote the m-th tensor power of $\mathbb{Z}_l(1)$ by $\mathbb{Z}_l(m)$.

We are often concerned with cohomology and sheaves up to torsion only. Then we also speak about $\mathbb{Q}_l$-sheaves, and denote cohomology by $H^q(X, \mathbb{Q}_l)$. We will show that such cohomology satisfy the formal properties of Weil's cohomology. As a rule, the required properties easily follow from similar properties for constructible sheaves.

7.2. Finiteness. The cohomology with coefficients $\mathbb{Z}/l^n\mathbb{Z}$ rather nicely depend on n, so we obtain reasonable objects in the limit as well.

Proposition. *Let $F = (F_n)$ be a sheaf of $\mathbb{Z}_l$-modules such that each F_n is locally free as a sheaf over $\mathbb{Z}/l^n\mathbb{Z}$. Then the $\mathbb{Z}_l$-modules $H^q(X, F)$ are finitely generated, and the universal coefficient formula holds:*

$$0 \to H^q(X, F) \otimes (\mathbb{Z}/l^n\mathbb{Z}) \to H^q(X, F_n) \to H^{q+1}(X, F)_{l^n} \to 0 .$$

A similar statement holds for the cohomology with compact support. We define the q-th *l-adic Betti number* as $\dim_{\mathbb{Q}_l} H^q(X, \mathbb{Q}_l)$, and denote it by $b^q(X)$.

Later, we shall see that the Betti numbers are independent of l (and equal to the topological Betti numbers for schemes over $\mathbb{C}$).

7.3. The Künneth Formula. *Let X and Y be complete varieties. Then*

$$H^*(X \times Y, \mathbb{Q}_l) = H^*(X, \mathbb{Q}_l) \underset{\mathbb{Q}_l}{\otimes} H^*(Y, \mathbb{Q}_l) .$$

A similar formula holds if instead of the coefficients $\mathbb{Q}_l$ we are given $\mathbb{Q}_l$-sheaves F on X and G on Y. We can also drop the assumption that X and Y are complete, and employ the cohomology with complact support.

Of course, this formula follows from a similar formula with coefficients $\Lambda = \mathbb{Z}/l^n\mathbb{Z}$. There are two essential steps in the proof. Let $p\colon X \times Y \to X$ be a projection; in order to apply the Leray spectral sequence, we have to calculate $R^q p_* \Lambda$. First, we deduce from the base change theorem of Sect. 6.2 that $R^q p_* \Lambda = \Lambda_X \otimes H^q(Y, \Lambda)$. Then we pass from the coefficients Λ to the coefficients $H^q(Y)$, which is a typical exercise on universal coefficients.

7.4. Poincaré Duality: Orientation. In the complex case, the duality on a smooth d-dimensional manifold X and even the fundamental class, i. e. the isomorphism $\mathbb{Z} \xrightarrow{\sim} H_{2d}^{\mathrm{BM}}(X)$ or $H_c^{2d}(X) \xrightarrow{\sim} \mathbb{Z}$, depend on an orientation of $\mathbb{C}$. The latter is determined by a choice of root $\sqrt{-1}$ or, in a more fancy terminology, a choice of an isomorphism between $\mathbb{Q}/\mathbb{Z}$ and the group μ_∞ of roots of unity.

As we have seen in Sect. 5.6, if X is a curve, the isomorphism between $H_c^2(X, \mathbb{Z}/n\mathbb{Z})$ and $\mathbb{Z}/n\mathbb{Z}$ becomes canonical after replacing the sheaf $\mathbb{Z}/n\mathbb{Z}$ by the sheaf μ_n. Therefore, in general, an isomorphism

$$H_c^{2d}(X, \mathbb{Z}_l(d)) \xrightarrow{\sim} \mathbb{Z}_l\,,$$

whose existence we are going to establish now, should be regarded as the *fundamental class*. The sheaf $\mathbb{Z}_l(d)$, $d = \dim X$, is said to be the *orientation* sheaf (compare with the dualizing sheaf in the coherent theory of Chap. 2, Sect. 5.6).

Theorem. *For every irreducible d-dimensional variety X, there is a canonical isomorphism*

$$H_c^{2d}(X, \mathbb{Z}_l(d)) \xrightarrow{\sim} \mathbb{Z}_t\,.$$

It is canonical in the sense that it is compatible with open embeddings and finite coverings. If U is an open subscheme of X, then $\dim X\backslash U < \dim X$, hence the map $H_c^{2d}(U, \mathbb{Z}_l(d)) \to H_c^{2d}(X, \mathbb{Z}_l(d))$ is an isomorphism by Sect. 6.3 and 6.4. Thus, in the definition of the fundamental class, we may replace X by any Zariski open piece. Recalling the construction of "nice" neighborhoods in Chap. 3, Sect. 4, we can assume the existence of a proper morphism $f\colon X \to Y$, where Y is a smooth variety of dimension $d-1$. Since $R^q f_* = 0$ for $q > 2$ and $H_c^q(Y) = 0$ for $q > 2d-2$, we get

$$H_c^{2d}(X, \mathbb{Z}_l(d)) = H_c^{2d-2}(Y, R^2 f_* \mathbb{Z}_l(d))\,.$$

It remains to observe that $R^2 f_* \mathbb{Z}_l(1)_X = \mathbb{Z}_{l,Y}$ by Sect. 6.

7.5. Poincaré Duality: Pairing. Again, let X be an irreducible variety of dimension d. The cup product in cohomology defines the pairing

$$H_c^q(X, F) \otimes H^{2d-q}(X, F^\vee) \to H_c^{2d}(X, \mathbb{Z}_l(d)) = \mathbb{Z}_l\,,$$

where $F^\vee = \operatorname{Hom}(F, \mathbb{Z}_l(d))$.

Theorem. *If X is smooth and the sheaf F of $\mathbb{Z}_l$-modules is locally free, then the pairing establishes a duality modulo torsion.*

The theorem is friequently applied to complete smooth varieties, where it gives a duality between $H^q(X, \mathbb{Q}_l)$ and $H^{2d-q}(X, \mathbb{Q}_l)$. However, it is more convenient to prove the "open" version of the theorem. Moreover, in course of the proof, we are forced to deal not only with locally free sheaves but constructible sheaves as well. Therefore, it is more convenient to prove the duality theorem in the form

$$H_c^q(X, F) \times \operatorname{Ext}_x^{2d-q}(F, \Lambda(d)) \to \Lambda\,.$$

Here $\Lambda = \mathbb{Z}/l^n\mathbb{Z}$. We omit the detailed description of Ext; note only that it is the derived functor of the functor $F \mapsto \operatorname{Hom}_X(F, \Lambda)$. In a sense, it is a homology of the sheaf F.

We will describe the scheme of the proof. The main point is to establish that the duality for X and its open subvarieties are equivalent statements (here we employ the induction on the support of the sheaf). Then, by shrinking X, we can assume that F is a locally free sheaf. Passing to an etale covering, we can assume that F is free and even isomorphic to Λ. Finally, employing "nice" neighborhoods, we can assume that X is a fibering of smooth complete curves over a smooth base Y. By induction, we have the duality on Y, as well as on the fibers (Sect. 5). Then it is not too hard to deduce the duality for X (for details, see (Grothendieck et al. (1972–1973), Grothendieck et al. (1977a), Milne (1980))).

7.6. The Gysin Homomorphism. As in the complex case, the duality allows us to furnish the cohomology, which are contravariant by nature, with some covariant features. Let $f\colon X \to Y$ be a proper morphism, where Y is smooth, $d = \dim X$, and $\dim Y = d + \delta$. We define the *Gysin homomorphisms*

$$f_*: H^q(X, \mathbb{Q}_l) \to H^{q+2\delta}(Y, \mathbb{Q}_l(\delta))$$

as Poincaré duals to the maps

$$f^*: H_c^{2d-q}(Y, \mathbb{Q}_l(d)) \to H_c^{2d-q}(X, \mathbb{Q}_l(d))\,.$$

The Gysin homomorphisms are functorial and satisfy the *projection formula*

$$f_*(\alpha \cup f^*(\beta)) = f_*(\alpha) \cup \beta$$

for $\alpha \in H^*(X)$ and $\beta \in H^*(Y)$.

In particular, if X is closed irreducible codimension δ subvariety of a smooth scheme Y, then the image of $1_X \in H^0(X, \mathbb{Q}_l)$ under the Gysin map gives the so-called *class of the subvariety* X, $\mathrm{cl}(X) \in H^{2\delta}(Y, \mathbb{Q}_l(\delta))$. As in the classical case, cl extends to a ring homomorphism which maps the class of a cycle of the Chow ring $A^*(X)$ to the cohomology ring $\oplus_{r \geq 0} H^{2r}(Y, \mathbb{Q}_l(r))$.

For divisors, we get the homomorphism

$$\operatorname{Pic} Y \to H^2(Y, \mathbb{Q}_l(1))\,,$$

which we have already encountered in Kummer's theory (Sect. 4.6). Under this homomorphism, the continuous part of the Picard group, $\operatorname{Pic}^0 Y$, goes to 0. The group $\operatorname{Pic}^0 Y$, in turn, is intimately related to the space $H^1(Y, \mathbb{Q}_l(1))$ (again see Kummer's theory). In particular, if $H^1(Y, \mathcal{O}_Y) = 0$ then $\operatorname{Pic}^0 Y = 0$, and consequently $H^1(Y, \mathbb{Q}_l(1)) = 0$. Conversely, if $H^1(Y, \mathbb{Q}_l(1)) = 0$ then $\operatorname{Pic}^0 Y$ is an infinitesimal group.

We will present a few more corollaries of the above formalism.

7.7. The Weak Lefschetz Theorem. Let X be an n-dimensional projective variety, and $Y \subset X$ a hyperplane section. Since $X \backslash Y$ is affine, $H^i(X \backslash Y) = 0$ for $i > n$ (see Remark in Sect. 6.4). We assume that $X \backslash Y$ is smooth. Then by Poincaré duality, $H_c^i(X \backslash Y) = 0$ for $i < n$, and the long exact sequence

$$\ldots \to H_c^i(X \backslash Y) \to H^i(X) \to H^i(Y) \to H_c^{i+1}(X \backslash Y) \to \ldots$$

(see Sect. 6.3) yields the isomorphism $H^i(X) \xrightarrow{\sim} H^i(Y)$ for $i < n - 1$, and the monomorphism $H^{n-1}(X) \hookrightarrow H^{n-1}(Y)$. If the varieties X and Y are smooth, then the Gysin maps dual to the above maps give the isomorphisms $H^i(Y) \xrightarrow{\sim} H^{i+2}(X)$ for $i \geq n$, and the epimorphism $H^{n-1}(Y) \to H^{n+1}(X)$.

7.8. The Lefschetz Trace Formula. Given an endomorphism $f\colon X \to X$, we denote by $f^*|H^r(X)$ the induced map in the vector space $H^r(X, \mathbb{Q}_l)$. If X is a smooth complete variety, the number of fixed points of f (see Chap. 3, Sect. 1.6) is given by the following Lefschetz formula:

$$(\Gamma_f.\Delta) = \sum_{r \geq 0} (-1)^r \operatorname{Tr}(f^*|H^r(X)).$$

Its proof is similar to the one given in Chap. 3, Sect. 1.6. Using the map cl, we translate the calculations of $(\Gamma_f.\Delta)$ into the ring $H^*(X \times X)$, and then repeat the argument given in Chap. 3, Sect. 1.6.

7.9. Applications to the Zeta Function. In Sect. 7.9 and 7.10, X is assumed to be an algebraic scheme over the finite field $\mathbb{F}_q$, $\overline{X} = X \otimes_{\mathbb{F}_q} \overline{F}_q$ its geometrization, and $\Phi\colon \overline{X} \to \overline{X}$ the Frobenius endomorphism over $\mathbb{F}_q$. Assume that X is smooth and complete of dimension d. As we mentioned in Sect. 1.7, the Lefschetz formula implies that the zeta function is rational

$$Z(X,t) = \prod_{r=0}^{2d} P_r(t)^{(-1)^{r+1}}, \qquad (*)$$

where $P_r(t) = \det(1 - t\Phi^*|H^r(\overline{X}, \mathbb{Q}_l))$. A priory the $P_r(t)$ are polynomials with coefficients in $\mathbb{Q}_l$. However, utilizing the fact that $Z(X,t)$ is a power series with integer coefficients, and the roots of the P_r do not intersect (see Sect. 8), one can prove that each P_r is a polynomial with integer coefficients which are independent of l. The Betti numbers $b^r(X)$ equal to the degrees of the P_r, so they are also independent of l.

It follows from Poincaré duality that $Z(X,t)$ satisfies a functional equation, namely:

$$Z(X, q^{-d}t^{-1}) = \pm q^{\frac{d\chi}{2}} t^{\chi} Z(X,t),$$

where $\chi = \sum(-1)^r b^r(X)$ is the topological Euler characteristic.

More generally, for any (not necessary complete or smooth) variety, we have a decomposition $(*)$, where

$$P_r(t) = \det(1 - t\Phi^*|H^r_c(\overline{X}, \mathbb{Q})).$$

It is convenient to prove an even more general assertion where the constant sheaf $\mathbb{Q}_l$ is replaced by an arbitrary $\mathbb{Q}_l$-sheaf F.

7.10. *L*-Functions. Let F be a $\mathbb{Q}_l$-sheaf on X, and $\overline{F}$ its lifting to $\overline{X}$. Then we have a canonical isomorphism $\overline{F} \simeq \Phi^*\overline{F}$, which allows us to define the Frobenius endomorphism

$$\Phi^* : H_c^r(\overline{X}, \overline{F}) \to H_c^r(\overline{X}, \overline{F}) .$$

Now, we will consider in detail the case when $X = \operatorname{Spec} \mathbb{F}_{q^n}$. Then X consists of a unique point x of degree n over $\mathbb{F}_q$, and $\overline{X}$ consists of n distinct points, $\overline{x}_1, \dots, \overline{x}_n$, that are cyclically permuted by the endomorphism Φ. The sheaf $\overline{F}$ has n fibers – the spaces $F_{\overline{x}_1}, \dots, F_{\overline{x}_n}$, that are permuted together with the points (Fig. 7)

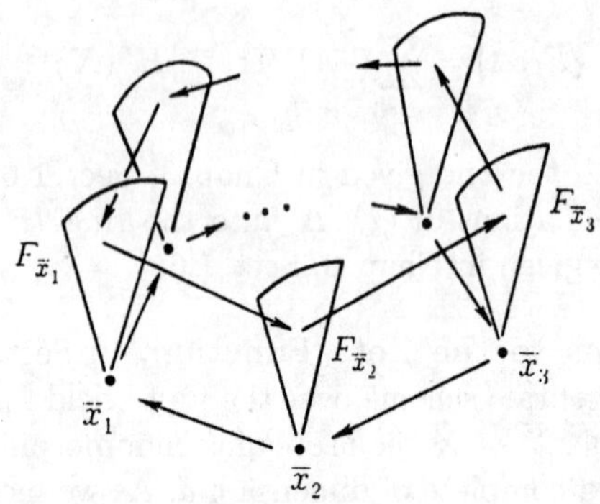

Fig. 7

By iterating this map n times, we obtain an endomorphis of $F_{\overline{x}}$, denoted by Φ_x: $F_{\overline{x}} \to F_{\overline{x}}$. One can verify that Φ_x is inverse to the Frobenius action as an element of the Galois group $\operatorname{Gal}(k(\overline{x})|k(x))$ in the geometric fiber of F (see Example in Sect. 3.2).

Given an arbitrary pair (X, F), where X is a scheme over $\mathbb{F}_q$ and F a $\mathbb{Q}_l$-sheaf on X, we consider the corresponding zeta function (also called an L-function):

$$Z(X, F; t) = \prod_{x \in X} \det(1 - \Phi_x t^{\deg(x)})^{-1} .$$

As in Sect. 1.3, x runs over all closed points of the scheme X, and Φ_x is the endomorphism of the $\mathbb{Q}_l$-space $F_{\overline{x}}$ described above ($\overline{x}$ denotes the geometric point over x). In particular, if $\mathbb{F} = \mathbb{Q}_l$, we get the usual zeta function. Generalizing the formula (*) of Sect. 7.9, Grothendieck obtained the following "integral" representation or cohomological interpretation of the zeta function

$$Z(X, F; t) = \prod_{r=0}^{2 \dim X} \det(1 - t\Phi^* | H_c^r(\overline{X}, \overline{F}))^{(-1)^{r+1}} .$$

This formula is just a formal consequence of the *generalized trace formula*

$$\sum_{\overline{x} \in \operatorname{Fix} \Phi} \operatorname{Tr}(\Phi_{\overline{x}} | F_{\overline{x}}) = \sum_r (-1)^r \operatorname{Tr}(\Phi^* | H_c^r(\overline{X}, \overline{F}))$$

(of course, there is a similar formula for Φ^n). (Its proof can be found in (Grothendieck et al. (1972–1973), Grothendieck et al. (1977a), Milne (1980)).)

For suitable X and F, it is possible to interpret the left-hand side of the previous formula as trigonometric sums (see Grothendieck et al. (1977a)). Concerning estimates of the right-hand side, see the next section.

§ 8. Deligne's Theorem

Let X be a variety over a finite field. Then $H^*(\overline{X}, \mathbb{Q}_l)$ are not just vector spaces over $\mathbb{Q}_l$. These cohomology admit the action of the Frobenius endomorphism. A theorem of Deligne provides a solution of the eigenvalue problem for this endomorphism, and, in particular, establishes the Riemann part of Weil's conjectures (Sect. 1.6). To state assertions, it is convenient to introduce a notion of weight. Then, roughly speaking, the r-dimensional cohomology have weight r, in complete analogy with the classical case (see Chap. 3, Sect. 3). Furthermore, many results on the classical cohomology may be presented as consequences of results on the cohomology of varieties over finite fields.

8.1. Weights. As for any endomorphism of a vector space over $\mathbb{Q}_l$, the eigenvalues of the Frobenius endomorphism are algebraic numbers over $\mathbb{Q}_l$, i. e. elements of the algebraic closure $\overline{\mathbb{Q}}_l$. A number $\alpha \in \overline{\mathbb{Q}}_l$ is said to be *pure*, if the absolute value of the complex number $\tau\alpha$, $|\tau\alpha|$, is independent of the choice of an embedding $\tau\colon \overline{\mathbb{Q}}_l \to \mathbb{C}$; we then denote this number by $|\alpha|$. Of course, pure numbers are very special; for instance, they are algebraic over $\mathbb{Q}$. Given a positive integer $N \in \mathbb{Z}$, the real number

$$w(\alpha) = w_N(\alpha) = 2\log_N |\alpha|$$

is said to be the *weight* of a (pure) number $\alpha \in \overline{\mathbb{Q}}_l$ with respect to the base N. Now, let X be a scheme over $\mathbb{F}_q$, and F an l-adic sheaf on X (l is prime to q). Then the Frobenius endomorphism Φ acts on the sheaf $\overline{F}$ over $\overline{X} = X \otimes_{\mathbb{F}_q} \overline{\mathbb{F}}_q$ (Sect. 7.10). At the (geometric) point $\overline{x}$ over $x \in X$, the endomorphism Φ_x: $F_{\overline{x}} \to F_{\overline{x}}$ is inverse to the action of the Frobenius element of the Galois group $\mathrm{Gal}(k(\overline{x})|k(x))$ (see Sect. 1.1 and 7.10).

Definition. The sheaf F is said to be *pure of weight* r, if for every point $x \in X$, all the eigenvalues of the Frobenius endomorphism $\Phi_x\colon F_{\overline{x}} \to F_{\overline{x}}$ are pure numbers of weight r with respect to the base $N(x) = \mathrm{Card}(k(x))$, the "absolute value" of the point x.

Example. Clearly the trivial sheaf $\mathbb{Q}_l$ is pure of weight 0, since the Frobenius acts trivially on its fibers.

The sheaves $\mathbb{Q}_l(1)$ and $\mathbb{Z}_l(1)$ are more interesting. The latter is the projective limit of the sheaves μ_{l^n} of roots of unity (Sect. 7.1). The Frobenius acts by the formula $\alpha \mapsto \alpha^q$ or, in additive notation, as multiplication by q.

Therefore, the endomorphism Φ_x, which is inverse to the Frobenius, acts on $\mathbb{Q}_l(1)_{\bar{x}}$ as multiplication by $N(x)^{-1}$. Hence $\mathbb{Q}_l(1)$ is a sheaf of weight -2.

It is clear how the weights behavior when we take the tensor product of pure sheaves, pass to the dual sheaf, etc. However, a sum of pure sheaves of different weights is not a pure sheaf. So, it is convenient to introduce the notion of *mixed sheaf* as a filtered sheaf whose factors are pure sheaves. Deligne assumes that any sheaf is mixed. The weights of pure factors of a mixed sheaf are said to be its weights.

8.2. Main Theorem. *Let $f\colon X \to C$ be a morphism of schemes over a finite field (or $\mathbb{Z}$), and F a mixed sheaf on X with weights $\leq n$. Then for every r, the sheaf $R^r f_!(F)$ on Y is mixed with weights $\leq n+r$.*

Corollary 1. *Let F be a mixed sheaf on X with weights $\leq n$. Then the eigenvalues of the Frobenius on $H^r_c(\bar{X}, \bar{F})$ are pure numbers of weights $\leq n+r$.*

Compare with Chap. 3, Sect. 3.3. By duality we obtain

Corollary 2. *Let X be a smooth variety, and F a mixed lisse sheaf with weights $\leq n$. Then the eigenvalues of the Frobenius on $H^r(\bar{X}, \bar{F})$ are pure numbers of weights $\leq n+r$.*

Compare with Chap. 3, Sect. 3.4. If X is a smooth complete variety, and $F = \mathbb{Q}_l$, this establishes the Riemann part of Weil's conjectures:

Theorem. *Let X be a smooth complete variety over $\mathbb{F}_q$. Then the Frobenius characteristic polynomial* $\det(1\cdot t - \Phi^* \,|\, H^r(X, \mathbb{Q}_l))$ *has integer coefficients, and the absolute values of all its roots equal $q^{r/2}$.*

In other words, as a sheaf over $\operatorname{Spec} \mathbb{F}_q$, $H^r(X, \mathbb{Q}_l)$ has weight r. In Sect. 1.3, we have described the significance of this statement and its applications to estimations of integral points. We will mention only two examples.

Example 1. Let $X \subset \mathbb{P}^{n+r}$ be a smooth complete intersection of dimension n over $\mathbb{F}_q$. Then

$$\| X(\mathbb{F}_q)| - |\mathbb{P}^n(\mathbb{F}_q)\| \leq b q^{n/2},$$

where b is the n-th Betti number of X.

Indeed, the cohomology of the complete intersection coincide with the cohomology of $\mathbb{P}^n$ in all dimensions but the middle one (see Sect. 7.7). The term $bq^{n/2}$ corresponds to the contribution of $H^n(X)$.

Example 2. Let Q be a polynomial of degree d in n variables over $\mathbb{F}_q$, and $\Psi\colon \mathbb{F}_q \to \mathbb{C}^*$ a nontrivial additive character. We assume that d is prime to q, and the hypersurface in $\mathbb{P}^{n-1}$ given by the equation $Q_d = 0$ is smooth, where Q_d is the homegeneous part of Q of degree d. Then the trigonometric sum

$$\Sigma = \sum_{x_1, \ldots, x_n \in F_q} \psi(Q(x_1, \ldots, x_n))$$

admits an estimate $|\Sigma| \leq (d-1)^n q^{n/2}$.

Again, the point is that the cohomology $H_c^*(\mathbb{A}^n)$ (with coefficients in a suitable sheaf F – see the trace formula in Sect. 7.10) are nontrivial in dimension n only. The space $H_c^n(\mathbb{A}^n F)$ has weight n and dimension $(d-1)^n$.

(Concerning p-adic estimates of the Frobenius action, see (Mazur (1975)).)

8.3. Outline of Proof. (**A**) Standard reductions, like fibering in curves, allows us to reduce the general statement of Sect. 8.2 to the following assertion about curves over $\mathbb{F}_q$:

Let X be a smooth complete curve, $j: U \to X$ an open inclusion, and F a lisse sheaf on U of weight n. Then the space $H^r(X, j_*F)$ has pure weight $n+r$.

By duality, the key case is that of H^1.

(**B**) We are interested in the sheaf j_*F but we have some information on the weight of F. An important point is that the weights of j_*F at the points $s \in X \backslash U$ are at most n (and even equal $n-k$, where k is a non-negative integer).

Here we employ a typical trick. It follows from general considerations that the weight of $(j_*F)_s$ (denoted by t) is at most $n+2$; moreover, this is true for any sheaf F. Applying this estimate to the sheaf $F^{\otimes k}$, we obtain the inequality $k \cdot t \leq k \cdot n + 2$, whence $t \leq n$.

(**C**) From now on, it is convenient to assume that F is pure of weight 0. Since the sheaf j_*F only slightly differ from $f_!F$, we are reduced to proving the following assertion: under the assumptions of (**A**), the weights of $H_c^1(U, F) = H^1(X, f_!F)$ are at most 1.

(**D**) Utilizing the ideas of Hadamard and de la Vallée Poussin, one can establish that the weights of $H_c^1(U, F)$ are *strictly less than* 2, which is the key point. We, however, must prove that the weights are less than or equal to 1. Again, we employ a trick similar to (**B**).

Suppose we know that the weights of the space $H_c^2(U \times U, F \boxtimes F)$ are less than $2+\varepsilon$, where $\varepsilon > 0$. Then by the Künneth formula, the weights of $H_c^1(U, F)$ are less than $1 + \varepsilon/2$.

(**E**) To investigate $H_c^2(U \times U, F \boxtimes F) = H^2(X \times X, j_!F \boxtimes j_!F)$, we consider a Lefschetz pencil on $X \times X$. Strictly speaking, we consider a fibering of the blowing-up of $X \times X$ along the axis of the pencil, however, this does not matter at all. Let $f: X \times X \to \mathbb{P}^1$ be our pencil, and $G = j_!F \boxtimes j_!F$.

Now, assume we have proved that the sheaf $R^1 f_*G$ has weights ≤ 1 (see (**F**) below). Then it follows from (**D**) that the weights of $H^1(\mathbb{P}^1, R^1 f_*G)$ are less than 3. Since this space essentially coincides with $H_c^2(U \times U, F \boxtimes F)$, the weights of $H_c^2(U \times U, F \boxtimes F)$ are less than 3. Then by a remark in (**D**), the weights of $H_c^1(U, F)$ are less than $1 + 1/2$, and this is true for any sheaf F of weight 0. Applying this improved estimate to the cohomology of $R^1 f_*G$ (whose weights are at most 1), we get that the weights of $H^1(\mathbb{P}^1, R^1 f_*G)$ (or $H_c^2(U \times U, F \boxtimes F)$) are less than $1 + 1/4$, and thereafter $1 + 1/8$, $1 + 1/16$, and so on. Finally, we conclude that the weights of $H_c^1(U, F)$ are at most 1.

(F) It remains to estimate the weights of $R^1 f_* G$ at general points of $\mathbb{P}^1$. An element of the fiber $(R^1 f_* G)_t$ over such a point $t \in \mathbb{P}^1$ (i. e., an element of $H^1(f^{-1}(t), G_t))$ either vanishes in some special fiber, or does not vanish at all. The latter elements form a lisse subsheaf of $R^1 f_* G$, and do not contribute to $H^1(\mathbb{P}^1, R^1 f_* G)$ because $\mathbb{P}^1$ is simply connected. As to the former elements, it follows from the local calculations in the fibers where they vanish that those elements have integer weights. Those weights are strictly less than 2 (see **(D)**), hence the weights are essentially at most 1, and we can argue as in **(E)**. (For a detailed proof, see Deligne (1980); a proof for a smooth complete X is given in (Deligne (1974c), Katz (1976)).)

8.4. Geometric Applications. In addition to natural applications to varieties over finite fields, Deligne gives several geometric corollaries of his Main Theorem for varieties over an arbitrary algebraically closed field k. In Chap. 2, Sect. 6.3, we discussed the principle of reduction to finite fields.

Semisimplicity Theorem. *Let $f: X \to Y$ be a smooth proper morphism of k-schemes, where Y is a normal variety. Then the sheaves $R^r f_*(\mathbb{Q}_l)$ on Y are semisimple.*

This means that the representation of $\pi_1(Y)$ in the fiber $H^r(X_Y, \mathbb{Q}_l)$ of that sheaf are semisimple (or completely reducible). This statement can be reduced to schemes over a finite field $\mathbb{F}_q$. By the theorem of Sect. 8.2, we know that $R^r f_*(\mathbb{Q}_l)$ is pure of weight r. Therefore, it will suffice to establish the following fact for varieties over $\mathbb{F}_q$: Given a smooth scheme Y over $\mathbb{F}_q$, and a pure lisse $\mathbb{Q}_l$-sheaf F on Y, then the sheaf $\overline{F}$ on $\overline{Y} = Y \otimes_{\mathbb{F}_q} \overline{\mathbb{F}}_q$ is semisimple.
To verify this, the $\overline{F}'$ denote the maximal semisimple subsheaf of $\overline{F}$ (a sum of simple subsheaves). It is invariant under the Frobenius action hence comes from a suitable subsheaf $F' \subset F$. Then F is an extension of $F'' = F/F'$ by F', given by an element of $\mathrm{Ext}^1(F'', F') = H^1(Y, \mathcal{H}om(F'', F'))$. Since F' and F'' have the same weights, the sheaf $\mathcal{H}om(F'', F')$ has weight 0, and the space $H^1(\overline{Y}, \mathcal{H}om)$ has weights ≥ 1 by Sect. 8.2. Therefore it does not contain invariant elements, the extension is trivial, and $F'' = 0$.

8.5. The Hard Lefschetz Theorem. This is yet another geometric corollary. Let X be a projective variety over k, $Y \subset X$ a hyperplane section, and η the class of Y in $H^2(X, \mathbb{Q}_l)$ (see Sect. 7.6; here we do not distinguish between $\mathbb{Q}_l$ and $\mathbb{Q}_l(1)$).

Theorem. *Let X be a smooth projective variety of dimension n. Then, for any $r \geq 0$, the multiplication by η^r*

$$\eta^r : H^{n-r}(X, \mathbb{Q}_l) \to H^{n+r}(X, \mathbb{Q}_l)$$

is an isomorphism.

One may obtain the multiplication by $\eta = \mathrm{cl}(Y)$ as a composition of i^*: $H^*(X) \to H^*(Y)$ and the Gysin homomorphism i_*: $H^*(Y) \to H^{*+2}(X)$.

Therefore, for the multiplication by η^r, we get a decomposition

$$H^{n-r}(X) \xrightarrow{i^*} H^{n-r}(Y) \xrightarrow{\eta^{r-1}} H^{n+r-2}(Y) \xrightarrow{i_*} H^{n+r}(X).$$

If $r \geq 2$, then i^* and i_* are isomorphisms by the weak Lefschetz theorem (Sect. 7.7), while η^{r-1} is an isomorphism by induction. So, it remains to consider the crucial case when $r = 1$. The composition of injection $H^{n-1}(X) \to H^{n-1}(Y)$ and the surjection $H^{n-1}(Y) \to H^{n+1}(X)$ is a bijection if and only if the pairing

$$H^{n-1}(Y) \times H^{n-1}(Y) \to H^{2n-2}(Y) = \mathbb{Q}_l$$

is nondegenerate on the image of $H^{n-1}(X)$.

For an interpretation of the image of $H^{n-1}(X)$ in $H^{n-1}(Y)$, we consider a Lefschetz pencil (Y_t), $t \in \mathbb{P}^1$, with Y as its member. Let $S \subset \mathbb{P}^1$ be a finite set of degenerate fibers. Then $H^{n-1}(Y_t)$ are fibers of the lisse sheaf $R^{n-1}f_*\mathbb{Q}_l$ on $\mathbb{P}^1\setminus S$, where $f\colon \widetilde{X} \to \mathbb{P}^1$ is the corresponding pencil. Applying the induction hypothesis to Y, one can show that the space $H^{n-1}(X)$ coincides with the subspace $H^{n-1}(Y)^{\pi_1(\mathbb{P}^1\setminus S)}$ of invariants in $H^{n-1}(Y)$ (see (Deligne-Katz (1973)) and Sect. 8.6).

According to Sect. 8.4, the representation of $\pi_1(\mathbb{P}^1\setminus S)$ on $H^{n-1}(Y)$ is completely reducible and compatible with the pairing. Let W denote a complement of $H^{n-1}(X)$ in $H^{n-1}(Y)$. It does not have trivial quotient spaces. The cup product with an arbitrary element of $H^{n-1}(X)$ determines an invariant linear functional on W, which is trivial by the preceding observation. It follows that W is orthogonal to $H^{n-1}(X)$, i. e., the restriction of the pairing to $H^{n-1}(X)$ is nondegenerate.

Corollary. *Let X be a smooth projective variety over the field k. Then the Betti numbers $b^r(X)$ are even for r odd, and positive for r even, $0 \leq r \leq 2\dim X$.*

This is probably true for arbitrary smooth complete varieties.

8.6. Theorem on Invariant Subspace. Let $f\colon X \to S$ be a smooth proper morphism of k-varieties. As in the classical case, it is possible to deduce the degeneration at E_2 of the Leray spectral sequence

$$E_2^{pq} = H^p(S, R^q f_*\mathbb{Q}_l) \Rightarrow H^{p+q}(X, \mathbb{Q}_l)$$

from the hard Lefschetz theorem (compare with Chap. 3, Sect. 3.5).

In particular, the following canonical map is surjective:

$$H^r(X, \mathbb{Q}_l) \to H^0(S, R^r f_*\mathbb{Q}_l) = H^r(X_s, \mathbb{Q}_l)^{\pi_1(S,s)},$$

where s is a closed point of S, and $X_s = f^{-1}(s)$ the fiber over s. Arguing as in the complex case and taking into account weights, we obtain the following

Theorem. *With the above notation and assumptions, let* $\overline{X}$ *be a smooth compactification of* X. *Then the image of* $H^r(\overline{X}, \mathbb{Q}_l)$ *in* $H^r(X, \mathbb{Q}_l)$ *coincides with the subspace* $H^r(X_s, \mathbb{Q}_l)^{\pi_1(S,s)}$ *of invariant elements.*

There is also a local version of the theorem (Deligne (1980)): Let $f: X \to S$ be a proper morphism, $s \in S$ a closed point, and $\eta \in S$ a generic point. Assume that f is smooth over η. Then, for every specialization of $\overline{\eta} \to s$ (Sect. 6.6), the specialization homomorphism

$$H^*(X_s) \to H^*(X_{\overline{\eta}})^{\mathrm{Gal}(\overline{\eta}|\eta)}$$

is surjective.

Thus, the l-adic cohomology of algebraic varieties over algebraically closed fields have the same properties as the classical cohomology of complex varieties. Furthermore, the results on classical cohomology mentioned before, including the theory of weights in Chap. 3, Sect. 3, can be deduced from the corresponding l-adic statements. As observed by Tate, the l-adic cohomology are even better than the classical cohomology in one instance, namely, they admit an action of the Galois group of the ground field. We will briefly consider one conjecture related to this phenomenon.

8.7. Tate's Conjecture. Now, we will leave the geometric setting (the ground field k is algebraically closed) and turn to the arithmetic one (the ground field K is finitely generated over its prime subfield). Let X be a complete smooth algebraic variety over K. Let $\overline{K}$ denote the algebraic closure of K. The Galois group $G = \mathrm{Gal}(\overline{K}|K)$ acts on the space $H^*(\overline{X}, \mathbb{Q}_l(m))$. Recalling the definition of the class of an algebraic cycle (see Sect. 7.6), we conclude that the class of any algebraic cycle of codimension r defined over K is invariant with respect to the action of G on the space $H^{2r}(\overline{X}, \mathbb{Q}_l(r))$. In other words, the subspace of $H^{2r}(\overline{X}, \mathbb{Q}_l(r))$ generated by algebraic cycles of codimension r lies in the subspace $H^{2r}(\overline{X}, \mathbb{Q}_l(r))^G$. The Tate conjecture states that both space, in fact, coincide (Tate (1965)).

This conjecture is, in a sense, analogous to the classical Hodge conjecture, though direct relations are yet to be found. More precise relations exist between those conjectures and Grothendieck's standard conjectures on algebraic cycles (Grothendieck (1969), Kleiman (1968)).

For the Tate conjectures on relations between the rank of the group of algebraic cycles and the orders of zeros and poles of the ζ-function of X, we refer to (Tate (1965)).

Bibliography

A general account of homological algebra is given in (Cartan-Eilenberg (1956), Grothendieck (1957)). For an up to date account, see (Gelfand-Manin (1988)). Unsurpassed introduction to the subject is the book (Godement (1958)). The cohomology of sheaves are discussed in (Gunning-Rossi (1965), Golovin (1986), Griffiths-Harris (1978), Hartshorne (1977), Onishchik (1986), Serre (1955), Shafarevich (1986)).

The cohomology of coherent sheaves were introduced in (Serre (1955)). The most accessible exposition of this theory is presented in (Hartshore (1977)), while the most comprehensive exposition in (Grothendieck-Dieudonné (1962–1963)). See also (Mumford (1966, 1970), Serre (1955)). More special topics dealing with the Riemann-Roch theorem are discussed in (Fulton (1984), Fulton-Lang (1985), Grothendieck et al. (1971), Hirzebruch (1966), Manin (1969)). There is no satisfactory exposition of duality, in part, because it requires the use of derived categories; however, see (Gelfand-Manin (1988), Golovin (1986), Griffiths-Harris (1978), Hartshorne (1966)).

There is no good exposition of the topology of complex algebraic varieties in spite of the classical nature of the subject. The exposition is often restricted to smooth varieties (see Griffiths-Harris (1978), Shafarevich (1988)); the general case is discussed in (Fulton (1984), Fulton-MacPherson (1981)). Complex algebraic varieties are studied in the context of analytic varieties in (Gunning-Rossi (1965), Griffiths-Harris (1978)), see also Onishchik (1986). A comparison of algebraic and analytic approaches was initiated in (Serre (1956)); see also (Grothendieck (1966, 1971)). The articles (Deligne (1971, 1974a)) are devoted to the Hodge structures in the cohomology of algebraic varieties; see also a survey article (Palamodov (1986)).

As we mentioned before, the etale cohomology theory was motivate by the Weil conjectures (Weil (1949)); for a historical account, see (Hartshorne (1977), Katz (1976), Milne (1980)). Finite fields and equations over finite fields are treated in (Lidl-Neiderreiter (1983)); this book also contains a large bibliography. An excellent survey on the arithmetic of algebraic curves is presented in (Mazur (1986)). The best exposition of the etale cohomology can be found in (Grothendieck et al. (1977a)); see also (Grothendieck et al. (1972–1973), Manin (1965), Milne (1980)). For a proof of the Weil conjectures, see (Deligne (1974c, 1980), Katz (1976)).

References[1]

Artin, M. [1969]: Algebraic Approximation of Structures over Complete Local Rings. Publ. Math. Inst. Haut. Étud. Sci. *36* (1969) 23–58, Zbl. 181, 488

Artin, M. [1971]: Algebraic Spaces. Yale Univ. Press: New Haven, 1971, Zbl. 226.14001

Beilinson, A. A. [1984]: Higher Regulators and Values of L-Functions. Itogi Nauki Tekh. VINITI, Ser. Sovrem. Probl. Mat. *24* (1984) 181–238. English transl.: J. Sov. Math. *30* (1985) 2036–2070, Zbl. 588.14013

Beilinson, A. A. [1986]: Notes on Absolute Hodge Cohomology. Contemp. Math. *55*, Part I (1986) 35–68, Zbl. 621.14011

Beilinson, A. A., MacPherson, R., Schechtman, V. [1987]: Notes on Motivic Cohomology. Duke Math. J. *54* (1987) 679–710, Zbl. 632.14010

Berthelot, P. Ogus, A. [1978]: Notes on Crystalline Cohomology. Princeton Univ. Press, Princeton, 1978, Zbl. 383.14010

Bloch, S. [1985]: Algebraic Cycles and Values of L-Functions II. Duke Math. J. *52* (1985) 379–397, Zbl. 628.14006

Borevich, Z. I., Shafarevich, I. R. [1985]: Number Theorey. Nauka: Moscow, 1985, 3rd compl. ed., Zbl. 592.12001. English transl.: Academic Press: New York 1966

Cartan, H., Eilenberg, S. [1956]: Homological Algebra. Princeton Univ. Press: Princeton, 1956, Zbl. 75, 243

Cassels, J. W. S., Fröhlich, A. (eds.) [1967]: Algebraic Number Theory. Academic Press: London, 1967, Zbl. 153, 74

Danilov, V. I. [1988]: Algebraic Varieties and Schemes. Itogi Nauki Tekh. (VINITI), Ser. Sovrem. Probl. Mat., Fundam. Napravleniya *23* (1988) 172–302. Engl. transl.: Encycl. Math. Sci. *23*. Springer-Verlag: Berlin Heidelberg New York 1994, 167–297, Zbl. 682.14002

Deligne, P. [1971]: Théorie de Hodge II. Publ. Math. Inst. Haut. Étud. Sci. *40* (1971) 5–57, Zbl. 223.14007

Deligne, P. [1974a]: Théorie de Hodge III. Publ. Math. Inst. Haut Étud. Sci. *44* (1974) 5–77, Zbl. 292.14005

Deligne, P. [1974b]: Poids dans la cohomologie des variétés algébriques. Actes Congrès Int. Math., Vancouver 1974, Vol. 1 (1975), 79–85, Zbl. 334.14011

Deligne, P. [1974c]: La conjecture de Weil. I. Publ. Math. Inst. Haut. Étud. Sci. *43* (1974) 273–307, Zbl. 287.14001

Deligne, P. [1980]: La conjecture de Weil. II. Publ. Math. Inst. Haut. Étud. Sci. *52* (1980) 137–252, Zbl. 456.14014

Deligne, P., Griffiths, Ph., Morgan, J., Sullivan, D. [1975]: Real homotopy theory of Kähler manifolds. Invent. Math. *29* (1975) 245–274, Zbl. 312.55011

Deligne, P., Illusie, L. [1987]: Relèvements modulo p^2 et décomposition de complex de De Rham. Invent. Math. *89* (1987) 247–270, Zbl. 632.14017

Deligne, P., Katz, N. [1973]: Groupes de monodromie en géométrie algébriques. Lecture Notes in Math. 340, Springer-Verlag: Berlin Heidelberg New York, 1973. 1–38 (Zbl. 272.14003), 39–61 (Zbl. 265.14007), 62–81 (Zbl. 266.14019), 82–115 (Zbl. 266.14008), 116–164 (Zbl. 266.14009), 165–196 (Zbl. 266.14010), 197–211 (Zbl. 266.14006), 212–253 (Zbl. 284.14006), 254–327 (Zbl. 284.14007), 328–340

[1] For the convenience of the reader, references to review in Zentralblatt für Mathematik (Zbl.), compiled using the MATH database, and Jahrbuch über die Fortschritte der Mathematik (Jbuch) have, as far as possible, been included in this bibliography.

(Zbl. 269.14019), 341–362 (Zbl. 271.14002), 363–400 (Zbl. 288.14001), 401–438 (Zbl. 275.14015)

Dold, A. [1972]: Lectures on Algebraic Topology. Springer-Verlag: Berlin Heidelberg New York, 1972, Zbl. 234.55001

Esnault, H., Viehweg, E. [1986]: Logarithmic De Rham Complexes and Vanishing Theorems. Invent. Math. *86* (1986) 161–194, Zbl. 603.32006

Faltings, G. [1983]: Endlichkeitssätze für abelsche Varietäten über Zahlenkörpern. Invent. Math. *73* (1983) 349–366. Erratum: 75 (1984), 381, Zbl. 588.14026

Fulton, W. [1984]: Intersection Theory. Springer-Verlag: Berlin Heidelberg New York, 1984. Ergebn. Math. Grenzgebiete 3. Folge, Bd. 2, Zbl. 541.14005

Fulton, W., Lang. S. [1985]: Riemann-Roch Algebra. Springer-Verlag: Berlin Heidelberg New York, 1985. Grundlehren Math. Wiss. 277, Zbl. 579.14011

Fulton, W., MacPherson, R. [1981]: Categorical Framework for the Study of Singular Spaces. Mem. Amer. Math. Soc. *243*. A.M.S.: Providence, RI, 1981, Zbl. 467.55005

Gabriel, P., Zisman, M. [1967]: Calculus of Fractions and Homotyp Theory. Springer-Verlag: Berlin Heidelberg New York, 1967, Zbl. 186, 568

Gelfand, S. I., Manin, Yu. I. [1988]: Methods of Homological Algebra. Vol. 1: Introduction to Cohomology Theory and Derived Categories. Nauka: Moscow, 1988 (Russian), Zbl. 668.18001

Godement, R. [1958]: Topologie algébrique et théorie des faisceaux. Hermann: Paris, 1958, Zbl. 80, 162

Golovin, V. D. [1986]: Homology of Analytic Sheaves and Duality Theorems. Nauka: Moscow, 1986, Zbl. 585.32005. English transl: Contemporary Soviet Mathematics. Consultants Bureau: New York, 1989

Goresky, M., MacPherson, R. [1980]: Intersection Homology Theory. Topology *19* (1980) 135–165, Zbl. 448.55004

Griffiths, Ph., Harris, J. [1978]: Principles of Algebraic Geometry. A Wiley-Interscience Publication: New York, 1978, Zbl. 408.14001

Grothendieck, A. [1957]: Sur quelques points d'algèbre homologique. Tohoku Math. J., II. Ser. *9* (1957) 119–221, Zbl. 118, 261

Grothendieck, A. [1966]: On the De Rham Cohomology of Algebraic Varieties. Publ. Math. Inst. Hautes Étud. Sci. *29* (1966) 351–359, Zbl. 145, 176

Grothendieck, A. [1968a]: Crystals and the de Rham cohomology of schemes. In: Dix Exposés sur la Cohomologie des Schémas. North-Holland: Amsterdam, 1968, Adv. Stud. Pure Math. *3*, 306–358, Zbl. 215, 371

Grothendieck, A. [1968b]: Cohomologie locale des faisceaux cohérents et Théorèmes de Lefschetz locaux et globaux. North-Holland: Amsterdam, 1968, Zbl. 197, 472

Grothendieck, A. [1969]: Standard Conjectures on Algebraic Cycles. In: Algebraic geometry (Bombay Colloquium 1968). Oxford University Press: London, 1969, 193–199, Zbl. 201, 233

Grothendieck, A. [1971]: Revêtements Étales et Groupe Fondemental. Lecture Notes in Math. 224, Springer-Verlag: Berlin Heidelberg New York, 1971, Zbl. 234.14002

Grothendieck, A. et al. [1971]: (with Berthelot, P., Illusie, L.) Théorie des Intersections et Théorème de Riemann-Roch. Lecture Notes in Math. 225, Springer-Verlag: Berlin Heidelberg New York, 1971, Zbl. 218.14001 (entire collection)

Grothendieck, A. et al. [1972–1973]: (with Artin, M., Verdier, J. L.) Théorie des Topos et Cohomologie Étale des Schémas. Lecture Notes in Math. 269, 270, 305, Springer-Verlag: Berlin Heidelberg New York, 1972–1973, Zbl. 234.00007; Zbl. 237.00012; Zbl. 245.00002

Grothendieck, A. et al. [1977a]: (with Deligne, P., ed., Boutot, J. F., Illusie, L., Verdier, J. L.) Cohomologie Étale. Lecture Notes in Math. 569, Springer-Verlag: Berlin Heidelberg New York, 1977, Zbl. 345.00010

Grothendieck, A. et al. [1977b]: (with Illusie, L., ed.) Cohomologie l-adique et Functions L. Lecture Notes in Math. 589, Springer-Verlag: Berlin Heidelberg New York, 1977, Zbl. 345.00011

Grothendieck, A., Dieudonné, J. [1962–1963]: Eléments de géométrie algébrique III. Étude cohomologique des faisceaux cohérents. Publ. Math., Inst. Haut. Étud. Sci. *11* (1962), 349–511 and *17* (1963), 137–223, Zbl. 118,362; Zbl. 122,161

Grothendieck, A., Dieudonné, J. [1967]: Eléments de géométrie algébrique IV. Étude locale des schémas et des morphismes de schémas. Publ. Math. Inst. Haut. Étud. Sci. *32* (1967) 361 pp., Zbl. 153,223

Gunning, R. C., Rossi, H. [1965]: Analytic Functions of Several Variables. Prentice-Hall: Englewood Cliffs, 1965 (Rev. Version, Wadsworth & Brooks: Florence/KY 1990), Zbl. 141,86

Hartshorne, R. [1966]: Residues and Duality. Lecture Notes in Math. 20, Springer-Verlag: Berlin Heidelberg New York, 1966, Zbl. 212,261

Hartshorne, R. [1976]: On the De Rham Cohomology of Algebraic Varieties. Publ. Math. Inst. Haut. Étud. Sci. *45* (1976) 5–99, Zbl. 326.14004

Hartshorne, R. [1977]: Algebraic Geometry. Graduate Texts in Mathematics *52*. Springer-Verlag: New York Berlin Heidelberg, 1977 (3rd print. 1983), Zbl. 367.14001

Hironaka, H. [1975]: Triangulation of Algebraic Sets. In: Algebraic Geometry (Arcata 1974). Amer. Math. Soc. Proc. Symp. Pure Math. *29* (1975) 165–185, Zbl. 332.14001

Hirzebruch, F. [1966]: Topological Methods in Algebraic Geometry. 3rd ed., Springer-Verlag: Berlin Heidelberg New York, 1966, Zbl. 138,420

Horrocks, G., Mumford, D. [1973]: A Rank 2 Vector Bundle on $\mathbb{P}^4$ with 15000 Symmetries. Toplogy *12* (1973) 63–81, Zbl. 255.14017

Illusie, L. [1979]: Complex de De Rham-Witt et cohomologie cristalline. Ann. Sci. Ec. Norm. Supér., IV. Sér. *12* (1979) 501–661, Zbl. 436.14007

Katz, N. [1976]: An Overview of Deligne's Proof of the Riemann Hypothesis for Varieties over Finite Fields (Hilbert's Problem 8). In: Math. Develop. Arising from Hilberg Probl., De Kalb 1974, Proc. Symp. Pure Math. *28* (1976) 275–306, Zbl. 339.14013

Kleiman, S. L. [1968]: Algebraic Cycles and the Weil Conjectures. In: Dix Exposés sur la Cohomologie des Schémas. North-Holland: Amsterdam, Adv. Stud. Pure Math. *3* (1968) 359–386, Zbl. 198,259

Kollár, J. [1986]: Higher Direct Images of Dualizing Sheaves. Ann. Math., II. Ser. *123* (1986) 11–40, Zbl. 598.14015

Lidl, R., Niederreiter, H. [1983]: Finite Fields. Encycl. Math. Appl. *20*, Addison-Wesley: London, 1983; Cambridge University Press, 1984, Zbl. 554.12010

Lubkin, S. [1967]: On a Conjecture of Andre Weil. Am. J. Math. *89* (1967) 443–548, Zbl. 208,484

Manin, Yu. I. [1965]: Algebraic Topology of Algebraic Varieties. Usp. Mat. Nauk *20*, No. 6 (1965) 3–12, Zbl. 141,182. English transl.: Russ. Math. Surv. *20* (1965), No. 5–6, 183–192

Manin, Yu. I. [1968]: Correspondence, Motifs, and Monoidal Transformations. Mat. Sb., Nov. Ser. *77* (1968) 475–507. English transl.: Math. USSR, Sb. *6* (1968) 439–470, Zbl. 199,248

Manin, Yu. I. [1969]: Lectures on the K-Functor in Algebraic Geometry. Usp. Mat. Nauk *24* (5) (1969) 3–86. English. transl.: Russ. Math. Surv. *24* (5) (1969) 1–89, Zbl. 199,555

Mazur, B. [1975]: Eigenvalues of Frobenius Acting on Algebraic Varieties over Finite Fields. In: Algebraic Geometry (Arcata 1974). Proc. Symp. Pure Math. *29* (1975) 231–261, Zbl. 306.14011

Mazur, B. [1986]: Arithmetic on Curves. Bull. Am. Math. Soc., New Ser. *14* (1986) 207–259, Zbl. 593.14021

Milne, J. [1980]: Étale Cohomology. Princeton Math. Ser. *33*, Princeton Univ. Press: Princeton, 1980, Zbl. 433.14012

Mumford, D. [1966]: Lectures on Curves on an Algebraic Surface. Princeton Univ. Press: Princeton, 1966, Zbl. 187, 427

Mumford, D. [1970]: Abelian Varieties. Oxford Univ. Press: London, 1970, Zbl. 223.14022 (1974 Oxford Zbl. 326.14012) (Repr.: Stud. Math. *5*, Tata Inst. Fund. Res., Oxford Univ. Press 1985)

Onishchik, A. L. [1986]: Methods of Sheaf Theory and Stein Spaces. Itogi Nauki Tekh. VINITI, Ser. Sovrem. Probl. Mat., Fundam. Napravleniya *10* (1986) 5–73. English transl.: Encycl. Math. Sci. *10*. Springer-Verlag: Berlin Heidelberg New York, 1990, 1–62, Zbl. 781.32017

Palamodov, V. P. [1986]: Deformations of Complex Structures. Itogi Nauki Tekh. VINITI, Ser. Sovrem. Probl. Mat., Fundam. Napravleniya *10* (1986) 123–221. English transl.: Encycl. Math. Sci. *10*, Springer-Verlag: Berlin Heidelberg New York, 1990, 105–194, Zbl. 781.32024

Raynaud, M. [1970]: Anneaux Local Henséliens. Lecture Notes in Math. 169, Springer-Verlag: Berlin Heidelberg New York, 1970, Zbl. 203, 51

Serre, J-P. [1955]: Faisceaux Algébriques Cohérents. Ann. Math., II. Ser. *61* (1955) 197–278, Zbl. 67, 162

Serre, J-P. [1956]: Géométrie Algébrique et Géométrie Analytique. Ann. Inst. Fourier *6* (1956) 1–42, Zbl. 75, 304

Serre, J-P. [1959]: Groupes Algébriques et Corps de Classes. Hermann: Paris, 1959, Zbl. 97, 356

Serre, J-P. [1964]: Cohomologie Galoisienne, 3rd ed. Lecture Notes in Math. 5, Springer-Verlag: Berlin Heidelberg New York, 1964 (5th ed. Springer 1994), Zbl. 128, 263

Shafarevich, I. R. [1986]: Basic Notions of Algebra. Itogi Nauki Tekh. VINITI, Ser. Sovrem. Probl. Mat., Fundam. Napravleniya *11* (1986) 258 pp. English transl.: Encycl. Math. Sci. *11*. Springer-Verlag: Berlin Heidelberg New York, 1990, Zbl. 711.16001 (Zbl. 655.00002)

Shafarevich, I. R. [1988]: Basic Algebraic Geometry. Vol. I, II; 2nd suppl. ed. Nauka: Moscow, 1988. English transl.: Springer-Verlag: Berlin Heidelberg New York, 1994, Zbl. 797.14001,2 (Zbl. 675.14001)

Shiffman, B., Sommese, A. J. [1985]: Vanishing Theorems on Complex Manifolds. Prog. Math. *56*. Birkhäuser: Boston, 1985, Zbl. 578.32055

Stepanov, S. A. [1972]: Congruences in Two Unknowns. Izv. Akad. Nauk SSSR, Ser. Mat. *36* (1972) 683–711. English transl.: Math. USSR, Izv. *6* (1972) 677–704, Zbl. 241.10002

Sullivan, D. [1970]: Geometric Topology. M.I.T. Press: Cambridge, 1970, Zbl. 366.57003

Tate, J. T. [1965]: Algebraic Cycles and Poles of the Zeta Function. In: Arithmetic Algebraic Geometry Proc. Conf. Purdue Univ. 1963 (Schilling, ed.). Harper & Row: New York, 1965, 93–110, Zbl. 213, 228

Weil, A. [1949]: Number of Solutions of Equations over Finite Fields. Bull. Am. Math. Soc. *55* (1949) 497–508, Zbl. 32, 394

Zarkhin, Yu. G., Parshin, A. I. [1986]: Finiteness Problems in Diophantine Geometry. Appendix in: Lang, S. Fundamentals of Diophantine Geometry. MIR: Moscow, 1986, 369–438 (Russian), Zbl. 644.14007

II. Algebraic Surfaces

V. A. Iskovskikh and I. R. Shafarevich

Translated from the Russian
by R. Treger

Contents

Introduction

The aim of this survey is to present a cohesive picture of the theory of algebraic surfaces, explain its problems, and describe its main methods. The proofs, when they are given, serve only to clarify the principal ideas employed in the field. For detailed proofs the reader is referred to the articles listed at the end of the survey.

The theory of algebraic surfaces is justifiably regarded as one of the most beautiful chapters in algebraic geometry. Its foundations were laid down more than 100 years ago by Clebsh who introduced the most important invariant of a surface, its *geometric genus*, which is an analog of the genus of a curve. He also proposed the problem of birational classification of surfaces and worked out several examples. His ideas were systematically developed by M. Noether. The most remarkable and beautiful achievement of the theory was the classification of algebraic surfaces obtained at the turn of the century by the Italian school of algebraic geometry – G. Castelnuovo, F. Enriques, and F. Severi. At the same time, Poincaré and Picard laid down the foundations of topological and analytic methods for investigating surfaces, and in the 1920s their ideas were fully developed by S. Lefschetz.

One may characterize the later development as a reconsideration of the theory in the spirit of new mathematical concepts, mainly in three directions: the theory of complex analytic varieties, the geometry over fields of positive charcteristic, and the investigation of surfaces as fibrations in curves. The latter direction was particularly fruitful for number theory because of applications of the intuition of "arithmetic surfaces" to the theory of algebraic curves with rational coefficients (i. e. Diophantine equations).

In spite of wealth of results, the theory of algebraic surfaces has a long way to go before it will reach the stage of maturity achieved by the theory of algebraic curves. One may describe the theory developed by the italian school as a birational classification of main types of surfaces according to their simplest numerical invariants. Here we have a comprehensive and clear picture. Far less is known about the values of those invariants. Finally, next to nothing is known about the continuous parameters on which the set of surfaces with given numerical invariants depend (i. e., moduli of algebraic surfaces).

The algebraic surfaces, i. e. algebraic varieties of dimension 2, are next to the algebraic curves by complexity. Here we first encounter "high-dimensional effects" that are typical for varieties of dimension > 1. However, several such effects first appear in dimension 3. Presumably, the varieties of arbitrary dimension are not fundamentally different from the varieties of dimension 3. So, in algebraic geometry, we have three cirtical dimensions: 1, 2, and 3.

The algebraic surfaces also occupy a similar intermediate place as far as topology goes. A smooth algebraic surface over the field of complex numbers is a 4-dimensional differential manifold, so the theory of algebraic surfaces is intimately related with 4-dimensional topology. On the other hand, the theory of algebraic surfaces (especially, the classification of algebraic surfaces – see Sect. 8) is very like the topology of 3-manifolds as the latter shaped recently up in the work of Thurston and others.

The above two connections show that the theory of algebraic surfaces is similar to the "low-dimensional topology" – the topology of manifolds of dimension 3 and 4; while the algebraic varieties of dimension 3 and higher are similar to the "stable case" – the theory of manifolds of dimension ≥ 5.

We assume the reader is familiar with basic algebraic geometry, namely: the theory of algebraic curves, divisors, intersection numbers, and differential forms on algebraic varieties, as well as with basic theorems on the cohomology of coherent sheaves. In dealing with the notions such as Betti numbers or Euler characteristic, the reader may restrict oneself to surfaces over the complex numbers in the standard topological sense; in the general case, they are introduced utilizing the l-adic cohomology.

We friequently refer to the surverys (Danilov (1988) and Shokurov (1988)) as well as to the first article of the present volume (which is referred to as "CAV"). However, we assume that the reader is familiar only with basic concepts and results described there. Moreover, in the present survey, we recall the most important facts needed in the sequel. To this end, we allow ourselves some liberty in the exposition: in studying examples, certain concepts from the above surveys are assumed to be known, although we recall them later when they appear in the systematic treatment of the theory.

The focal point of this survey is the theory of algebraic surfaces over an algebraically closed field k of caracteristic 0; to illustrate the theory, we often restrict ourselves to the case when k is the field $\mathbb{C}$ of complex numbers. However, we will also try to describe, rather briefly, two large adjacent fields – the theory of algebraic surfaces over fields of finite characteristic and the theory of 2-dimensional compact complex analytic manifolds. As for algebraic surfaces over non-algebraically closed field or 2-dimensional schemes of arithmetic nature (e. g., over $\mathbb{Z}$), they are completely left outside the scope of this survey.

The manuscript was carefully read by V. I. Danilov. He made so many remarks and suggestions that it is impossible to recount them here. In particular, the exposition of the theory of minimals models of algebraic surfaces in Sect. 7 is based on the manuscript that he gave us. We are happy to express our warm thanks to him.

§ 1. Main Invariants

In the survey, by an *algebraic surface* we mean an algebraic surface (almost always without singular points) in a projective space, or its open (in the Zariski topology) subset, i. e. a quasiprojective surface. This is justified because any abstract complete nonsingular algebraic variety of dimension 2 is projective, though in general this is no longer holds for the varieties with singularities. (See, e. g. (Hartshorne (1977) and Zariski (1958)).)

In studying algebraic surfaces, it is always helpful to have in mind the simplest type of algebraic varieties – curves. In case of curves different definitions lead to a sole discrete invariant, the genus. Surface, on the other hand, posses several numerical invariants connected by several relations.

First of all, there are topological invariants, namely the Betti numbers b_i, $i = 0, 1, 2, 3, 4$ (for an algebraic surface over $\mathbb{C}$, we consider the Betti numbers of $X(\mathbb{C})$; in the general case, we consider l-adic Betti numbers). By Poincaré duality, we get essentially two numbers, b_1 and b_2. We denote by $e(X)$ the Euler characteristic

$$\sum_{i=0}^{4}(-1)^i b_i = 2 - 2b_1 + b_2 .$$

Another kind of invariants are defined using cohomology of coherent sheaves. We set $h^{p,q} = \dim H^q(X, \Omega_X^p)$, $0 \leq p + q \leq 2$, where Ω_X^p is the sheaf of p-dimensional differential forms on X. We have the following relation

$$h^{p,q} = h^{2-p,2-q} \tag{1}$$

(which is a consequence of the duality theorem, e. g., see CAV, Chap. II, Sect. 5).

Over the complex numbers, by the Hodge theory, we get (Griffiths-Harris (1978)):

$$h^{p,q} = h^{q,p} , \tag{2}$$

$$b_1 = h^{1,0} + h^{0,1} , \tag{3}$$

$$b_2 = h^{2,0} + h^{1,1} + h^{0,2} . \tag{4}$$

So, the same relations hold over an arbitrary field of characteristic 0.

Finally, there is yet another invariant: the *canonical class* $K_X = c_1(\Omega_X^2)$ of an algebraic surface X. It is the divisor class on X containing the divisors of differential 2-forms (Danilov (1988), Chapt. I, Sect. 7 and 3). Its self-intersection number, $(K, K) = (K^2)$, gives another important numerical invariant (for details on intersection numbers on surfaces, see Sect. 4). By the Riemann-Roch theorem, we get the *Noether formula*

$$\frac{(K^2) + e(X)}{12} = 1 - h^{0,1} + h^{0,2} . \tag{5}$$

The number $1 - h^{0,1} + h^{0,2} = \sum(-1)^i \dim H^q(X, \mathcal{O}_X)$ is denoted by $\chi(\mathcal{O}_X)$ (or $p_a(X)$) and is called the *arithmetic genus* of X. The number $h^{2,0} = \dim H^0(\Omega_X^2)$ is denoted by p_g (or p) and is called the *geometric genus.* Thus p_g is the number of linearly independent regular differential 2-forms. Finally, $h^{0,1} = \dim H^1(X, \mathcal{O}_X)$ is denoted by q and is called the *irregularity.* It follows from (2) and (3) that $q = b_1/2$ over the field of characteristic 0.

Henceforth, we denote the dimension of $H^i(X, \mathcal{F})$ by $h^i(X, \mathcal{F})$ or $h^i(\mathcal{F})$.

The number $P_n = l(nK) = h^0(nK_X)$ is called the *n-genus of a surface.* It is the number of linearly independent regular 2-forms ω of degree n; locally $\omega = f(dx \wedge dy)^n$, where x, y are local parameters and f a regular function.

The numbers $h^{1,0}$, $h^{2,0}$, and P_n are birational invariants. This follows from their interpretation via differential forms (Danilov (1988), Chap. II, Sect. 7). It follows fom (1) that the same holds for $h^{0,2}$ (in characteristic 0, the same holds for b_1 and $h^{0,1}$ by (2) and (3)). The numbers $h^{1,1}$, b_2, and (K^2) are *not* invariant under birational equivalences of smooth projective surfaces.

We have three fundamentally different kind of curves, namely: curves of genus $g = 0$, $g = 1$, and $g > 1$. For instance, consider the n-canonical map φ_{nK} given by the n-canonical class nK (Shokurov (1988)). If $g = 0$, then $H^0(nK) = 0$ and φ_{nK} is not defined. If $g = 1$, then $H^0(nK) = 1$ hence φ_{nK} maps the curve to a point for every $n \geq 1$. But if $g > 1$, then φ_{nK} is an embedding for $n \geq 3$.

Analogously, we consider an important birational invariant κ, the *canonical* (or *Kodaira*) *dimension*, which makes sense for arbitrary complete nonsingular variety X. It is defined as follows

$$\kappa = \kappa(X) = \begin{cases} \max_n \dim \varphi_{nK}(X), & \text{if } H^0(nK) \neq 0 \text{ for at least} \\ & \text{one } n \geq 1, \\ -\infty, & \text{otherwise}, \end{cases}$$

where K is the canonical divisor on X, and $\varphi_{nK}(X)$ is the image of X under the n-canonical rational map. It follows at once from the definition that $\kappa(X)$ may take the following values: $-\infty, 0, 1, \ldots, \dim X$. Thus, in the case of curves: $\kappa = -\infty \Leftrightarrow g = 0$, $\kappa = 0 \Leftrightarrow g = 1$, and $\kappa = 1 \Leftrightarrow g > 1$.

For surfaces: $\kappa = -\infty \Leftrightarrow (\forall n \geq 1)\,[P_n = 0]$, $\kappa = 0 \Leftrightarrow (\forall n \geq 1)\,[P_n \leq 1]$ and $(\exists n_0)\,[P_{n_0} = 1]$, while for $\kappa = 1$ or 2 the numbers P_n grow, as we will see in Sect. 8–10, as polynomials in n of degree 1 or 2, respectively, starting from a suitable number n_0. So, one may say that κ expresses the asymptotic behavior of the birational invariants P_n as $n \to \infty$.

The birational classification of surfaces is the description of geometric properties of surfaces according to the values of the discrete invariants P_n, $h^{p,q}$, b_i, (K^2), and foremost the description of surfaces according to their canonical dimensions $\kappa = -\infty, 0, 1, 2$ (for more details, see Sect. 8).

For instance, the most simple to formulate and at the same time a highly nontrivial question related to the classification is the so-called "rationality problem", namely, *how to characterize the rational surfaces by the above invariants* (by a rational surface we mean a surface birationally equivalent to $\mathbb{P}^2$). In other words, for which surfaces X, the field $k(X)$ is isomorphic to the field $k(x, y)$ of rational functions in two variables. The answer is obtained in Sect. 13.

We observe that in the 1-dimensional case, the rational curves are characterized by a single condition, namely the vanishing of its genus.

§ 2. Examples

We will present several important examples of algebraic surfaces and describe their invariants introduced in Sect. 1. Most invariants can be calculated with the help of (1)–(5), or by cohomological methods described in CAV.

Example 1. A smooth surface X_d of degree d in $\mathbb{P}^3$. It is given by a single irreducible equation $F(x_0, x_1, x_2, x_3) = 0$ of degree d. We get

$$b_1 = h^{1,0} = h^{0,1} = 0\,, \quad b_2 = d^3 - 4d^2 + 6d - 2\,, \quad e = d^3 - 4d^2 + 6d\,, \quad q = 0\,,$$

$K = (d-4)H$, where H is a hyperplane section, (1)

$$(K^2) = d(d-4)^2\,, \quad p_g = \frac{(d-1)(d-2)(d-3)}{6}\,.$$

In particular, for $d = 1$, $X = \mathbb{P}^2$ and $K_{\mathbb{P}^2} = -3H$, where H is a line, and $K^2_{\mathbb{P}^2} = 9$. We can verify it by hand. By definition, K is the divisor class of any differential 2-form. We take affine coordinates x, y in $\mathbb{P}^2$, and consider $\omega = dx \wedge dy$. Since $(x-a, y-b)$ are local coordinates at a point (a, b) of $\mathbb{A}^2 \subset \mathbb{P}^2$, and clearly $d(x-a) \wedge d(y-b) = dx \wedge dy$, the form ω has no zeroes and poles on $\mathbb{A}^2$. Next, we will describe its behavior at infinity. If $(x_0 : x_1 : x_2)$ are homogeneous coordinates in $\mathbb{P}^2$ and $x = x_1/x_0$, $y = x_2/x_0$, the line $\mathbb{P}^2 \backslash \mathbb{A}^2$ is given by the equation $x_0 = 0$. Consider the affine chart $\{(u, v)\} = \mathbb{A}^2$ with center $(0 : 0 : 1)$, where $u = x_0/x_2$, $v = x_1/x_2$ are affine coordinates. Then $x = v/u$, $y = u^{-1}$, and

$$\omega = dx \wedge dy = \begin{vmatrix} \dfrac{\partial x}{\partial u} & \dfrac{\partial x}{\partial v} \\ \dfrac{\partial y}{\partial u} & \dfrac{\partial y}{\partial v} \end{vmatrix} du \wedge dv = \begin{vmatrix} -\dfrac{v}{u^2} & \dfrac{1}{u} \\ -\dfrac{1}{u^2} & 0 \end{vmatrix} du \wedge dv = \frac{1}{u^2} du \wedge dv\,.$$

Therefore, in the affine chart $\{(u, v)\}$, the form ω has a pole of order 3 along the line $u = 0$. Since the two charts cover $\mathbb{P}^2 \backslash (0 : 1 : 0)$, the divisor K of ω equals $-3H$, where H is the line at infinity $x_0 = 0$.

Turning to the general case, we will describe explicitly the space $H^0(\Omega^2_{X_d})$ of regular 2-forms. If in nonhomogeneous coordinates $x = x_1/x_0$, $y = x_2/x_0$, $z = x_3/x_0$, the equation of the surface takes the form $f(x, y, z) = 0$, and

$$\omega_0 = \frac{dx \wedge dy}{f'_z} = \frac{dy \wedge dz}{f'_x} = \frac{dz \wedge dx}{f'_y}\,,$$

then

$$H^0(\Omega^0_{X_d}) = \{\varphi\omega_0\}\,. \qquad (2)$$

where $\varphi(x, y, z)$ is a polynomial of degree $\leq d-4$ ($H^0(\Omega^2_{x_d}) = 0$ for $d < 4$).

The surface X_2 is called a *quadric*. Its equation can be written in the form $x_0x_1 - x_2x_3 = 0$. The map $x_1 = u_1v_1$, $x_2 = u_0v_1$, $x_3 = u_1v_0$, $x_0 = u_0v_0$

identifies X_2 with $\mathbb{P}^1 \times \mathbb{P}^1$, where $(u_0 : u_1)$ and $(v_0 : v_1)$ are homogeneous coordinates on factors. Then $s \times \mathbb{P}^1$ and $\mathbb{P}^1 \times t$, $s, t \in \mathbb{P}^1$, correspond to two families of line generators (rulings) on X_2. Clearly $\mathbb{P}^1 \times \mathbb{P}^1$ contains an open set $\mathbb{A}^1 \times \mathbb{A}^1 = \mathbb{A}^2$, hence the quadric is rational. Geometrically, the birational equivalence is given by stereographic projection from the point $x_0 \in X_2$. It maps a point $x \in X_2$ to the point of intersection of the line l_x, through x_0 and x, with a fixed plane $\mathbb{P}^2 \subset \mathbb{P}^3$ (Fig. 1).

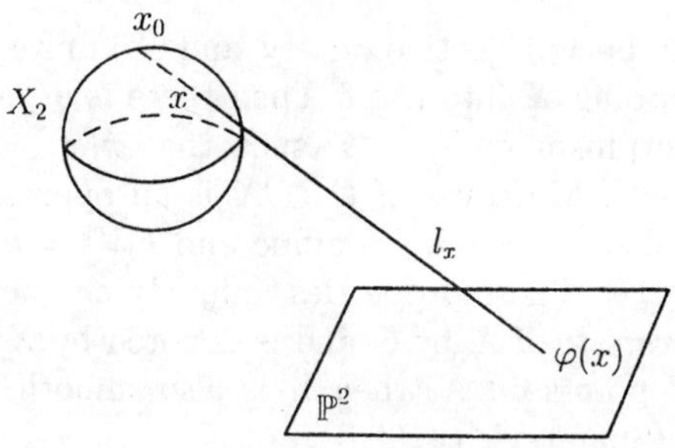

Fig. 1

The surface X_3 is called a *cubic*. It contains 27 distinct lines (see Sect. 13). The cubic X_3 is also rational (i. e., birationally equivalent to $\mathbb{P}^2$). A birational equivalence φ: $X_3 \dashrightarrow \mathbb{P}^2$ is given by the formula $\varphi(x) = l_x \cdot L$, where $L = \mathbb{P}^2 \subset \mathbb{P}^3$, and l_x is a line through the point $x \in X_3$ that intersects two fixed skew lines on X_3, m and m'.

In the simplest case, when $d = 1$, we have $X_1 = \mathbb{P}^2$ and $K = -3H$, where H is a line. Hence $p_g(X_1) = 0$ as well as $P_n(X_1) = 0$ for $n \geq 1$. Moreover $q(X_1) = 0$.

Since those numbers are birational invariants, $P_n(X) = q(X) = 0$ for any rational surface. Thus X_d is not rational if $d > 3$ since $p_g > 0$. A characterization of rational surfaces by the equalities $P_n(X) = q(X) = 0$ (and even a part of the equalities) is just the rationality criterion for surface (see Sect. 13).

Example 2. A surface $X \subset \mathbb{P}^{r+2}$ is said to be a *complete intersection* if it is a transversal intersection of r hypersurfaces $Y_1, \ldots, Y_r$ that are smooth at the points of intersections. If $\deg Y_i = d_i$, $i = 1, \ldots, r$, then $\delta = (d_1, \ldots, d_r)$ is said to be the type of X, and we denote X by $X_{(\delta)}$. For $X_{(\delta)}$, we get

$$q(X_{(\delta)}) = 0\,, \quad h^{1,0}(X_{(\delta)}) = 0\,, \quad b_1(X_{(\delta)}) = 0\,, \quad b_2(X_{(b)}) = e(X_{(\delta)}) - 2\,,$$

$$e(X_{(\delta)}) = \left[\binom{r+3}{2} - (r+3)\sum d_i + \sum d_i^2 + \sum_{i \neq j} d_i d_j\right] \prod_{i=1}^{r} d_i\,, \tag{3}$$

$$K_{X_{(\delta)}} = \left(\sum d_i - (r+3)\right) H\,,$$

where H is a hyperplane section, and

$$K^2_{X_{(\delta)}} = \left(\sum d_i - (r+3)\right)^2 \prod_{i=1}^{r} d_i .$$

We derive the formulas of Example 1 by setting $r = 1$. For $r = 2$ and $\delta = (2,2)$, we get a new kind of rational surfaces, namely an intersection of two quadrics in $\mathbb{P}^4$ (see Sect. 13). All the remaining complete intersections (with $r > 1$ and $d_i > 1$) are not rational.

Example 3. Let X be a projective variety, and G a finite group of projective transformations mapping X into itself. Then there is a projective variety Y, and a surjective morphism $\pi\colon X \to Y$ such that $\pi(x) = \pi(x')$ if and only if $x' = g(x)$ for $g \in G$. Moreover, if $U \subset X$ is an open affine set, invariant with respect to G, then $V = \pi(U)$ is affine and $k[V] = k[U]^G$ is the ring of invariants of G in $k[U]$. These properties uniquely characterize Y, and Y is called the *quotient variety* of X by G and is denoted by X/G. If X is smooth and G has no fixed points on X, then Y is also smooth; otherwise, Y may have singular pints (Shafarevich (1988)).

The following surface is our first application of the above construction. Let $X \subset \mathbb{P}^3$ be a smooth surface over a field of characteristic 0 given by the equation $x_0^5 + x_1^5 + x_2^5 + x_3^5 = 0$. Let $G = \{g\}$ be a cyclic group of order 5 with a generator g such that $gx_i = \varepsilon^i x_i$, $i = 0, 1, 2, 3$, $\varepsilon^5 = 1$, $\varepsilon \neq 1$, and $Y = X/G$. It is easy to see that if $\eta \in H^0(\Omega^1_Y)$ is a regular 1-form, then $\pi^*\eta \in H^0(\Omega^1_X)$. According to Example 1, $H^0(\Omega^1_X) = 0$ hence $H^0(\Omega^1_Y) = 0$, i.e. $q(Y) = 0$. If $\omega \in H^0(\Omega^2_Y)$, then $\pi^*\omega \in H^0(\Omega^2_X)^G$, i.e., $\pi^*\omega$ is a G-invariant regular form. With the notation of Example 1, $H^0(\Omega^2_X) = \{\varphi\omega_0\}$ where $\varphi(x,y,z)$ is a polynomial of order at most 1. Clearly G has no fixed vectors in the space $\{\varphi\omega_0\}$ except 0, so $p_g(Y) = 0$. On the other hand, the quadratic differential

$$\omega_0^2 = z^{-7}(dx \wedge dy)^2 = \omega$$

is G-invariant and regular. One can easily verify that $\omega = \pi^*\omega'$, where ω' is a regular quadratic differential on Y, hence $P_2(Y) \neq 0$. Therefore Y is not a rational surface, so the condition $q = p_g = 0$ is not sufficient to characterize rational surfaces; Y is called a *Godeaux surface.*

The algebraic curves fall into two classes: the curves of genus $g > 1$ are of "general type", and the curves of genus 1 and 0 are "specical". The same holds for algebraic surfaces. It is related to the notion of canonical dimension κ: "general type" corresponds to $\kappa = 2$, and the remaining surfaces are "special". Furthermore, "special surfaces" with $\kappa < 2$ can be viewed as generalizations of curves of genus 1 or 0. However, curves of genus 1 as well as curves of genus 0 can be generalized in several different ways which we will now describe (Examples 4, 5, 6).

The general principle governing such constructions is that an analog of curves of a certain type are surfaces with a pencil of curves of that type.

Definition. A *fibration in algebraic curves* is a smooth surface X and a proper morphism $f\colon X \to B$ onto a smooth curve B with connected fibers. Then B is said to be a *base*, and the curves $X_b = f^{-1}(b)$, $b \in B$, are said to be *fibers*. Sometimes X is called a *pencil of curves*.

The well-known *Bertini theorem* says that if $\operatorname{char} k = 0$, then all the fibers but a finite number are smooth curves (in $\operatorname{char} k = p > 0$, this fails in general – see Sect. 15). This result is an algebraic analog of *Sard's theorem* in the theory of differential manifolds, namely: for a proper smooth mapping $f\colon X \to Y$ of differential manifolds, the set of points $y \in Y$ such that the corresponding map of tangent spaces $d_x\colon T_x \to T_y$ fails to be an epimorphism for at least one $x \in f^{-1}(y)$ has measure 0. In other words, for a set $U \subset Y$ whose complement has measure 0, the mapping $f^{-1}(U) \to U \cap f(X)$ defines a locally trivial smooth fibration of manifolds (de Rham (1955)).

A morphism has connected fibers if and only if it is not a composition $X \to B' \to B$, where $B' \to B$ is a nontrivial finite covering. Such a *factorization* is said to be *Stein* if the fibers of $X \to B'$ are connected (Danilov (1988)). This, in turn, means that the field $k(B)$ is algebraically closed in $k(X)$. Then all the fibers, but finitely many, are irreducible. A general fiber is irreducible and smooth; its genus g is called the *genus of pencil.* All the smooth fibers have the same genus g. The singular fibers are said to be *degenerate.*

If $k = \mathbb{C}$, and the fibration $f\colon X \to B$ has no singular fibers, then f is topologically locally trivial, and we have a well-known relation $e(X) = e(B)e(F)$, where F is a general (any nondegenerate) fiber. One can easily generalize this relation to arbitrary fibrations.

Proposition. *For a fibration* $f\colon X \to B$

$$e(X) = e(B)e(F) + \sum_{b \in B} (e(F_b) - e(F)) \tag{4}$$

where $F_b = f^{-1}(b)$ *and* e *is the topological Euler characteristic. The formal sum in* (4) *runs over all the fibers; however, only the terms corresponding to singular fibers are different from* 0.

The relation (4) holds not only for $k = \mathbb{C}$ but for an arbitrary field of characteristic 0, and even for an arbitrary field if we use the l-adic cohomology (with a few modifications in characteristic $p > 0$).

The relation (4) is especially useful since $e(F_b) \geq e(F)$ for all $b \in B$, as we will see later.

The notion of fibration in curves reflects a general principle in algebraic geometry that an object fully reveals itself only "in dynamics", when we view it varying, i. e., as an object over a sufficiently general base B. Thus, one should think about a fibration in curves as a "curve" over a base B. The generic fiber $F_\xi = f^{-1}(\xi)$ of the fibration is actually a curve, where ξ is the generic point. It is important, however, that F_ξ is a curve over a non-algebraically closed field $k(\xi) = k(B)$, the function field of B.

An analog of a point of a curve is a section of a morphism $f\colon X \to B$, i. e. a choice of a point in each fiber F_b of the fibration (which vary "algebraically" with b); precisely, we are given a morphism $\varphi\colon B \to X$ such that $\varphi(b) \in F_b$ or, in other words, $f\varphi = 1$. Such a morphism is said to be a *section* of the fibration. The image of the morphism, namely the curve $\varphi(B)$, is also called a section. Clearly the sections are in a one-to-one correspondence with the points of F_ξ that are rational over $k(B)$ (i. e. with coordinates in $k(B)$).

Example 4. The above discussion suggests that, in addition to the trivial surface $\mathbb{P}^2$, various fibrations in curves of genus 0 are analogs of curves of genus 0 (i. e. $\mathbb{P}^1$).

Definition. A surface X is said to be *ruled* if there is a morphism π: $X \to B$ to a smooth curve B (called a *base*) whose fibers are isomorphic to $\mathbb{P}^1$. One can prove that the fibration π is locally trivial, i. e., there exixts a covering $\cup U_i = B$, where the U_i's are open sets, such that $\pi^{-1}(U_i) \cong \mathbb{P}^1 \times U_i$ (see Sect. 13).

The ruled surfaces with $B = \mathbb{P}^1$ can be described (as abstract varieties) as follows.

Let $\mathbb{P}^1 = U_0 \cup U_1$, where $U_0 = \{t \in \mathbb{P}^1 | t \neq 0\}$ and $U_1 = \{t \in \mathbb{P}^1 | t \neq \infty\}$ (t is an affine coordinate in $\mathbb{P}^1$). We set

$$X = (\mathbb{P}^1 \times U_0) \cup (\mathbb{P}^1 \times U_1) .$$

and on $U_0 \cap U_1$, we identify $(x, t) \in \mathbb{P}^1 \times U_0$ with $(y, t) \in \mathbb{P}^1 \times U_1$ if $y = xt^n$, $n \geq 0$. We denote the corresponding surface by $\mathbf{F}_n$.

Under the glueing of $\mathbb{P}^1 \times U_0$ with $\mathbb{P}^1 \times U_1$, the section $(0) \times U_0$ is glued with the section $(0) \times U_1$, and we get a section S_0. Similarly, $(\infty) \times U_0$ gives a section S_∞. Clearly, $S_0 \cap S_\infty = \emptyset$. Consider the function y which is a coordinate along $\mathbb{P}^1$ in $\mathbb{P}^1 \times U_0$. Its polar divisor $(y)_\infty$ equals S_∞, and its zero divisor $(y)_0$ equals $(xt^n) = S_0 + nF$, where F is a fiber (for a discussion of divisors, see (Danilov (1988), Chap. III, Sect. 3) and Sect. 3 of the present article). It follows that $S_\infty \sim S_0 + nF$, and since $(S_0, F) = (S_\infty, F) = 1$, we get $(S_\infty^2) = n$ and $(S_0^2) = -n$ (for a discussion of intersection numbers, see Sect. 4). One can show that S_0 is a unique curve on $X = \mathbf{F}_n$, $n > 0$, with negative self-intersection number. Consequently $\mathbf{F}_n$ and $\mathbf{F}_m$ are not isomorphic if $n \neq m$.

Since over a suitable open set $U \subset B$, we have $\pi^{-1}(U) \simeq \mathbb{P}^1 \times U$, the ruled surfaces with base $\mathbb{P}^1$ are rational. They are called *rational ruled surfaces.* For such surfaces

$$b_1 = h^{1,0} = h^{0,1} = 0 , \qquad p_g = 0 .$$

For $\mathbf{F}_n$, $K = -2S_0 - (n+2)F$ where F is a fiber of the projection $\mathbf{F}_n \to \mathbb{P}^1$.

For more details on ruled surfaces, see Sect. 13. In particular, we will prove there that any rational ruled surface is isomorphic to one of the surfaces $\mathbf{F}_n$.

Example 5. Similarly, one can construct analogs of curves of genus 1, namely pencils of genus 1, or *elliptic pencils* (fibrations in elliptic curves).

A surface X is said to be *elliptic* if it admits a morphism $f\colon X \to B$ whose generic fiber is a curve of genus 1. One can construct elliptic surfaces over $B = \mathbb{P}^1$ as follows (we assume that $\operatorname{char} k \neq 2, 3$).

Consider a covering of $\mathbb{P}^1$ by two open sets: $\mathbb{A}^1 = \mathbb{P}^1 \backslash \{\infty\}$ with coordinate t, and $\mathbb{A}^1 = \mathbb{P}^1 \backslash \{0\}$ with coordinate $s = t^{-1}$. Consider the surface X' in $\mathbb{P}^2 \times \mathbb{A}^1$ given by the equation

$$x_0 x_2^2 = x_1^3 + \alpha(t) x_0^2 x_1 + \beta(t) x_0^3$$

where $\alpha(t)$ and $\beta(t)$ are polynominals, $\deg \alpha(t) = n$ and $\deg \beta(t) = m$. In nonhomogeneous coordinates $x = x_1/x_0$, $y = x_2/x_0$, the equation takes the form

$$y^2 = x^3 + \alpha(t) x + \beta(t)\,. \tag{5}$$

The projection $\mathbb{P}^2 \times \mathbb{A}^1 \to \mathbb{A}^1$ defines a morphism $p'\colon X' \to \mathbb{A}^1$. Similarly, consider the surface X'' in $\mathbb{P}^2 \times \mathbb{A}^1$ with an equation

$$v^2 = u^3 + \gamma(s) u + \delta(s)\,,$$

where $\gamma(s) = \alpha\left(\frac{1}{s}\right) s^{4k}$, $\delta(s) = \beta\left(\frac{1}{s}\right) s^{6k}$, $k \geq \max\left(\frac{n}{4}, \frac{m}{6}\right)$.

It admits the projection $p''\colon X'' \to \mathbb{A}^1$. We identify the open sets $t \neq 0$ in X' and $s \neq 0$ in X'' by the formulas

$$s = \frac{1}{t}\,, \quad u = \frac{x}{t^{2k}}\,, \quad v = \frac{y}{t^{3k}}\,.$$

Then X' and X'' produce a surface $X = X' \cup X''$ with a projection $p\colon X \to \mathbb{P}^1$. The morphism p is smooth in a neighborhood of a fiber $p^{-1}(b)$, provided $D(b) \neq 0$ where $D(t) = 4\alpha(t)^3 + 27\beta(t)^2$. If we assume that D is not identically zero, then X is an elliptic surface. We constructed it as an abstract variety, though it is easy to define its projectie embedding. However, X is not smooth in general. The fibers $p^{-1}(b)$ with $D(b) \neq 0$ are always nonsingular. But if $D(b) = 0$, then $p^{-1}(b)$ has a singular point, which is a singular point of X if and only if $D'(b) = 0$ and $\beta(b) \neq 0$, or $\beta(b) = \beta'(b) = 0$ ($'$ stands for a derivative). Those singular points are not very complicated, and can be easily resolved by simple birational transformations (for details, see Sect. 6). Thus we get smooth elliptic surfaces. In general elliptic surfaces are not locally trivial. If the absolute invariant

$$J(t) = \frac{\alpha(t)^3}{4\alpha(t)^3 + 27\beta(t)^2}$$

in the equation (5) is not constant, then the fibers vary with t. In fact, not all curves of genus 1 are isomorphic, while all curves of genus 0 are isomorphic. Therefore elliptic fibrations usually have degenerate fibers. In (5), they correspond to the roots of $D(t)$.

As a generalization of curves of genus 0, we have obtained two types of surfaces: $\mathbb{P}^2$ and ruled surfaces. In trying to generalize curves of genus 1, we come across even a wider class of surfaces. Next we will construct an analog that is different from elliptic surfaces.

Example 6. Recall that an *Abelian variety* is a projective variety with a group structure such that the addition map

$$X \times X \to X\,, \qquad (x,y) \mapsto x+y\,,$$

and the inverse map

$$X \to X\,, \qquad x \mapsto -x$$

are morphism (Mumford (1970a), Shafarevich (1988)).

A 1-dimensional Abelian variety is an elliptic curve. Therefore its natural generalization is a 2-dimensional Abelian variety. An example of such a variety is a product $E_1 \times E_2$, where E_1 and E_2 are elliptic curves. However, this is a very special case. A more general example is the Jacobian $J(C)$ of a curve C of genus 2 (Shokurov (1988), Chap. II).

In affine coordinates, the curve C can be given by an equation $y^2 = f(x)$, where f is a polynomial of degree 5 without multiple roots. The variety $J(C)$ parametrizes the divisor classes of degree 0 on C. By the Riemann-Roch theorem, such a divisor class contains a representative of the form $D - D_0$, where $D > 0$, $D_0 > 0$, $\deg D = \deg D_0 = 2$, and D_0 is a fixed divisor. We take the divisor $2P$ as D_0, where P is a unique point where the function x has a pole. Such a representative is unique in the class $D - D_0$, except for the case when $D \in K_C$, where K_C denotes the canonical class of C (clearly $D_0 = 2P \in K_C$). In the latter case, $D = D_0$ or $D = (a,b) + (a,b')$, where b and b' are two solutions of $y^2 = f(x)$, and $D - D_0 = (x-a) \sim 0$ (Fig. 2).

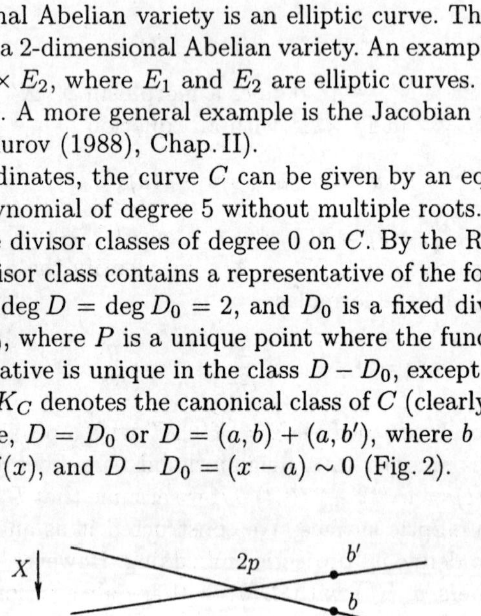

Fig. 2

Thus, in order to construct $J(C)$, we must first construct the set Y of divisors D of degree 2, i. e., unordered pairs of points $(P,Q) \in C \times C$; then we must contract the set of divisors $(x-a)_0$ to a point. To begin with, consider the set of unordered pairs, i. e. the surface $Y = (C \times C)/G$, where $G = \{1, g\}$ and $g(P,Q) = (Q,P)$ (one can easily verify that g is a projective transformation for a certain projective embedding of $C \times C$). The surface Y contains a curve E that parametrizes the set of divisors $(x-a)_0$ and D_0. Obviously $E \cong \mathbb{P}^1$. One can show that there is a surface X and a morphism $\varphi\colon Y \to X$ such that $\varphi(E)$ is a point on X, and φ is an isomorphism between $Y \backslash E$ and $X \backslash \varphi(E)$ (see Sect. 6, Example 4). The surface X is the required $J(C)$.

The Abelian variety $E_1 \times E_2$ has structures of elliptic fibrations: $E_1 \times E_2 \to E_1$ and $E_1 \times E_2 \to E_2$. The Jacobian $J(C)$, on the other hand, has no such a structure in general. So, $J(C)$ is an analog of a curve of genus 1 that is different from elliptic surfaces.

Given an Abelian variety A, the translation by a point defines a canonical isomorphism of the tangent spaces T_0 and T_a, $a \in A$. Thus for A, as for any group variety, the tangent bundle is isomorphic to $T_0 \times A$. Hence $\Omega_A^1 \cong \mathcal{O}_A \oplus \mathcal{O}_A$. It follows that $h^0(\Omega_A^1) = 2$, $\Omega_A^2 = \wedge^2 \Omega_A^1 \cong \mathcal{O}_A$, and $K_A = 0$. In addition

$$q(A) = h^{0,1} = 2\,, \quad b_1 = 4\,, \quad b_2 = 6\,, \quad e(A) = 0$$

(in characteristic 0 this follows from (1)–(5) of Sect. 1).

Finally, we will describe the third class of surfaces that can be regarded as an analog of curves of genus 1.

The elliptic curves are characterized by the condition $K = 0$, so it is natural to regard the surfaces X with $K_X = 0$ as one of their analogs. In general, $K \neq 0$ for surfaces with a pencil of elliptic curves, so we get two different analogs of elliptic curves. On the other hand, $K_A = 0$ for 2-dimensional Abelian varieties, as we have seen before. Now we will construct another example of a surface with this property – yet another analog of an elliptic curve.

Example 7. Let A be a 2-dimensional Abelian variety (the reader may assume that $A = E_1 \times E_2$, where E_1 and E_2 are elliptic curves). We will assume that $\operatorname{char} k \neq 2$. Let σ denote the involution $\sigma(a) = -a$, and set $Y = A/G$, where $G = \{1, \sigma\}$. One can show that σ has 16 fixed points on A, i. e. points of order 2 (that is clear for $A = E_1 \times E_2$, since a elliptic curve has 4 points of order 2). Consequently Y has 16 singular points. These singularities are very nice, so its is easy to construct their resolution, i. e. a smooth surface X and a birational morphism $\varphi\colon X \to Y$ that is an isomorphism away from those points. It will suffice to apply σ-processes at those points of Y (σ-processes are described in (Danilov (1988), Chap. II, Sect. 1) and Sect. 6 of the present article). It turns out that $\varphi^{-1}(y_i)$ are curves isomorphic to $\mathbb{P}^1$, where y_i are singular points of Y ($i = 1, \ldots, 16$). Such an X is said to be a *Kummer surface.*

The main property of X is that $K_X = 0$, as for an Abelian variety A. On the other hand, $q(X) = 0$ while for Abelian varieties $q \neq 0$. The reason is that a regular 2-form ω on A is invariant with respect to σ (e. g., for $A = E_1 \times E_2$, we get $\omega = \pi_1^*(\eta_1) \wedge \pi_2^*(\eta_2)$, where $\pi_i\colon E_1 \times E_2 \to E_i$ are the projections, η_i is a regular 1-form on E_i, and $\sigma^*\eta_i = -\eta_i$). It follows that ω induces a form on X, and it is easy to see, using transformation formulas for φ, that that form (on X) has trivial divisor. On the other hand, if $\eta \in H^0(\Omega_A^1)$ then $\sigma^*\eta = -\eta$, and one easily derives that $H^0(\Omega_X^1) = 0$. Consequently $q(X) = 0$ in characteristic 0.

The surfaces X with $q(X) = 0$ and $K_X = 0$ are called $K3$ *surfaces.* The Kummer surfaces are not the only examples of $K3$ surfaces. Clearly $\kappa = 0$ for every $K3$ surface. Among smooth surfaces in $\mathbb{P}^3$, the surfaces of degree 4

(Example 1) are $K3$ surfaces. Among complete intersections, the surfaces of type $(2,3)$ or $(2,2,2)$ are $K3$ surfaces (Example 2).

Example 8. Let A be a 2-dimensional Abelian variety (e. g., over a field of characteristic 0). Let G be a fine group of automorphisms acting freely on A such that not all of its elements act as translations. Set $X = A/G$. If ω is an invariant differential 2-form on A, then $g^*\omega = \chi(g)\omega$, $\forall g \in G$, where χ is character of G such that $\chi \neq 1$. It follows that $H^0(\Omega_X^2) = 0$ but $nK_X = 0$, so $P_n(X) = 1$ where $n = |G|$ is the order of G. One can easily verify that $q(X) = 1$ and $p_g(X) = 0$.

One can specialize the above construction as follows. Let E and B be elliptic curves, and G a finite group of automorphisms of $E \times B$ that commute with the projection $E \times B \to E$ and act as translations on E (in other words, for $g \in G$ and $(e,b) \in E \times B$,

$$g(e,b) = (e + \gamma_g, \varphi_g(e,b)),$$

where $\gamma_g \in E$ and $\varphi_g(e,b)$ is a morphism $E \times B \to E$). It follows that G acts freely on $E \times B$. Assume that not every element of G is a translation (i. e., $b \mapsto \varphi_g(e,b)$ is not a translation of B for a certain $g \in G$ and $e \in E$). If $\operatorname{char} k \neq 2$ or 3, then one can easily show that a 2-form $\omega \in H^0(\Omega^2_{E\times B})$ is not g-invariant, hence for $X = (E \times B)/G$, we get $p_g(X) = 0$, $q(X) = 1$, $K_X \neq 0$, and $nK_X = 0$, where $n = |G|$. Such surfaces are called *bi-elliptic*. (The classical term *hyperelliptic* is more often used. In (Beauville (1978)), Beauville proposed the nice term "bi-elliptic", which allows to avoid unfounded associations with hyperelliptic curves.)

Example 9. Let Y be a $K3$ surface with an automorphism g of order 2 without fixed points. Let $G = \{1, g\}$ and $X = Y/G$. As in Example 8, one can easily show that $q(X) = p_g(X) = 0$ but $P_2(X) \neq 0$. The surfaces X with $q = 0$, $H^0(\Omega_X^2) = 0$, and $2K_X \sim 0$ (so that $P_2(X) = 1$) are called the *Enriques surfaces.*

To obtain an example of a $K3$ surface with an antomorphism g as above over a ground field k with $\operatorname{char} k \neq 2$, we consider a complete intersection of type $(2,2,2)$ in $\mathbb{P}^5$ given by equations

$$f_i(x_0,x_1,x_2) + h_i(x_3,x_4,x_5) = 0 \qquad (i = 1,2,3),$$

where f_i and h_i are quadratic forms; moreover, the conics $f_1 = f_2 = f_3 = 0$ as well as $h_1 = h_2 = h_3 = 0$ have no common points. Then

$$g(x_0,x_1,x_2,x_3,x_4,x_5) = (-x_0,-x_1,-x_2,x_3,x_4,x_5).$$

§ 3. Curves on an Algebraic Surface

3.1. Divisors. By a *curve* on an algebraic surface we mean a closed reduced irreducible 1-dimensional subvariety. By a *divisor* we mean a formal sum $D = \sum n_i C_i$, where $n_i \in \mathbb{Z}$ and C_i are curves. The set $\cup C_i$ is denoted by $\operatorname{Supp} D$. The divisors form a group: $\sum n_i C_i + \sum m_i C_i = \sum (n_i + m_i) C_i$. We denote this group by $\operatorname{Div} X$. A divisor is said to be *effective* if all $n_i \geq 0$. We then write $D \geq 0$.

For divisors on arbitrary algebraic varieties, see, e. g. (Danilov (1988), Chap. III, Sect. 3). We recall their main properties.

Every rational function $f \in k(X)$, $f \neq 0$, on a smooth variety X defines two effective divisors, namely the zero divisor $(f)_0$ and the polar divisor $(f)_\infty$. The divisor $(f)_0 - (f)_\infty$ is denoted by (f) and is called the *divisor of the function* f. The divisors of the form (f), $f \in k(X)^* = k(X) \backslash \{0\}$, are said to be *principle*. The correspondence $f \mapsto (f)$ is a homomorphism of the group $k(X)^*$ to $\operatorname{Div} X$. If X is a complete variety, its kernel coincides with the group of constants. The divisors D_1 and D_2 are said to be *equivalent* if the divisor $D_1 - D_2$ is principle; we then write $D_1 \sim D_2$.

Given a divisor D on X, we consider the linear subspace $L(D) \subset k(X)$ consisting of 0 and the functions f with $(f) + D \geq 0$. Its dimension is denoted by $l(D)$.

Given a divisor D, we consider the sheaf $\mathcal{O}(D)$ whose sections over an open set $U \subset X$ are the functions f such that $(f) + D \geq 0$ on U. Clearly $\mathcal{O}(D)$ is a sheaf of $\mathcal{O}$-modules (where $\mathcal{O} = \mathcal{O}_X$ is the structure sheaf of X). It is a coherent sheaf and $L(D) = H^0(X, \mathcal{O}(D))$. Hence $l(D)$ is finite if X is a complete variety. The sheaves $\mathcal{O}(D_1)$ and $\mathcal{O}(D_2)$ are isomorphic if and only if $D_1 \sim D_2$.

Every divisor on a smooth variety is locally principle, i. e., in a sufficiently small neighborhood U of any point, it has the form (f_U). The function f_U is called a *local equation* of the divisor on U. Let $X = \cup U_\alpha$ be an open covering such that D is principle on each U_α. If f_α are corresponding local equations, then D is uniquely determined by these equations. The functions $\varphi_{\alpha,\beta} = f_\alpha f_\beta^{-1}$ are regular, and $\varphi_{\alpha,\beta} \in \mathcal{O}_X(U_\alpha \cap U_\beta)^*$. Conversely, every such collection of f_α's defines a divisor D. One may view the $\varphi_{\alpha,\beta}$'s as transition functions of the line bundle L with base X, namely: $L = \cup(\mathbb{A}^1 \times U_\alpha)$, and (a, x) is identified with (b, x) ($x \in U_\alpha \cap U_\beta$, $a, b \in \mathbb{A}^1$) if $b = \varphi_{\alpha,\beta}(x) a$ (Danilov (1988), Chap. I, Sect. 5). The sheaf of sections of L coincides with $\mathcal{O}(D)$.

For example, if $X = \mathbb{P}^2$, then any curve $C \subset X$ is given by an equation $F = 0$, where F is an irreducible form. The degree of this form is called the *degree of the curve* and is denoted by $\deg C$. The divisor $D = \sum n_i C_i$ on $\mathbb{P}^2$ is principle if and only if $\sum n_i \deg C_i = 0$. In this case $D = (f)$, where $f = \prod F_i^{n_i}$ (F_i denotes the equation of C_i). If x_0, x_1, x_2 are homogeneous coordinates in $\mathbb{P}^2$, then the function

$$\left(\prod F_i^{n_i}\right) x_\alpha^{-m} \qquad \left(m = \sum n_i \deg C_i\right)$$

is a local equation of D on $U_\alpha = \{x_\alpha \neq 0\}$.

If $f\colon X \to B$ is a morphism of a smooth surface X to a smooth curve B, then $f^{-1}(b)$, $b \in B$ is an effective divisor, and we can take $f^*(\varphi)$ as its local equation, where $\varphi \in k(B)$ is any function with a zero of order 1 at b.

The behavior of divisors under a morphism $\varphi\colon X \to Y$ is determined by two maps.

We assume $\varphi(X) \not\subset \operatorname{Supp} D$, and D is given an open sets $U_i \cup U_i = Y$, by local equations f_i. Then the functions $\varphi^*(f_i)$ on $\varphi^{-1}(U_i)$ are local equations (in the covering $\varphi^{-1}(U_i)$) of a certain divisor on X, which is denoted by $\varphi^*(D)$. Clearly $\mathcal{O}(\varphi^*(D)) = \varphi^*\mathcal{O}(D)$. Further, $\varphi^*((f)) = (\varphi^*(f))$ hence $D_1 \sim D_2$ implies $\varphi^*(D_1) \sim \varphi^*(D_2)$.

The second transformation maps divisors on X to divisors on Y. It is denoted by φ_*. By definition, $\varphi_* = 0$ if $\dim \varphi(X) < \dim X$. Assume that $\dim X = \dim Y$. If $C \subset X$ is an irreducible curve and $\dim \varphi(C) = \dim C$, i. e. φ induces a morphism of finite degree, then $\varphi_*(C) = r\varphi(C)$, where r is the degree of that morphism. We extend the map φ_* by additivity to the whole $\operatorname{Div} X$. If n denote the degree of $X \to \varphi(X)$, then

$$\varphi_*\varphi^*(D) = nD\,, \quad D \in \operatorname{Div} Y\,, \quad Y = \varphi(X)\,. \tag{1}$$

In particular, if $\varphi\colon X \to Y$ is etale (i. e. unramified), then clearly $K_X = \varphi^* K_Y$, and it follows from (1) that $\varphi_* K_X = nK_Y$. But if $K_X \sim 0$ then $nK_Y \sim 0$.

Finally, recall a connection between divisors and rational maps. The projectivization $\mathbb{P}L(D)$ of the space $L(D)$ is denoted by $|D|$. It consists of the effective divisors $D' \geq 0$ with $D' \sim D$, and is called the *complete linear system* of divisors associated with D. Any subspace of that space is called a *linear system*. Each linear system Δ of divisors defines a rational map φ_Δ from X to a projective space. If $\Delta = \mathbb{P}(M)$, $M \subset L(D)$, and $f_0, \ldots, f_n$ is a basis of M, then $\varphi_\Delta = (f_0, \ldots, f_n)$.

In particular, if $\Delta = |D|$ then we write φ_D in place of φ_Δ.

Set $D_i = (f_i) + D$. It may happen that all the D_i's have a common part. Then $D_i = D_i' + D^0$, where the D_i''s have no common components. The divisor D^0 is called the *fixed part* of Δ, and $\Delta' = \Delta - D^0 \subset |D - D^0|$ is called a system *without fixed components*.

If Δ has no fixed components, then the map φ_Δ is not regular (i. e. not a morphism) at the points of the set $\cap \operatorname{Supp} D_i$. The map φ_D is regular only at the points $x \in X$ where the stalk of $\mathcal{O}(D)$ is generated (as $\mathcal{O}$-module) by its sections over X.

3.2. Algebraic Equivalence. For $k = \mathbb{C}$, the method employed in studying curves and divisors on surfaces comes from topology. The set $X(\mathbb{C})$ of complex points of a smooth surface X is a 4-dimensional manifold with a canonical orientation (Griffiths-Harris (1978)). Given a curve $C \subset X$, its normalization C^ν is a smooth projective curve, hence $C^\nu(\mathbb{C})$ is a compact oriented surface (Shokurov (1988)). Therefore the normalization map ν: $C^\nu(\mathbb{C}) \to C(\mathbb{C})$ makes C into a 2-dimensional cycle on $X(\mathbb{C})$. We extend this map by linearlity to arbitrary divisors; then D corresponds to a homology class $\langle D \rangle = H_2(X(\mathbb{C}), \mathbb{Z})$. In case X is an algebraic curve, a divisor D is a collection of points, i. e. a 0-dimensional cycle $\sum n_i P_i$, and its homology class $\langle D \rangle \in H_0(X(\mathbb{C}), \mathbb{Z}) = \mathbb{Z}$ is determined by its degree $\sum n_i$. So, the correspondence $D \mapsto \langle D \rangle \in H_2(X(\mathbb{C}), \mathbb{Z})$ generalizes the notion of degree of divisor on a curve. However, $\langle D \rangle$ is not a number but an element of a finitely generated group $H_2(X(\mathbb{C}), \mathbb{Z})$.

Now, we will describe an algebraic analog of the homological equivalence. The idea is that divisors are homologically equivalent if they fit in a continuous family. Let S be an algebraic variety. A locally principle divisor ϑ on $X \times S$ whose support $\operatorname{Supp} \vartheta$ does not contain $X \times s$, $s \in S$, is said to be an *algebraic family* of divisors with base S. Then the inclusion $i_s : X \xrightarrow{\sim} X \times s \to X \times S$ defines a divisor $i_s^*(\vartheta) = D_s$ on X. In this sense, the divisors D_s form a "family" parametrized by the points $s \in S$.

In certain questions (e. g., in studying infinitesimal deformations) it is important to consider families whose base is an arbitrary, not necessary reduced, scheme. This is, however, irrelevant in our presentation.

A divisor $D \in \operatorname{Div} X$ is said to be *algebraically equivalent to zero* if it has the form $D_{s_1} - D_{s_2}$, where D_{s_1} and D_{s_2} are elements of an algebraic family with connected base. One can show that the divisors algebraically equivalent to zero form a subgroup of $\operatorname{Div} X$; it is denoted by $\operatorname{Div}_a X$. For $k = \mathbb{C}$, one can prove that the cycle $\langle D \rangle$ corresponding to a divisor D on an algebraic surface is homologically equivalent to zero if and only if D is algebraically equivalent to zero (Griffiths-Harris (1978)). So, for $k = \mathbb{C}$, the group $\operatorname{Div} X / \operatorname{Div}_a X$ is isomorphic to the image of $\operatorname{Div} X$ in $H_2(X(\mathbb{C}), \mathbb{Z})$.

In general, $\operatorname{Div} X / \operatorname{Div}_a X$ is called the *Severi* (or *Neron-Severi*) *group* of X and is denoted by S_X. It has been established that this group has a finite number of generators. The number of free generators (rank) of this group is denoted by ϱ and is called the *Picard number* of the surface. As in the case $k = \mathbb{C}$, we get $\varrho \leq b_2$ (where b_2 is the 2-dimensional l-adic Betti number).

So, by passing to $S_X = \operatorname{Div} X / \operatorname{Div}_a X$, we ignore all continuous deformation of curves, and obtain a discrete group.

Example 1. For $X = \mathbb{P}^1$, we get $S_X \cong \mathbb{Z}$ and the image of $D = \sum n_i C_i$ in $S_X = \operatorname{Div} X / \operatorname{Div}_a X$ is determined by the integer $\sum n_i \deg C_i$.

Example 2. If X is a quadric then $X \cong \mathbb{P}^1 \times \mathbb{P}^1$. Each curve is given by an irreducible form $F(x_0 : x_1; y_0 : y_1)$ homogeneous in $(x_0 : x_1)$ as well as $(y_0 : y_1)$ (here $(x_0 : x_1)$ are homogeneous coordinates of a point on the first factor, and $(y_0 : y_1)$ on the second factor). Thus a curve C has two "degrees", namely $\deg' C$ and $\deg'' C$, degree of the form F in $x_0 : x_1$ and $y_0 : y_1$, respectively. The divisor $D = \sum n_i C_i$ is algebraically equivalent to zero if and only if $\sum n_i \deg' C_i = \sum n_i \deg'' C_1 = 0$. Clearly $\operatorname{Div} X / \operatorname{Div}_a X \cong \mathbb{Z}l_1 \oplus \mathbb{Z}l_2$, where l_1 and l_2 are the images of generators of the two different systems.

Example 3. Let $X = E_1 \times E_2$, where E_1 and E_2 are elliptic curves. In the "general case" (e. g., if the absolute invariants J_1 and J_2 of E_1 and E_2 are algebraically independent over $\mathbb{Q}$) $\varrho_X = 2$ and $S_X \cong \mathbb{Z} \oplus \mathbb{Z}$ and is generated by $E_1 \times e_2$ and $e_1 \times E_2$. In certain special cases $\varrho_X > 2$, and the new classes in S_X are generated by the graphs of surjective homomorphisms $\varphi\colon E_1 \to E_2$. If $k = \mathbb{C}$, $E_1 = \mathbb{C}/\Omega_1$, and $E_2 = \mathbb{C}/\Omega_2$, where $\Omega_1, \Omega_2 \subset \mathbb{C}$ are lattices (Shokurov (1988)), then those homomorphisms correspond to the complex numbers $\alpha \neq 0$ such that $\alpha\Omega_1 \subset \Omega_2$, i. e. to sublattices of Ω_2 congruent to Ω_1.

In general, for a connected family of surfaces, the number ϱ takes a certain minimal value, which jumps if the parameters defining the surfaces of that family ("moduli") satisfy certain additional conditions. As a rule, it is a very delicate task to find all possible values of those "jumps". Several special cases are discussed in Sect. 11 and 12.

Example 4. We assume that a surface X is embedded in $\mathbb{P}^n$. We obtain a divisor on X by intersecting X with a hyperplane not containing it. Since the hyperplanes in $\mathbb{P}^n$ form an algebraic family, the divisors obtained this way are all algebraically equivalent and define an element of S_X. An element $H \in S_X$ corresponding to the projective embedding $X \hookrightarrow \mathbb{P}^n$ is said to be *very ample.* Any positive multiple of a very ample element is also very ample – use a Veronese map. An element $E \in S_X$ is said to be *ample* if nE is very ample for a certain $n > 0$.

Since X is a projective space, S_X always contains very ample algebraic equivalence classes. Sometimes S_X has no other elements, i. e. $\varrho = 1$. In a certain very vague sense, this is a "general case". For example, Noether's theorem states that for $n \geq 4$, there is a subset in the space of forms of degree n in 4 variables, which is a union of a countably many algebraic subvarieties, such that a form, not in that subset, defines a surface $X \subset \mathbb{P}^3$ with $S_X = \mathbb{Z}H$, where H is a plane section.

On the other hand, for the "Fermat surface" in $\mathbb{P}^3$ with an equation $x_0^n + x_1^n + x_2^n + x_3^n = 0$, we get $\varrho = 3(n-1)(n-2) + 1 > 1$ if $(n, 6) = 1$, and ϱ is even larger if $(n, 6) > 1$ (Aoki-Shioda (1983)).

3.3. Linear Equivalence. How to describe the set of all divisors algebraically equivalent to a given one? The situation becomes more accesible if we consider only effective divisors algebraically equivalent to a divisor D. In Example 1, the effective divisors equivalent to a curve of degree n are described by nontrivial forms of degree n up to constant factors. Thus they are parametrized by the points of the projective space $\mathbb{P}^{n(n+3)/2}$. On the other hand, in Example 3, the effective divisors algebraically equivalent to $E_1 \times e_2$

have the form $E_1 \times a$, $a \in E_2$, so they are parametrized by points of the curve E_2. Further, the effective divisors algebraically equivalent $(E_1 \times e_2)+(e_1 \times E_2)$ have the form $(E_1 \times a_2) + (a_1 + E_2)$, so they are parametrized by the points (a_1, a_2) of the variety $E_1 \times E_2$. These two examples point on two typical cases, namely, the set of effective divisors algebraically equivalent to a given one are parametrized by the points of rational (as in Example 1) or irrational (as in Example 3) variety.

A *rational family* of divisors on a surface X is defined exactly as an algebraic family – only a base S must be rational variety. A divisor is said to be *rationally* (or *linearly*) *equivalent to zero* if it has the form $D_{s_1} - D_{s_2}$, where D_{s_1} and D_{s_2} are members of a rational family of divisors. The divisors rationally equivalent to zero form a subgroup of $\operatorname{Div} X$, denoted by $\operatorname{Div}_l X$. Clearly $\operatorname{Div}_l X \subset \operatorname{Div}_a X$. Divisors D_1 and D_2 are said to be *rationally* (or *linearly*) *equivalent to zero* if the divisor $D_1 - D_2$ is rationally equivalent to zero.

The simplest case is the one when the base S coincides with $\mathbb{P}^1$. Then the family is given by a rational function $f \in k(X)$, and the divisor ϑ has a local equation $f - t$, where t is a coordinate in $\mathbb{P}^1$. In this case, D_s are the "level curves" of the function f, i. e., $D_a = (f - a)$ for $a \neq \infty$ and $D_\infty = (f)_\infty$. If $D = D_0 - D_\infty$ then $D = (f)$. One can show that the general case is reduced to this special case, i. e., D is linearly equivalent to zero if and only if it is a divisor of a function, namely $D = (f)$.

The group $\operatorname{Div} X / \operatorname{Div}_l X$ is called the *Picard group* of X and is denoted by $\operatorname{Pic} X$. Since a line bundle corresponds to a divisor, one can verify that this group is isomorphic to $H^1(X, \mathcal{O}_X^*)$ (in the Zariski topology). One can make the group $\operatorname{Pic} X$ into a scheme compatible with multiplication. Then the group $\operatorname{Div}_a X / \operatorname{Div}_l X$ is the connected component of unity and is denoted by $\operatorname{Pic}^0 X$. Over the field of characteristic 0, it has a structure of an Abelian variety. In general, it is a group scheme, perhaps nonreduced. Clearly $\operatorname{Pic} X / \operatorname{Pic}^0 X \cong S_X$.

To clarify the picture, we consider the case when $k = \mathbb{C}$. Then we may view X as a complex analytic variety, and there is a one-to-one correspondence between analytic and algebraic line bundles over $X(\mathbb{C})$ (Griffiths-Harris (1978)). Therefore $\operatorname{Pic} X \cong H^1(X, (\mathcal{O}^*)^{\mathrm{an}})$, where the cohomology are understood in the complex sense. The map exp: $f \mapsto e^{2\pi i f}$ defines a sheaf homomorphism exp: $\mathcal{O}^{\mathrm{an}} \to (\mathcal{O}^*)^{\mathrm{an}}$, and the exact sequence

$$0 \to \mathbb{Z} \xrightarrow{\alpha} \mathcal{O}^{\mathrm{an}} \xrightarrow{\exp} (\mathcal{O}^*)^{\mathrm{an}} \to 1\,, \tag{2}$$

is called the *exponential sequence* (CAV and (Griffiths-Harris (1978))). It yields an exact cohomology sequence

$$\begin{aligned} 0 \to H^1(X,\mathbb{Z}) \xrightarrow{\alpha_1} H^1(X,\mathcal{O}^{\mathrm{an}}) \xrightarrow{\exp} H^1(X,(\mathcal{O}^*)^{\mathrm{an}}) \xrightarrow{\delta_1} H^2(X,\mathbb{Z}) \xrightarrow{\alpha_2} \\ \xrightarrow{\alpha_2} H^2(X,\mathcal{O}^{\mathrm{an}})\,. \end{aligned} \tag{3}$$

Utilizing the Hodge theory (see (Griffiths-Harris (1978)) or Sect. 14), one may interpret the homomorphisms of (3) as follows. For $i = 1, 2$, the homomorphism α_i maps a cycle $c \in H^i(X, \mathbb{Z})$ to its component of type $(0, i)$ (in particular,

$$H^1(X, \mathcal{O}^{\mathrm{an}}) \simeq \mathbb{C}^q\,, \quad H^1(X, \mathbb{Z}) \simeq \mathbb{Z}^{2q}\,, \quad H^1(X, \mathcal{O}^{\mathrm{an}})/\alpha_1 H^1(X, \mathbb{Z}) \simeq \mathbb{C}^q/\mathbb{Z}^{2q}$$

and the latter is a q-dimensional complex torus). On the other hand, δ_1 maps a line bundle (as an element of $H^1(X, (\mathcal{O}^*)^{\mathrm{an}})$) to its Chern class or, what is the same, maps a divisor defining the bundle to its Poincaré dual cohomology class. Thus $\operatorname{Ker}\delta_1$ consists of the divisor classes homologically equivalent to zero, i. e. algebraically equivalent to zero, as we have see before. In other words, $\operatorname{Ker}\delta_1 = \operatorname{Pic}^0 X \cong \mathbb{C}^q/\mathbb{Z}^{2q}$.

The sequence (3) yields two more important corollaries.

First, we have seen that $\operatorname{Pic}^0 X = \operatorname{Ker}\delta_1$ consists of line bundles with trivial Chern class, which means they are trivial as topological bundles. Thus $\operatorname{Pic}^0 X$ measures the difference between topological isomorphisms of line bundles and analytic (or algebraic) isomorphisms.

Secondly, $\operatorname{Im}\delta_1 = \operatorname{Ker}\alpha_2$ consists of integral cocycles c whose components of type $(0, 2)$ are trivial, i. e. $c^{0,2} = 0$. Since c is integral hence real, we get also $c^{2,0} = 0$, i. e., c is a cocycle of type $(1, 1)$. In other words, the image of the Severi group S_X in $H^2(X, \mathbb{Z})$ coincides with the cycles of type $(1, 1)$. The condition that a real cocycle c has type $(1, 1)$ means that $c \cup x = 0$ for all cocycles x of type $(0, 2)$. Since $H^{0,2} \cong H^0(\Omega^2_X)$ ($H^{0,2}$ coincides with the space of regular differential 2-forms on X), the last condition means that $\int_d \omega = 0$ for all $\omega \in H^0(\Omega^2_X)$, where d is a cycle dual to c, i. e. a homology class of a divisor $D \in S_X$. This result is called the *Lefschetz theorem*.

Example 5. For a surface X with $p_g = 0$, $H^0(\Omega^2_X) = 0$ so every cycle is algebraic, i. e. $S_X \cong H^2(X, \mathbb{Z})$. If, in addition, $q = 0$ then $\operatorname{Pic}^0 X = 0$ and $\operatorname{Pic} X \cong S_X \cong H^2(X, \mathbb{Z})$. In particular, this holds for rational surfaces.

Example 6. Let A be a 2-dimensional Abelian variety (Sect. 2, Example 6). Then $q = 2$, $b_1 = 4$, $b_2 = 6$, and $A \cong \mathbb{C}^2/\oplus_{i=1}^4 \mathbb{Z}e_i$ (as complex varieties). The e_i's, $i = 1, \ldots, 4$, correspond to a basis $\sigma_1, \ldots, \sigma_4$ of the group $H_1(A, \mathbb{Z})$, and coordinates in $\mathbb{C}^2$ to a basis φ_1, φ_2 in $H^0(A, \Omega^1_A)$. The isomorphism $A \cong \mathbb{C}^2/\oplus_{i=1}^4 \mathbb{Z}e_i$ is given as follows:

$$e \mapsto \left(\int_{a_0}^{a} \varphi_1, \int_{a_0}^{a} \varphi_2\right),$$

where the integrals are taken along a path from a fixed point a_0 to a. In addition, $H^0(A, \Omega^2_A) \cong \omega\mathbb{C}$ where $\omega = \varphi_1 \wedge \varphi_2$. The cycles $\sigma_{ij} = \sigma_i \cup \sigma_j$, $i < j$, $i, j = 1, \ldots, 4$, form a basis of $H_2(A, \mathbb{Z})$, and

$$\int_{\sigma_{ij}} \omega = \int_{\sigma_i} \varphi_1 \int_{\sigma_j} \varphi_2 - \int_{\sigma_i} \varphi_2 \int_{\sigma_j} \varphi_1 = c_i c'_j - c_j c'_i\,, \quad \text{where } (c_i, c'_i) = e_i\,.$$

It follows that $\sum n_{ij}\sigma_{ij}$ is algebraic if and only if $\sum n_{ij}(c_i c'_j - c_j c'_i) = 0$ and the Picard number ϱ $(= \operatorname{rk} S_X)$ equals to the number of linearly independent integral relations between the numbers $c_i c'_j - c_j c'_i$, $1 \le i < j \le 4$. For "general" c_i and c_i', there are no relations at all – this corresponds to the case when the torus $\mathbb{C}^2/\bigoplus_{i=1}^4 \mathbb{Z}e_i$ is not an algebraic variety. But if the torus is algebraic, i. e. an Abelian variety, we get, in the "general case", a unique relation, corresponding to the ample divisor classes. In more special cases (e. g., if $A = E_1 \times E_2$ where E_1 and E_2 are elliptic curves), we can have 2, 3, or 4 independent relations, corresponding to $\varrho = 2, 3, 4$. By the same Lefschetz theorem, $\varrho \leq h^{1,1} = b_2 - 2p_g$, and $\varrho \leq 4$ in our case.

Finally, we will briefly describe the algebraic counterpart of this theory. It is based on the fact that for $k = \mathbb{C}$, the variety $\operatorname{Pic}^0 X$ not only set-theoretically parametrizes the divisor classes algebraically equivalent to zero, but allows to construct all algebraic families of such classes as well. In fact, there is a line bundle L on $X \times \operatorname{Pic}^0 X$ such that given an arbitrary variety S, we get a one-to-one correspondence between the maps $f\colon S \to \operatorname{Pic}^0 X$ and the elements of the group $P^0(X \times S)/\pi^* P(S)$, namely the map f corresponds to the sheaf $(1 \times f)^* L$ on $X \times S$. Here $P^0(X \times S)$ is the group of line bundles on $X \times S$ such that the corresponding divisors on $X \times x$, $s \in S$, are algebraically equivalent to zero, $P(S)$ is the group of all line bundles on S, and $\pi\colon X \times S \to S$ the projection.

This universal property characterizes $\operatorname{Pic}^0 X$ up to isomorphism. We have described the universal property algebraically, and, for algebraic surfaces over an arbitrary field, one may take it as a definition; in this case it is natural to take arbitrary schemes of finite type over k as S. It is not clear a priori that $\operatorname{Pic}^0 X$ exists. The main existence theorem states that $\operatorname{Pic}^0 X$ (defined by the universal property) exists over an arbitrary field k, though only as a (not necessary reduced) connected projective group scheme. However, according to a well-known simple result, in characteristic 0 any group scheme is reduced, so $\operatorname{Pic}^0 X$ is an Abelian variety provided $\operatorname{char} k = 0$. For examples of nonreduced $\operatorname{Pic}^0 X$ over fields of finite characteristic, see Sect. 15.

3.4. Picard and Albanese Varieties. The group scheme $\operatorname{Pic}^0 X$ always contains a maximal reduced subscheme $(\operatorname{Pic}^0 X)_{\mathrm{red}}$ which is an Abelian variety. It is called the *Picard variety.* There is a connection between this variety and another very important variety, namely the *Albanese variety* $\operatorname{Alb} X$. The latter is defined as an Abelian variety A together with a morphism $\varphi\colon X \to A$ (such that $\varphi(X)$ generates A as an abstract group) that satisfy the following universal property. For any morphism $\psi\colon X \to B$, there exists a unique morphism $f\colon A \to B$ such that the diagram

$$\begin{array}{ccc} X & \xrightarrow{\ \varphi\ } & A \\ & {\scriptstyle\psi}\searrow \quad \swarrow{\scriptstyle f} & \\ & B & \end{array}$$

is commutative. It has been established that the Albanese variety, $\operatorname{Alb} X$, exists and is unique. Moreover, the morphism φ is unique up to a translation by a point of $\operatorname{Alb} X$.

For $k = \mathbb{C}$, the map φ is defined as follows

$$\varphi(x) = \left(\int_{x_0}^{x} \omega_1, \ldots, \int_{x_0}^{x} \omega_q\right),$$

where $x \in X$, $x_0 \in X$ is a fixed point, $\omega_1, \ldots, \omega_q$ form a basis of $H^0(\Omega^1_X)$, and all the integrals are taken along a path in X from x_0 to x. The $\varphi(X)$ is clearly defined only modulo the vectors

$$\left(\int_\sigma \omega_1, \ldots, \int_\sigma \omega_q\right) \in \mathbb{C}^q, \qquad \sigma \in H_1(X, \mathbb{Z}),$$

which form a lattice $\Omega = \mathbb{Z}^{2q}$ in $\mathbb{C}^q$, and $\operatorname{Alb} X = \mathbb{C}^q/\Omega$.

If X is a curve, we get a well-known definition of the Jacobian of a curve: $J(X) \cong \operatorname{Pic}^0 X$. If X is a surface, we get two distinct Abelian varieties, $\operatorname{Pic}^0 X$ and $\operatorname{Alb} X$, which are dual to each other. In particular, $(\operatorname{Pic}^0 X)_{\text{red}} \cong (\operatorname{Alb} X)/\Gamma$ where $\Gamma \subset \operatorname{Alb} X$ is a finite group subscheme (Mumford (1970a)).

One can easily calculate the tangent space to $\operatorname{Pic}^0 X$ at its zero element. It is isomorphic to $H^1(X, \mathcal{O}_X)$ and has dimension $h^{0,1}$. So $\dim \operatorname{Pic}^0 X \leq h^{0,1}$ with equality if and only if $\operatorname{Pic}^0 X$ is reduced (e. g., over a field of characteristic 0).

The properties of $\operatorname{Pic}^0 X$ and $\operatorname{Alb} X$, described above, are not specific for algebraic surfaces. They can be generalized to arbitrary smooth projective varieties (and certain more general schemes). However, in contrast with algebraic curves, one can obtain any Abelian variety as $\operatorname{Alb} X$ or $\operatorname{Pic}^0 X$ for a suitable algebraic surface X – there is no "Schottky problem" for algebraic surfaces.

3.5. Divisors on Fibrations. In case of fibrations in curves $f\colon X \to B$, we get a "relative" version of the main notions of divisor theory. The restriction to the generic fiber F_ξ defines a homomorphism of the group $\operatorname{Div} X$ onto the group $\operatorname{Div} F_\xi$ of divisors on the curve F_ξ over $k(B)$ defined over this field. The kernel is the group Φ of divisors generated by components of fibers, including the fibers itself – only they do not intersect the generic fiber. The homomorphism $\operatorname{Div} X \to \operatorname{Div} F_\xi$ yields a homomorphism $\operatorname{Pic} X \to \operatorname{Pic} F_\xi$, where $\operatorname{Pic} F_\xi$ is the divisor class group of F_ξ over $k(B)$. The degree homomorphism $\operatorname{Pic} F_\xi \to \mathbb{Z}$ is compatible with the homomorphism $D \mapsto (D, F)$, where F is a fiber and (D, F) is the degree of the restriction of D to F (it equals the intersection number of D and F defined in Sect. 4, below). So, we get an exact sequence

$$0 \to \Phi \to (\operatorname{Pic} X)_F \to \operatorname{Pic}^0 F_\xi \to 0, \tag{3}$$

where $(\operatorname{Pic} X)_F$ is the kernel of the homomorphism

$$\operatorname{Pic} X \to \mathbb{Z}, \qquad D \mapsto (D, F).$$

§ 4. Intersection Numbers

4.1. Main Properties. As before, we first assume that $k = \mathbb{C}$ and consider the homological interpretation of the divisor classes. As we mentioned in Sect. 3, each divisor $D \in \operatorname{Div} X$ defines a cycle and a homology class $\langle D \rangle \in H_2(X(\mathbb{C}), \mathbb{Z})$. Given two classes $\alpha, \beta \in H_2(X(\mathbb{C}), \mathbb{Z})$, the intersection number $(\alpha, \beta) \in \mathbb{Z}$ is defined, hence we can define the intersection number (D_1, D_2) for every $D_1, D_2 \in \operatorname{Div} X$. The same construction can be also described as follows. By the Poincaré duality $H_2(X(\mathbb{C}), \mathbb{Z}) \cong H^2(X(\mathbb{C}), \mathbb{Z})$, to the cycle $\langle C \rangle$ there corresponds the dual class $[D] \in H^2(X(\mathbb{C}), \mathbb{Z})$. Given $\sigma, \eta \in H^2(X(\mathbb{C}), \mathbb{Z})$, we have a cup-product $\xi \cup \eta \in H^4(X(\mathbb{C}), \mathbb{Z})$. The value of this class on the fundamental cohomology class, $\langle X(\mathbb{C}) \rangle$, defines the same intersection number (recall that $X(\mathbb{C})$ has a canonical orientation). So, the intersection number of divisors is defined via the morphism $\operatorname{Div} X \to H^2(X(\mathbb{C}), \mathbb{Z})$ and the pairing $H^2(X(\mathbb{C}), \mathbb{Z}) \times H^2(X(\mathbb{C}), \mathbb{Z}) \to \mathbb{Z}$.

Now, we will give an algebraic description of the intersection number, which will serve as the definition for algebraic surfaces over an arbitrary field. Two divisors $D_1, D_2 \in \operatorname{Div} X$ are said to be in *general position* if their supports have no common components, i. e., $\operatorname{Supp} D_1 \cap \operatorname{Supp} D_2$ consists of finitely many points. Assuming this, let $x \in X$ be an arbitrary point, and f_1 and f_2 are local equations of the effective divisors D_1 and D_2 in a neighborhood of x. Then the space $\mathcal{O}_x/(f_1, f_2)$ has finite dimension over k (here (f_1, f_2) denotes the ideal generated by $f_1, f_2 \in \mathcal{O}_X$). The number

$$(D_1, D_2)_x = \dim_k \mathcal{O}_x/(f_1, f_2) \tag{1}$$

is said to be the *local intersection number* of the effective divisors D_1 and D_2 at x.

Given two effective divisors D_1 and D_2 in general position, their *intersection number* (D_1, D_2) is defined by the formula

$$(D_1, D_2) = \sum_{x \in \operatorname{Supp} D_1 \cap \operatorname{Supp} D_2} (D_1, D_2)_x \,. \tag{2}$$

This definition can be extended to arbitrary (not necessary effective) divisors by linearity: if

$$D_1 = D_1' - D_1'' , \quad D_2 = D_2' - D_2'' ,$$
$$D_1' \geq 0 , \quad D_1'' \geq 0 , \quad D_2' \geq 0 , \quad D_2'' \geq 0 ,$$

then

$$(D_1, D_2) = (D_1', D_2') - (D_1'', D_2') - (D_1', D_2'') + (D_1'', D_2'') .$$

The intersection numbers have the following properties (assuming all the divisors are in general position):

Symmetry.

$$(D_1, D_2) = (D_2, D_1) . \tag{3}$$

Bilinearity.

$$\begin{aligned}(D_1' + D_1'', D_2) &= (D_1', D_2) + (D_1'', D_2), \\ (D_1, D_2' + D_2'') &= (D_1, D_2') + (D_1, D_2'').\end{aligned} \tag{4}$$

Invariance. If D_1' and D_1'' are algebraically equivalent (in particular, linearly equivalent), then

$$(D_1', D_2) = (D_1'', D_2). \tag{5}$$

The last property allows us to extend the definition of intersection number to arbitrary (not necessary in general position) divisors. It is easy to show that given any two divisors D_1 and D_2, there exists a divisor D_1' linearly equivalent to D_1 such that D_1' and D_2 are in general position (e. g., see (Shafarevich (1988))). Then set $(D_1, D_2) = (D_1', D_2)$; in view of (5), the result is independent of the choice of an auxiliary divisor D_1'. Clearly this intersection number satisfies (3)–(5) as well.

In particular, the number (D, D) is well defined and is denoted by (D^2).

The simplest but typical application of the properties of the intersection numbers is the Bezout theorem on $\mathbb{P}^2$. We know that $\operatorname{Pic} \mathbb{P}^2 = S_{\mathbb{P}^2} = \mathbb{Z}H$ and $(H^2) = 1$, where H is the class of a line. Given two curves $C_1, C_2 \subset \mathbb{P}^2$, we get $C_i \sim (\deg C_i)H$. Therefore

$$(C_1, C_2) = \sum (C_1, C_2)_x = ((\deg C_1)H, (\deg C_2)H) = \deg C_1 \cdot \deg C_2 .$$

This is just the *Bezout theorem: the number of intersections of two curves, counted with appropriate multiplicities, equals the product of the degrees of curves.*

The most important property, specific for the intersection numbers of divisors, is that $(D_1, D_2) \geq 0$ for two divisors in general position. This follows at once from (2). For $k = \mathbb{C}$, we can illustrate this property by an example of two curves, C_1 and C_2, intersecting transversely. In this case

$$(\langle C_1 \rangle, \langle C_2 \rangle) = \sum_{x \in C_1 \cap C_2} (C_1, C_2)_x$$

and the local topological index $(\langle C_1 \rangle, \langle C_2 \rangle)_x$ equals the intersection number of the tangent lines T_{C_1} and T_{C_2} in the tangent plane. Moreover, T_{C_1}, T_{C_2}, and T_X have a natural orientation, and $T_{C_1} \oplus T_{C_2}$ and T_X have the same orientation, because an isomorphism between them is a complex linear map, say φ, such that the determinant of the corresponding real transformation equals $|\varphi|^2$. Hence $(\langle C_1 \rangle, \langle C_2 \rangle) \geq 0$.

However, it is important to keep in mind that we can have $(D^2) < 0$ for an effective divisor D (and even a curve).

Example 1. Let X be a smooth surface of degree n in $\mathbb{P}^3$, and let $L \subset X$ be a line. We will calculate (L^2). Take a plane $\mathbb{P}^2_L$ through L. Let $H = X \cap \mathbb{P}^2_L$. On one hand, H has degree n hence

$$H = L + C$$

where C is a divisor corresponding to a complementary curve of degree $n-1$. On the other hand, H is equivalent to any other hyperplane section H'. Hence

$$(H, L) = (H', L) = (L^2) + (L, C).$$

Since H is a hyperplane section and L is a line, $(H, L) = 1$ and $(C, L) = n-1$ by the Bezout theorem. It follows that $(L^2) < 0$ if $n \geq 3$. The "Fermat surface" with an equation $x_0^n + x_1^n + x_2^n + x_3^n = 0$ provides an example of a surface containing a line (lines: $x_0 = \xi x_1$, $x_2 = \eta x_3$, $\xi^n = \eta^n = -1$).

It follows from the above discussion that a curve C with $(C^2) < 0$ have a remarkable property: there are no other curves algebraically equivalent to C, i. e., one cannot "move" C. The curves C with $(C^2) < 0$ are said to be *exceptional*.

Now, we will list several important properties of the intersection numbers. First, we consider the local intersection numbers (we assume D_1 and D_2 are in general position):

1. $(D_1, D_2)_x \geq 0$, and $(D_1, D_2)_x > 0$ if $x \in \operatorname{Supp} D_1 \cap \operatorname{Supp} D_2$.
2. $(D_1, D_2)_x = 1$ if and only if D_1 and D_2 intersect transversely at x, i. e., only one component with multiplicity 1 of D_1 and D_2 is passing through x, the point is simple on both components, and the tangent spaces of the components at x are distinct.
3. Let D_1 be an irreducible curve, and x its singular point with multiplicity r, i. e., if f is the local equation of D_1 at x and $\mathfrak{m}_x \subset \mathcal{O}_x$ a maximal ideal, then $f \in \mathfrak{m}_x^r$ and $f \notin \mathfrak{m}_x^{r+1}$. Then

$$\min_{D_2}(D_1, D_2) = r.$$

These properties imply the following properties of global intersection numbers.

4. If D_1 and D_2 are effective and in general position, then $(D_1, D_2)_x \geq 0$ with $(D_1, D_2)_x > 0$ provided $\operatorname{Supp} D_1 \cap \operatorname{Supp} D_2 \neq \emptyset$.
5. If $\operatorname{Supp} D_1 \cap \operatorname{Supp} D_2 = \{x_1, \ldots, x_r\}$, then $(D_1, D_2) \geq r$ with $(D_1, D_2) = r$ if and only if D_1 and D_2 intersect transversely.
6. If H is a very ample class, then $(H, H) > 0$ for the corresponding embedding $X \hookrightarrow \mathbb{P}^N$. The number (H^2) is called the *degree* of X in $\mathbb{P}^N$. It is equal to the maximal number of intersection points of X with a subspace $L \subset \mathbb{P}^N$ of dimension $N-2$ such that $X \cap L$ has only finitely many points.
7. For a curve C, we get $(C, H) > 0$, and (C, H) is the maximal number of intersection points of C with a hyperplane not containing C, i. e. the degree of C in $\mathbb{P}^N$.
8. For an effective divisor $D \neq 0$, we get $(D, H) > 0$.
9. If $\varphi: X \to Y$ is a morphism, then the maps φ^* and φ_* are adjoint, i. e., we have the following *projection formula* for $D_1 \in \operatorname{Div} X$ and $D_2 \in \operatorname{Div} Y$:

$$(\varphi_* D_1, D_2)_Y = (D_1, \varphi^* D_2)_X. \tag{6}$$

(Since φ^* and φ_* are compatible with equivalence of divisors, the formula (6) now makes sense for arbitrary D_1 and D_2.)

If φ has finite degree, we can apply (6) to divisors $D, D' \in \operatorname{Div} Y$ by setting $D_1 = \varphi^*(D')$ and $D_2 = D$. Applying the formula (1) of Sect. 3, we then get

$$(\varphi^* D', \varphi^* D)_X = (\deg \varphi)(D, D')_Y \,. \tag{7}$$

Example 2. In Example 6 of Sect. 2, we considered a degree 2 morphism φ: $X \to Y$, an irreducible curve $C_1 \subset Y$, and $C = \varphi^*(C_1)$. In this case, (7) gives

$$(C_1^2)_Y = \tfrac{1}{2}(C^2)_X \,.$$

4.2. Adjunction Formula. Intersection numbers enter in the genus formula for a curve on surface ("adjunction formula"). If $C \subset X$ is a smooth curve, its genus is given by the formula

$$g = \frac{(C^2) + (C, K_X)}{2} + 1 \,. \tag{8}$$

If C is not smooth, and g is the genus of its normalization, then

$$g = \frac{(C^2) + (C, K_X)}{2} + 1 - \delta \,, \tag{9}$$

where $\delta > 0$ is the sum of positive multiplicities of all singular points of C (for details, see Sect. 6).

For a smooth curve C, (8) is equivalent to

$$(C^2) + (C, K) = 2g - 2 \,,$$

and the latter follows at once by considering the restriction of T_X to C, the tangent bundle T_C, and the quotient $N_{C/X} = T_X/T_C$ called the *normal bundle* of C. Then the previous formula is just the assertion: $\det T_X = N_{C/X} \otimes T_C$ (as bundles), provided we take into account $(C^2) = \deg N_{C/X}$. The latter follows by simple calculations. In other words, the restriction of the sheaf $\mathcal{O}_X(C + K_X)$ to the curve C gives the canonical sheaf $\mathcal{O}_C(K_C)$.

We can rewrite (9) in the form

$$\frac{(C^2) + (C, K)}{2} = h^1(C, \mathcal{O}_C) - h^0(C, \mathcal{O}_C) \,. \tag{10}$$

Indeed, consider the normalization ν: $C^\nu \to C$, and the exact sequence

$$0 \to \mathcal{O}_C \to \nu_* \mathcal{O}_{C^\nu} \to \nu_* \mathcal{O}_{C^\nu} / \mathcal{O}_C \to 0 \,.$$

Then

$$h^1(C, \mathcal{O}_C) = g + \delta \,, \quad \delta = h^0(C, \nu_* \mathcal{O}_{C^\nu} / \mathcal{O}_C) \,. \tag{11}$$

We can rewrite (10) in the form $(C^2) + (C, K) = -2\chi(C, \mathcal{O}_C)$, and the exact sequence $0 \to \mathcal{O}_X(-C) \to \mathcal{O}_X \to \mathcal{O}_C \to 0$ shows that $\chi(C, \mathcal{O}_C) =$

$\chi(X, \mathcal{O}_X) - \chi(X, \mathcal{O}_X(-C))$, hence $\chi(C, \mathcal{O}_C)$ depends only on the equivalence class of C. This allows us to reduce the proof to the nonsingular case, i. e. (8), with a help of a few additional arguments.

Example 3. Let $X = \mathbb{P}^2$ and $\deg C = n$. Then we get $C = nH$ in S_X, where H is the class of a line and $K_X = -3H$ (see Example 1 of Sect. 2). Hence $g = (n-1)(n-2)/2$.

Example 4. Let $X = C_1 \times C_2$, where C_1 and C_2 are curves. One can easily verify that

$$K_X = (K_{C_1} \times C_2) + (C_1 \times K_{C_2}) = (2g_1 - 2)\widetilde{C}_2 + (2g_2 - 2)\widetilde{C}_1 ,$$

where g_1 and g_2 are the genera of C_1 and C_2, and $\widetilde{C}_1$ and $\widetilde{C}_2$ are the images of curves $C_1 \times c_2$ and $c_1 \times C_2$ ($c_1 \in C_1$ and $c_2 \in C_2$). Let $C_1 = C_2 = \mathbb{P}^1$ (so X is a quadric) and $C = n_1\widetilde{C}_1 + n_2\widetilde{C}_2$. Then $g = (n_1 - 1)(n_2 - 1)$ by (8). Next, let $C_1 = C_2 = \Gamma$, and $C = \Delta$ be the diagonal in $\Gamma \times \Gamma$. Then $\Delta \cong \Gamma$, and by (8)

$$(\Delta^2) = 2 - 2g , \tag{12}$$

where g is the genus of Γ and Δ. In particular $(\Delta^2) < 0$ if $g > 1$.

Example 5. Let C be a hyperelliptic curve of genus g, given by an equation $y^2 = f(x)$. Let i be an automorphism: $i(x, y) = (x, -y)$. Let $\Gamma \subset C \times C$ be the graph of i, i. e. $\Gamma = \{(c, i(c)) | c \in C\}$. The same argument as in Example 4 shows that $(\Gamma^2) = 2 - 2g$.

Consider the surface

$$Y = (C \times C)/G , \quad G = \{1, g\} , \quad g(c_1, c_2) = (c_2, c_1) .$$

It is easy to verify that Y is a smooth surface, though g has fixed points. The transformation g induces the automorphism i of Γ, $\Gamma/G = L \cong \mathbb{P}^1$, and the natural map $\Gamma \to L$ coincides with $(x, y) \mapsto x$. By (8), we get on Y:

$$(L^2) = 1 - g . \tag{13}$$

Yet another application of (9) and (10) is connected with the proposition of Sect. 2.

Proposition. *Let $X \to B$ be a fibration in curves, and $F_b = f^{-1}(b)$ a fiber. Then*

$$e(F_b) \geq e(F) ,$$

where F is a general fiber, with equality only if $F_b = mE$, where E is a smooth curve, and the genus of a general fiber of the family equals 1.

We will give a proof in case $F_b = \sum_{i=1}^m C_i$ has no multiple components. Consider the normalization $F_b^\nu = \sum C_i^\nu$ (disjoint sum), and the morphism ν: $F_b^\nu \to F_b$. Then

$$\chi(F_b, \mathcal{O}_{F_b}) = \chi(F_b^\nu, \mathcal{O}_{F_b^\nu}) - \delta\,,$$
$$e(F_b) = e(F_b^\nu) - \varepsilon\,, \tag{14}$$

where $\varepsilon = \sum_{x \in F_b}(r_x - 1)$, and r_x is the number of points in the inverse image $\nu^{-1}(x)$ of $x \in F_b$. The first relation of (14) follows from (1), and the second from the definition of the Euler characteristic via a triangulation.

Finally, $e(B) = 2\chi(B, \mathcal{O}_B)$ for a smooth complete curve B. Hence $e(F_b) = 2\chi(\mathcal{O}_{F_b}) + 2\delta - \varepsilon$ by (14). In particular $e(F) = 2\chi(\mathcal{O}_F)$ for a smooth fiber F. According to (10), $\chi(C, \mathcal{O}_C)$ is the same for algebraically equivalent curves C. Therefore

$$e(F_b) - e(F) = 2\delta - \varepsilon\,.$$

It remains to verify that $\varepsilon < 2\delta$. In fact, one can easily verify that $\varepsilon \le \delta$. In Sect. 6, we will show that $\delta = \sum_{x \in F_b} \delta_x$, as in the definition of ε; moreover, $\delta_x \ge r_x(r_x - 1)/2 \ge r_x - 1$ for $r_x \ge 2$ (formula (11)). It follows that $e(F_b) = e(F)$ only if F_b is a smooth curve (under the above assumptions).

We get the same result in the general case, provided by $e(F_b)$ we mean $e(\operatorname{Supp} F_b)$. So, we get the equality if $\operatorname{Supp} F_b$ is smooth, i. e. $F_b = mF$ where E is a smooth curve. The same argument shows that $\chi(\mathcal{O}(F)) = \chi(\mathcal{O}(F_b)) = m\chi(\mathcal{O}(E))$ by (10) since $(E^2) = 0$; furthermore, since $2\chi(\mathcal{O}(F)) = e(F)$ and $2\chi(\mathcal{O}(E)) = e(E)$, the equality $e(F) = e(E)$ is possible for $m > 1$ only if $e(E) = 0$, i. e. E is an elliptic curve.

§ 5. Numerical Equivalence of Divisors

5.1. Riemann-Roch Theorem. The Riemann-Roch theorem is the main numerical relation on a surface involving intersection numbers. We will use its special case,namely, we will apply it to sheaves $\mathcal{O}(D)$ corresponding to divisors. (For a general statement, see CAV, Chap. II, Sect. 4.) The number

$$\chi(\mathcal{O}(D)) = \chi(D) = h^0(X, \mathcal{O}(D)) - h^1(X, \mathcal{O}(D)) + h^2(X, \mathcal{O}(D))$$

is called the *Euler characteristic* of a divisor D. The *Riemann-Roch theorem* states that

$$\chi(D) = \frac{(D^2) - (D, K)}{2} + \chi(\mathcal{O}_X)\,, \tag{1}$$

$$\chi(\mathcal{O}_X) = \frac{(K^2) + e(X)}{12}\,, \tag{2}$$

where $e(X) = b_0 - b_1 + b_2 - b_3 + b_4 = 2b_0 - 2b_1 + b_2$ is the topological Euler characteristic of the surface, and the b_i's are Betti numbers, topological (if $k = \mathbb{C}$) or l-adic. The formula (2) is also called *Noether's formula.*

In practice, we apply the Riemann-Roch theorem to calculate $h^0(X, \mathcal{O}(D)) = l(D)$. By the *duality theorem* (CAV, Chap. II, Sect. 5), the spaces $H^2(X, \mathcal{O}(D))$ and $H^0(X, \mathcal{O}(K - D))$ are dual, so $h^2(X, \mathcal{O}(D)) = l(K - D)$. The term $h^1(X, \mathcal{O}(D))$ is the most difficult to control. One is able to prove the vanishing of this term only in certain special cases. In particular, we have the following result, which holds in characteristic 0 only.

The Kodaira Vanishing Theorem. *If D is ample, then $h^1(X, \mathcal{O}(-D)) = 0$. By duality $h^1(X, \mathcal{O}(K+D)) = 0$* (CAV, Chap. II, Sect. 6).

The following is an improvement.

The Ramanujam Theorem. *If $(D^2) > 0$ and $(D, C) \geq 0$ for every effective divisor C, then $h^1(X, \mathcal{O}(-D)) = 0$* (Barth-Peters-Van de Ven (1984), Ramanujam (1978)).

This statement is stronger than Kodaira's theorem, since for an ample divisor D, we get $(D^2) > 0$ and $(D, C) > 0$ if $C \neq 0$, by the properties 6 and 8 of the intersection number. We will see below that these inequalities characterize ample divisors.

However, the most simple and frequently used approach is to drop the term $h^1(X, \mathcal{O}(D))$ in (1), and replace (1) by a more crude *Riemann-Roch inequality*:

$$l(D) + l(K - D) \geq \frac{(D^2) - (D, K)}{2} + \chi(\mathcal{O}_X). \tag{3}$$

Here is a classical application of (3).

Proposition. *If $(D^2) > 0$, then one of the divisors, nD or $-nD$, is equivalent to an effective divisor for n sufficiently large.*

Indeed, for a divisor nD, $n \in \mathbb{Z}$, the left hand side of (3) is a quadratic function of n with the highest coefficient $(D^2)/2$. Writing (3) for nD and $-nD$, we conclude that there are three possibilities: (i) $l(D) \to \infty$, (ii) $l(-D) \to \infty$, or (iii) $l(K - nD) \to \infty$ and $l(K + nD) \to \infty$ as $n \to \infty$. We have to show that the latter is impossible.

Suppose we have (iii). Then for sufficiently large n, we get $K - nD \sim D_0 > 0$, and for $D' > 0$ and $D' \sim K + nD$, the map $D' \to D' + D_0$ defines an inclusion

$$L(K + nD) \to L(K - nD + K + nD) = L(2K).$$

We get $l(2K) \geq l(K + nD) \to \infty$, a contradiction.

Clearly whether nD or $-nD$ is equivalent to an effective divisor for $n \gg 0$ depends on the sign of (D, H), where H is an ample divisor.

5.2. The Cone of Effective Classes of Curves. The preceding proposition is closely related to a very important invariant of algebraic surfaces, which we will now describe.

In view of (5) of Sect. 4.1, the intersection number defines a bilinear function $(x, y) \in \mathbb{Z}$ $(x, y \in S_X)$ on the Severi group S_X. Clearly, if x is a torsion element in S_X, then $(x, y) = 0$ for all $y \in S_X$. The converse also holds by the Riemann-Roch inequality (Milne (1980), Chap. V, Sect. 3). In other words, if $t(S_X)$ denotes the torsion subgroup of S_X and $N_X = S_X/t(S_X)$, then (x, y) is defined on N_X and is nondegenerate (if $(x, y) = 0$ for all $y \in N_X$ then $x = 0$). The divisors that have the same image in N_X are said to be *numerically equivalent.* So, D_1 and D_2 are numerically equivalent if $(D_1, D) = (D_2, D)$ for every divisor D. We then write $D_1 \approx D_2$.

The group N_X is isomorphic to $\mathbb{Z}^\varrho$, where ϱ is the Picard number. Such a group N (i. e. a free $\mathbb{Z}$-module) equipped with a symmetric bilinear form $(x, y) \in \mathbb{Z}$ is said to be a *lattice.* To give a lattice is the same as to give an integral quadratic form: If $N = \oplus_{i=1}^n \mathbb{Z}e_i$ with $(e_i, e_j) = c_{ij}$, then $F(x_1, \ldots, x_\varrho) = \sum c_{ij}x_ix_j$, i. e. $F(x_1, \ldots, x_\varrho) = (x, x)$ where $x = \sum x_ie_i$. When we pass to another basis of N, the form F is replaced by an equivalent integral form, since the transformation is in fact a linear integral transformation of the variables $x_1, \ldots, x_\varrho$ with determinant ± 1. By the above discussion, $\det(c_{ij}) \neq 0$.

Thus the analog of the group $\mathbb{Z}$, which is the range of the function $\deg D$ of divisors on an algebraic curve, is a much more delicate object, namely the lattice N_X or the corresponding integral quadratic form. It turns out that the theory of algebraic surfaces is related to the theory of quadratic forms via N_X.

Example 1. For $X = \mathbb{P}^2$, we get $N_X = \mathbb{Z}l$, where l is the class of a line $\mathbb{P}^1 \subset \mathbb{P}^2$, and $(l^2) = 1$.

Example 2. If X is a quadric $\mathbb{P}^1 \times \mathbb{P}^1 \subset \mathbb{P}^3$ then $N_X = \mathbb{Z}l_1 + \mathbb{Z}l_2$, where l_1 corresponds to the line $\mathbb{P}^1 \times x$ and l_2 to the line $y \times \mathbb{P}^1$. Then

$$\begin{pmatrix} 0 & 1 \\ 1 & 0 \end{pmatrix}$$

is the matrix of the form (x, y) in the basis l_1, l_2.

Example 3. If X is a smooth cubic surface in $\mathbb{P}^3$, then $\varrho = 7$, and the quadratic form corresponding to the $\mathbb{Z}$-module N_X can be written in the form

$$x_1^2 - x_2^2 - \ldots - x_7^2$$

in a certain basis (see Sect. 13).

Example 4. Let a surface X' be the blowing-up of a surface X at a point (for a definition of this notion, see (Shafarevich (1988)); we will recall the definition in the next section). Then

$$N_{X'} \cong N_X \oplus \mathbb{Z}e\,, \quad (e^2) = -1\,, \quad (x, e) = 0 \qquad (x \in N_X)\,.$$

A much more crude invariant is the space $N_X \otimes \mathbb{R}$ with the induced form (x, y). In general, $N_X \otimes \mathbb{R}$ is a pseodoeuclidean space, since (x^2) is not necessary positive (see Example 1 of Sect. 4.1, and Examples 2 and 3 above). It is given by its *type* (r, s), where r is the number of positive and s is the number of negative coefficients c_i in $F(x_1, \ldots, x_\varrho) = \sum c_i x_i^2$ (provided the form is reduced to a sum of squares). The type of $N_X \otimes \mathbb{R}$ is determined by the *Hodge index theorem.*

Theorem 1. *The space* $N_X \otimes \mathbb{R}$ *is of type* $(1, \varrho - 1)$.

Proof. We know that N_X contains elements whose squares are positive, e. g., the element H corresponding to a hyperplane section. It will suffice to show that if $(D, H) = 0$ then $(D^2) \leq 0$. Suppose $(D, H) = 0$ and $(D^2) > 0$. Then, by the proposition of Sect. 5.1, $nD \sim D_0 > 0$ for a suitable $n \neq 0$. In view of property (8) of intersection numbers (see Sect. 4.1), it follows that $(D_0, H) > 0$. Hence $(D, H) \neq 0$, a contradiction.

An important property of pseodoeuclidean spaces of type (r, s) with $r > 0$ and $s > 0$ is the existence of isotropic vectors $x \neq 0$ with $(x, x) = 0$. In the special case when $r = 1$ and $s > 0$, we encounter another phenomenon: the cone Ω of vectors with $(x, x) > 0$ is disconnected, and has two components or *halves.* Indeed, for any vector $x_0 \in \Omega$, the linear function (x_0, y) does not vanish on Ω, since otherwise $x_0\mathbb{R} + y\mathbb{R}$ would have been a subspace of type $(2, 0)$ in $N_X \otimes \mathbb{R}$, a contradiction. Therefore Ω is a union of at least 2 components: $(x_0, y) > 0$ and $(x_0, y) < 0$. It is easy to verify that both components are connected. We see that the vectors x and y from the same component are characterized by the condition $(x, y) > 0$ (see Fig. 3).

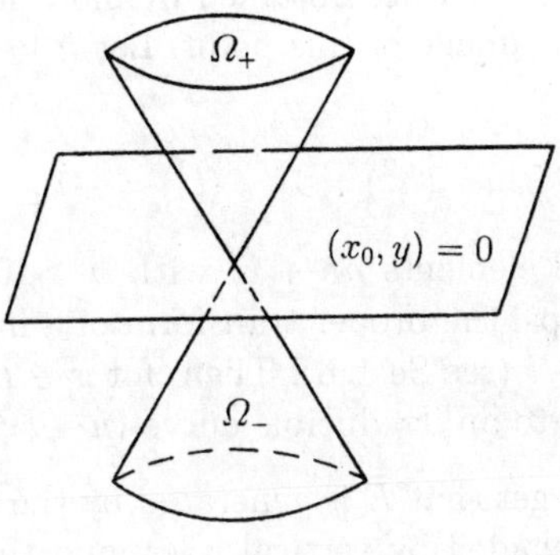

Fig. 3

In the case of $N_X \otimes \mathbb{R}$, one of the components of the cone Ω contains all the ample divisors. It is denoted by Ω_+ and is called the *positive half.*

The most important invariant of a surface X is the set of classes of N_X that contain effective divisors. A cruder but more accessible invariant is the closure in $N_X \otimes \mathbb{R}$ of the convex hull of those classes. We denote this convex cone by $\overline{E}$. By the proposition of Sect. 5.1, we get $\Omega_+ \subset \overline{E}$. So, every algebraic surface X has the following invariants: a) lattice N_X, b) pseodoeuclidean space $N_X \otimes \mathbb{R}$ of type $(1, \varrho - 1)$ that contains N_X, c) convex cone $\overline{E}$ in this space, and d) vector $K_X \in N_X$ (sometimes it is more convinient to draw the hyperplane $(K_X, x) = 0$ in $N_X \otimes \mathbb{R}$).

Example 5. Let $X = \mathbb{P}^1 \times \mathbb{P}^1$ be a quadric, and $N_X = \mathbb{Z}l_1 + \mathbb{Z}l_2$ in the notation of Example 2 of Sect. 3.2. If $D = a_1L_1 + a_2L_2$ where L_i is the divisor class of l_i $(i = 1, 2)$, and $D > 0$, then $(D, L_1) \geq 0$ and $(D, L_2) \geq 0$ since the curves L_i are mobile. It follows that $a_1 \geq 0$ and $a_2 \geq 0$, so $\overline{E} = \Omega_+$ coincides with the positive quadrant (Fig. 4).

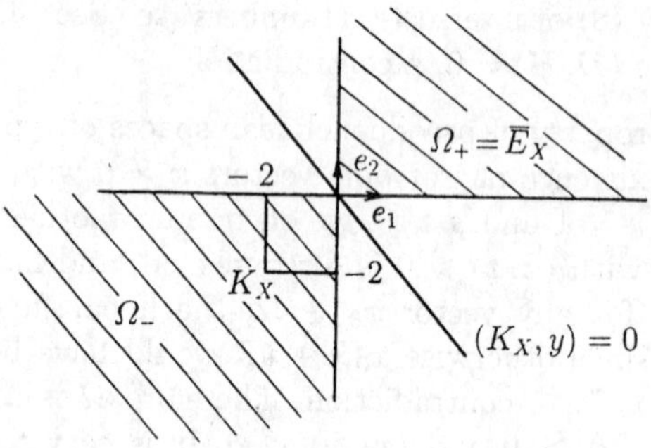

Fig. 4

Example 6. Let X be a surface obtained by blowing up $\mathbb{P}^2$ in a point, and l the class of the inverse image of this point. Let h be the total transform of a line. Then

$$N_X = \mathbb{Z}h + \mathbb{Z}l\,, \quad (h^2) = 1\,, \quad (l^2) = -1\,, \quad (h, l) = 0\,.$$

So, Ω_+ consists of the elements $ah + bl$ with $a > 0$ and $|b| < a$. By the properties of blowing-ups, the proper transform of a line through the point is effective and equals $h - l$ (see Sect. 6). Then, for $x \in \overline{E}$, we get $(h - l, x) \geq 0$ since $h - l$ corresponds to an irreducible curve $(h - l)^2 = 0$.

As in Example 5, we get that $\overline{E}$ is generated by the vectors l and $h - l$ (see Fig. 5; the cone Ω_+ is shaded by vertical lines, and the cone $\overline{E}$ is shaded by horizontal lines).

Example 7. Let X be a surface obtained by blowing up $\mathbb{P}^2$ in two points. As in Example 6,

$$N_X = \mathbb{Z}h + \mathbb{Z}l_1 + \mathbb{Z}l_2\,, \quad (h^2) = 1\,,$$
$$(h, l_i) = 0\,, \quad (l_1, l_2) = 0\,, \quad (l_i)^2 = -1\,, \qquad i = 1, 2\,.$$

Figure 6 contains a 2-dimensional section of the cones $\overline{E}$ and Ω_+, or if one prefers, their images in the projective plane $\mathbb{P}(N_X \otimes \mathbb{R})$; the plane $(K_X, y) = 0$ intersects $\overline{E}$ along an edge (a point in Fig. 6).

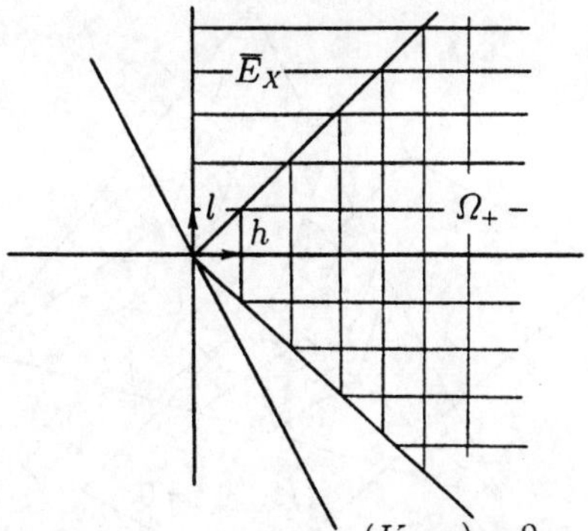

Fig. 5

Example 8. Let X be a surface obtained by blowing up $\mathbb{P}^2$ in three points on a line. Now a figure similar to Fig. 6 is 3-dimensional (see Fig. 7); the ball Ω_+ is tangent to three faces of the tetrahedron $\overline{E}$.

Example 9. If X is an Abelian variety, then any curve $C \subset X$ is mobile, namely it can be moved by a point of X. Therefore $(C^2) \geq 0$ and $\overline{E} = \overline{\Omega}_+$ (e. g., if $\varrho = 3$ we get a cone over a circle).

In general, $\overline{E}$ can have a very complicated structure – it can have infinitely many extremal rays saturated near Ω_+ (Fig. 8). The detailed study of it will play a major role in Sect. 7.

The lattice N contains yet another important set, namely the set of classes containing ample divisors (ample classes). They can be described with a help of the *Nakai-Moisheson ampleness criterion* (Hartshorne (1977)).

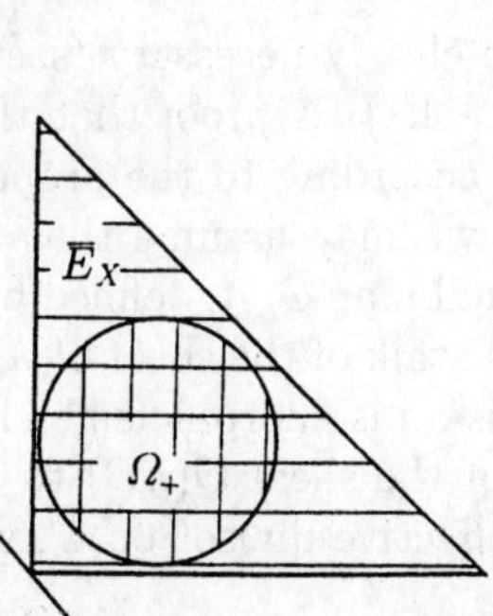

Fig. 6

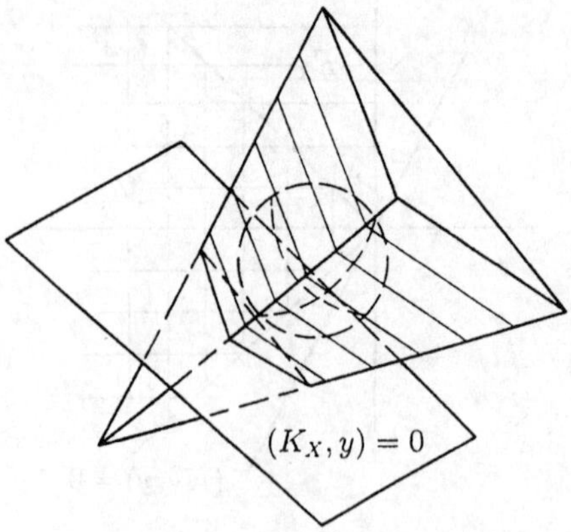

Fig. 7

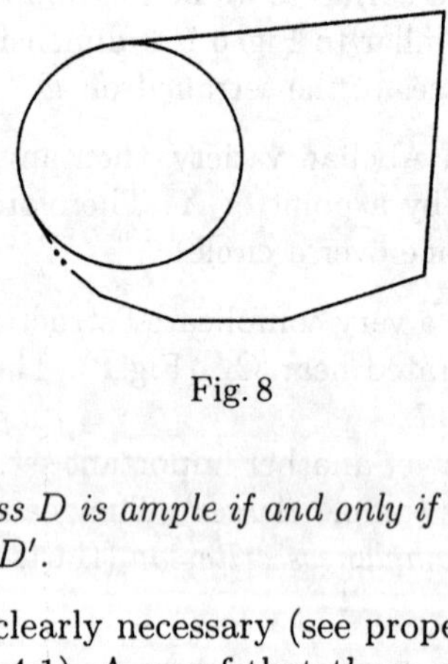

Fig. 8

Theorem 2. *The class D is ample if and only if $(D^2) > 0$ and $(D, D') > 0$ for any effective divisor D'.*

These conditions are clearly necessary (see properties 6 and 8 of the intesection numbers in Sect. 4.1). A proof that these conditions are sufficient is based on the fact that according to the proposition of Sect. 5.1, $nD > 0$ for a suitable $n > 0$, and we may assume that already $D > 0$. Then one can establish that the rational map φ_{nD}, defined by nD, is a morphism for a sufficiently large n (i. e., the stalk of the sheaf $\mathcal{O}(nD)$ at any point is generated by sections); moreover, it is an isomorphic embedding into $\mathbb{P}^N$. The main reason is that the restriction of the sheaf $\mathcal{O}(nD)$ to any curve $C \subset X$ corresponds, by assumption, to an effective divisor on C, so it posses those properties on C.

As in case of effective divisors, consider a closed convex cone $\overline{A}$ generated in the space $N_X \otimes \mathbb{R}$ by the set of ample divisors. It is given by the conditions $(x^2) \geq 0$, and $(x, y) \geq 0$ for all $y \in \overline{E}$. We will prove that the condition $(x^2) \geq 0$ is superfluous. Indeed, the set A', given by the condition "$(x, y) \geq 0$ for all $y \in \overline{E}$", is the cone $\overline{E}^*$ dual to $\overline{E}$. We know that $\overline{E} \supset \Omega_+$, and an elementary linear algebra argument establishes that Ω is self-dual, precisely

$\Omega_+^* = \overline{\Omega}_+$, where Ω_+^* is the dual cone and $\overline{\Omega}_+$ is the closure of Ω_+. It follows from $\overline{E} \supset \Omega_+$ that $\overline{E}^* \subset \Omega_+^* = \overline{\Omega}_+$, i. e. $A' \subset \overline{\Omega}_+$, which means that $A' = \overline{A}$. Thus $\overline{A} = \overline{E}^*$ is a cone dual to the cone $\overline{E}$.

One can deduce from the above discussion the *Kleiman ampleness criterion*: a divisor H is ample if and only if the linear function (H, y) is positive on $E\backslash\{0\}$.

Remark. A specific property of surfaces is that both curves and divisors have the same dimension, namely 1. In dimension 3 and higher, we get two spaces which are dual with respect to the intersection form: $N_1(X) = \{$1-cycles (curves)/ mod $\approx\} \otimes \mathbb{R}$ and $N^1(X) = \{(n-1)$-cycles (divisors)/ mod $\approx\} \otimes \mathbb{R}$. Each space has dimension ϱ, the rank of the Neron-Severi group (i. e. the Picard number). The space $N^1(X)$ contains the cones of effective, ample, and numerically effective divisor classes, while the space $N_1(X)$ contains the cone $NE(X)$ generated by the effective 1-cycles. Its closure is called the *Mori cone* of the variety X (in our case it coincides with $\overline{E}$). It is of utmost importance in the theory of minimal models of algebraic varieties of dimension $n \geq 3$ (see a remark at the end of Sect. 7).

§ 6. Birational Maps

6.1. σ-Process. One of the basic properties of algebraic curves is that a rational map $\varphi: X \to \mathbb{P}^N$ of an algebraic curve to a projective space is regular at nonsingular points of X. In particular, it follows that a birational equivalence of smooth projective curves is an isomorphism – the birational uniqueness of smooth projective models. Both properties do not hold for algebraic varieties of dimension > 1, and we first encounter this "higher-dimensional phenomenon" in case of algebraic surfaces. A classical example of this phenomenon is a birational transformation called a *σ-process* (it is also goes by the name *dilation, monoidal transformation*, and *blowing-up*). One can find the definition of σ-process in (Danilov. (1988a); see also (Hartshorne (1977))). We will recall the definition in the case of surfaces.

First, we consider X and a point $p \in X$ such that there are functions x and y, regular on X, that form a system of local parameters at p, and the equations $x = 0$ and $y = 0$ have a unique solution on X, namely the point p. (Of course, this implies that X is not projective. However, every point p on any smooth projective surface has such a neighborhood: one has to remove from the surface the polar divisor of functions x and y that form a local system of parameters at p, as well as the solutions of the system $x = y = 0$ distinct from p.) Consider the surface $Y \subset X \times \mathbb{P}^1$ given by the equation

$$x\xi_1 - y\xi_0 = 0 \tag{1}$$

where $(\xi_0 : \xi_1)$ are coordinates in $\mathbb{P}^1$. The projection $X \times \mathbb{P}^1 \to X$ defines the map $\sigma: Y \to X$, which is called the *σ-process at a point p*. Clearly $Y \supset p \times \mathbb{P}^1$.

Indeed, if $x = y = 0$, then (1) is satisfied by any ξ_0 and ξ_1. On the other hand, if at a point $q \in X$, either $x(q) \neq 0$ or $y(q) \neq 0$, then ξ_0 and ξ_1 are uniquely determined by (1). Therefore σ defines an isomorphism between $Y \backslash L$ and $X \backslash p$ (L is the curve $p \times \mathbb{P}^1$ isomorphic to $\mathbb{P}^1$), and it maps L to p, i.e., σ is a birational morphism contracting L to a point.

The variety Y is not affine even if X is affine, because Y contains the projective line L. However, it can be covered by two charts: $Y = U_0 \cup U_1$ ($U_0 = \{\xi_0 \neq 0\}$ and $U_1 = \{\xi_1 \neq 0\}$), which are affine provided X is affine. Indeed, setting $t = \xi_1/\xi_0$ in U_0, we get $k[U_0] = k[X][t]$ and $y = xt$, and setting $s = \xi_0/\xi_1$ in U_1, we get $k[U_1] = k[X][s]$ and $x = ys$. Thus, in each chart, the maximal ideal $\mathfrak{m}_p = (x, y)$ of the point p becomes a principle ideal. The line L has an equation $x = 0$ in U_0, and $y = 0$ in U_1.

Let $C \subset X$ be a curve. Then the curve $C \backslash p$ (which coincides with C if $p \notin C$) has an isomorphic inverse image $\sigma^{-1}(C \backslash p) \subset Y \backslash L$. We denote its closure in Y by $\sigma'(C)$. This closure is called the *proper transform of the curve* C.

One can easily deduce from the above description of the σ-process in charts U_0 and U_1 that if $p \in C$ and C is smooth at p, then the intersection point of $\sigma'(C)$ and L is uniquely determined by the tangent line to C at p (one should write the local equation f of C at p in the form $f = \alpha x + \beta y + g$, $g \in \mathfrak{m}_p^2$). Thus the line $L \cong \mathbb{P}^1$ is identified with the set of directions in the tangent plane T_p of X at p, i.e. with the projectivization, $\mathbb{P}(T_p)$, of this plane. We now see the geometric meaning of a σ-process: Y is obtained from X by "blowing up" a point p, which is replaced by the line $L = \mathbb{P}(T_p)$ (see Fig. 9).

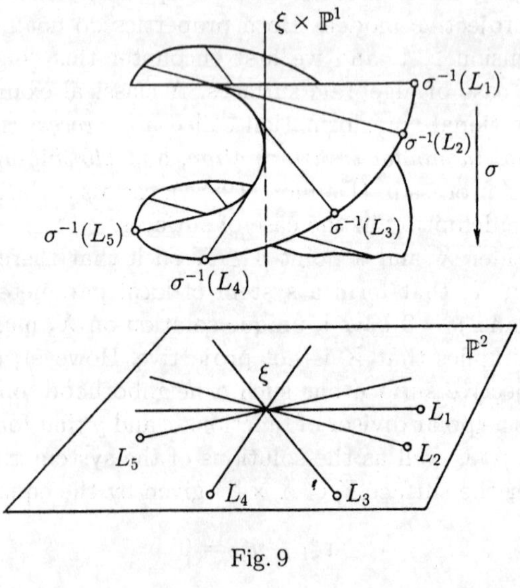

Fig. 9

Given an arbitrary surface and its simple point $p \in X$, the σ-process at p can be defined as an abstract variety. We cover X by two open sets: $X = X_0 \cup X_1$, where $p \in X_0$ and X_0 satisfies the listed at the beginning conditions, and $p \notin X_1$. Then let $Y = Y_0 \cup X_1$, where $\sigma\colon Y_0 \to X_0$ is the σ-process defined above, and Y_0 and X_1 are glued via an isomorphism

$$X_1 \supset X_0 \cap X_1 \simeq \sigma^{-1}(X_0 \cap X_1) \subset Y_0 .$$

Then the map $Y \to X$, which coincides with σ for $q \in Y_0$ and is an identity for $q \in X_1$, is said to be the *σ-process at the point* p, and is denoted by σ. Clearly $\sigma^{-1}(p) = L \simeq \mathbb{P}^1$, and $\sigma\colon Y \backslash L \to X \backslash p$ is an isomorphism.

One can easily construct a projective embedding of Y. If $X \subset \mathbb{P}^N$ then Y can be embedded into $\mathbb{P}^N \times \mathbb{P}^1$. If X is projective then Y is also projective. For a projective X, a σ-process provides the simplest example of a birational equivalence of smooth projective surfaces which is not an isomorphism. The rational map $\sigma^{-1}\colon X \dashrightarrow Y$ is an example of a rational map of a smooth surface into a projective space that is not regular.

Henceforth, we will assume that the surface X is smooth and projective, and $\sigma\colon Y \to X$ is the σ-process at the point $p \in X$. The invariants of X and Y are related as follows:

$$\operatorname{Pic} Y = \sigma^* \operatorname{Pic} X \oplus \mathbb{Z} l \quad (\text{where } l \text{ is the class of } L) \tag{2}$$

$$\begin{cases} S_Y = \sigma^* S_X \oplus \mathbb{Z} l\,, \quad N_Y = \sigma^* N_X \oplus \mathbb{Z} l\,, \\ \sigma^* N_X \simeq N_X \text{ (as lattices)}\,, \quad (\sigma^* N_X, l) = 0\,, \quad (l^2) = -1\,, \end{cases} \tag{3}$$

$$\varrho_Y = \varrho_X + 1\,, \tag{4}$$

$$b_2(Y) = b_2(X) + 1\,, \tag{5}$$

$$\begin{cases} K_Y = \sigma^* K_X + l\,, \\ (K_Y^2) = (K_X^2) - 1\,. \end{cases} \tag{6}$$

The behavior of a curve under a σ-process is determined by the relations

$$\sigma^*(C) = \sigma'(C) + rL \tag{7}$$

(where r is the multiplicity of p on C, i.e., if f is the local equation of C at p, then $f \in \mathfrak{m}_p^r$ and $f \notin \mathfrak{m}_p^{r+1}$);

$$(\sigma'(C), L) = r\,; \tag{8}$$

$$(\sigma'(C_1), \sigma'(C_2)) = (C_1, C_2) - r_1 r_2 \tag{9}$$

(where r_1 and r_2 are the multiplicities of C_1 and C_2 at p); and

$$(\sigma'(C), K_Y) = (C, K_X) + r\,. \tag{10}$$

All the relations are easily established by considering local equations of C and utilizing the simplest properties of intersection numbers.

By (7), (8), and the trivial relation $(\sigma^* D, L) = 0$ (suffice to move D away from the point p), we get an important relation $(l^2) = -1$ (compare (3)).

6.2. Birational Transformations. The above properties of a σ-process allows us to establish basic properties of curves on algebraic surfaces as well as properties of rational maps of surfaces.

A. **Properties of Curves.** We will say that a map $\varphi\colon Y \to X$ of smooth surfaces is a *product of σ-process* if there is a sequence σ_i, $i = 1, \ldots, n$, of σ-processes:

$$Y \xrightarrow{\sigma_n} X_{n-1} \xrightarrow{\sigma_{n-1}} X_{n-2} \to \ldots \xrightarrow{\sigma_2} X_1 \xrightarrow{\sigma_1} X$$

such that $\varphi = \sigma_1 \cdot \ldots \cdot \sigma_n$. Clearly φ then defines an isomorphism $Y \setminus \cup L_i \cong X \setminus \cup p_i$, where $L_i \subset Y$ are curves and $p_i \in X$ are points. If $C \subset X$ is a curve, then the closure of the curve $\varphi^{-1}(C \setminus \cup p_i)$ in Y is a curve, denoted by $\varphi'(C)$.

Theorem 1. *Given any smooth surface X and a curve $C \subset X$, there is a smooth surface Y and a map $\varphi\colon Y \to X$ which is a product of σ-processes such that the curve $\varphi'(C)$ is smooth.*

The theorem reflects an almost trivial fact: σ-processes simplify singularities. For instance, consider a curve C with the simplest singularity at p. Under the σ-process the two tangents to the branches at p correspond to distinct points on the line L, so the singularity is resolved (see Fig. 10).

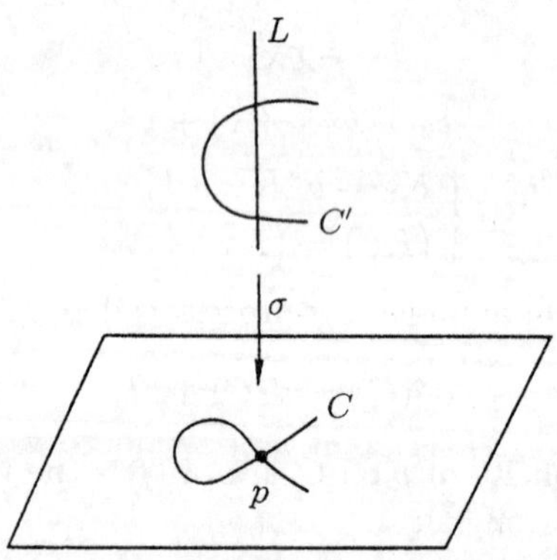

Fig. 10

The proof of the theorem is also almost trivial. First, apply σ-process at the singular points of C; next, at the singular points of the proper transforms, and so on. This process will eventually terminate, since according to (9) and (10), the number $(C^2) + (C, K)$ will decrease by $r^2 - r$ under a σ-process, while by the adjunction formula (see (9) of Sect. 4.2), $(C^2) + (C, K) \geq -2$.

Clearly the curve $\varphi'(C)$ is birationally equivalent to C. Theorem 1 provides an explicit construction of a smooth model, as well as a tool for a more detailed analysis of singular points. The points of each of the curves $(\sigma_1 \cdot \ldots \cdot \sigma_i)'(C)$ that go to the point $p \in C$ are said to be *infinitely near points of order* i lying over p.

Given a singular point p, we consider a tree of infinitely near points: Over p, we draw the infinitely near points of order 1; over each of these points, we draw the infinitely near points of order 2 that lie over them; and so on until we arrive at simple points. Next to each point we write its multiplicity as a singular point of the curve $(\sigma_1 \cdot \ldots \cdot \sigma_i)'(C)$ on the surface X_i. The tree of infinitely near points of the point $(0,0)$ on the curve $y^2 = x^2y^4 + x^4$ looks as follows:

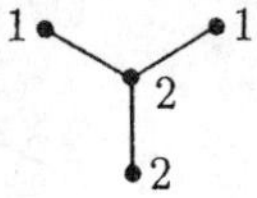

One can easily deduce from (7) the following relation:

$$(C_1, C_2) = \sum_i r_1^{(i)} r_2^{(i)}, \tag{11}$$

where $r_1^{(i)}$ and $r_2^{(i)}$ are the multiplicities of the curves C_1 and C_2 at all their common infinitely near points.

Applying (8) of Sect. 4.2 and property (6) of a σ-process in Sect. 6.1, we easily obtain the following formula for a curve C on a smooth surface:

$$g_C = \frac{(C^2) + (C, K_X)}{2} + 1 - \sum_i \frac{r_i(r_i - 1)}{2}, \tag{12}$$

where g_C is the genus of the normalization of C, and r_i are the multiplicities of all infinitely near points lying over the singular points of C. In particular, for the genus of the normalization of a plane curve of degree n, we get

$$g = \frac{(n-1)(n-2)}{2} - \sum \frac{r_i(r_i - 1)}{2}. \tag{13}$$

B. **Properties of Maps.** A point p of a smooth variety X is said to be a *point of indeterminacy* of a rational function f, if it belongs to the support of the zero divisor and polar divisor of f, namely $p \in \operatorname{Supp}(f)_0 \cap \operatorname{Supp}(f)_\infty$. For example, $(0,0)$ is the point of indeterminacy of the function y/x. Only varieties of dimension > 1 may have points of indeterminacy – it is a higher-dimensional phenomenon too. If X is a variety over the field of real or complex numbers, and p is a points of indeterminacy of a function f, then $\lim_{b \to p} f(q)$ may take any value, depending on how q tends to p.

Theorem 2 (resolution of points of indeterminacy). *Given a rational function $f\cdot$ on a smooth surface X, there are a smooth surface Y and a map φ: $Y \to X$, which is a product of σ-processes, such that $\varphi^*(f)$ has no points of indeterminacy on Y.*

For example, for the function y/x, one may take the σ-process at $(0,0)$ as φ.

The proof follows at once from the fact that the intersection number of a component of $(f)_0$ with a component of $(f)_\infty$ drops under a σ-process by (9) of Sect. 6.1, so a finite sequence of σ-processes will make it 0.

The last σ-process in the sequence will "separate" the fibers of the map f, making f a morphism (see Fig. 11).

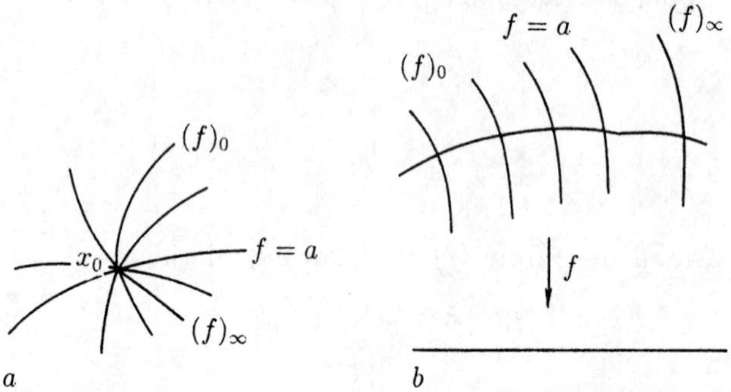

Fig. 11

Since every rational map $X \dashrightarrow \mathbb{P}^N$ is given by a collection of functions, the following corollary follows at once from Theorem 2.

Corollary 1. *Let $f\colon X \dashrightarrow \mathbb{P}^N$ be a rational map of a smooth surface X to a projective space. Then there exist smooth surface Y and a map $\varphi\colon Y \to X$, which is a product of σ-processes, such that the map $f \circ \varphi\colon Y \to \mathbb{P}^N$ is regular.*

For example, if $X = \mathbb{A}^2$ and $f(x,y) = (x : y) \in \mathbb{P}^1$, one can take the σ-process at $(0,0)$ as φ.

The following is a useful application of the above discussion.

Corollary 2. *A rational map $f\colon X \to B$ of a surface X to a curve B of genus $g(B) > 0$ is a morphism.*

Indeed, suppose f is not a morphism. Then, by Fig. 11(b), we get a morphism $f\colon L \to B$ where $L \cong \mathbb{P}^1$, a contradiction since $g(B) > 0$ (Lüroth's theorem for curves!).

The most precise result concerning the structure of birational equivalences is the following theorem.

Theorem 3. *Let $f\colon X \to B$ be any birational equivalence of smooth projective surfaces. Then there exist a smooth projective surface Z and maps $\varphi\colon Z \to X$ and $\psi\colon Z \to Y$, which are products of σ-processes, such that the following diagram is commutative:*

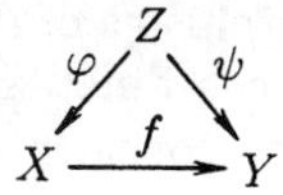

In other words, any birational equivalence can be factored into a finite sequence of σ-processes and their inverses.

In view of Corollary 1 of Theorem 1, it will suffice to show that any birational morphism $f\colon X \to Y$ is a product of σ-processes. The proof relies on the Zariski Main Theorem (Danilov (1988)): If f is not an isomorphism, then it contracts a curve $E \subset X$ to a point $p \in Y$. Let $\sigma\colon Y' \to Y$ be the σ-process at p, and $L \subset Y'$ the line contracted to the point. We have to show that $f = \sigma \circ g$, where g is a morphism; then clearly g contracts less curves than f, and the theorem follows from Zariski's theorem.

Suppose g is not a morphism. Then the rational map g^{-1} contracts a curve to a point – this follows from an easy generalization of Zariski's theorem to rational maps. Clearly g^{-1} may contract only L. Let $g^{-1}(L) = q \in E \subset X$. We get a diagram

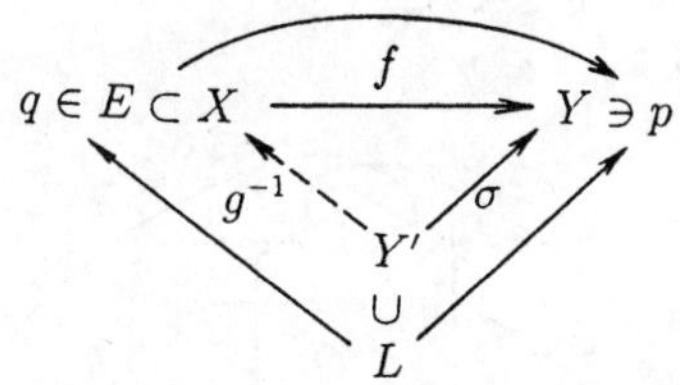

Clearly the diagram is absurd and the differential $d_q f\colon T_q \to T_p$ has a kernel, since f contracts the curve E; therefore its image lies in a line $l \subset T_p$ and, for all $y' \in L$ where g^{-1} is defined, $T_{y'}$ goes to l under $d_{y'} \cdot \sigma$. However, it follows at once from Fig. 9 that $(d_{y'}\sigma)T_{y'}$ is the whole T_p.

Remark. Theorems 1 and 2 hold for varieties of arbitrary dimension, although their proofs are much more complicated in the general case (clearly σ-processes should be replaced by blowing-ups with arbitrary smooth centers; see (Danilov (1988a), Chap. II, Sect. 1)). An analog of Theorem 3, however, does not hold even for 3-dimensional varieties – it belongs to the results specific for varieties of dimension 2.

Example 1. The stereographic projection f from a point p of a nondegenerate quadric $X \subset \mathbb{P}^3$, is a birational equivalence between X and $\mathbb{P}^2$ (see Fig. 1 in Sect. 2). The map f is not regular at p, and contracts to $x_1, x_2 \in \mathbb{P}^2$ two rulings, l_1 and l_2, of X through p (they are imaginary on the sphere in Fig. 1). The inverse map f^{-1} is not regular at x_1 and x_2, and contracts the line through x_1 and x_2 to p. Clearly $f = \tau_2 \circ \tau_1 \circ \sigma^{-1}$, where σ is the σ-process at p, and τ_1 and τ_2 are the σ-processes at x_1 and x_2.

Example 2 (birational automorphisms of $\mathbb{P}^2$). Let $p_0, p_1, p_2 \in \mathbb{P}^2$ be three points not on a line. We choose a coordinate system such that $p_0 = (1:0:0)$, $p_1 = (0:1:0)$, $p_2 = (0:0:1)$, and consider the transformation

$$f: \mathbb{P}^2 \to \mathbb{P}^2, \qquad f(x_0 : x_1 : x_2) = (x_1 x_2 : x_0 x_2 : x_0 x_1).$$

It is called a *standard quadratic* transformation. Since

$$f^2 = (x_0^2 x_1 x_2 : x_0 x_1^2 x_2 : x_0 x_1 x_2^2) = (x_0 : x_1 : x_2),$$

we get $f^2 = 1$ (identity transformation). Hence $f^{-1} = f$ and f is a birational automorphism. It is not regular at p_0, p_1, p_2, and contracts the line $x_0 = 0$ to p_0, the line $x_1 = 0$ to p_1, and the line $x_2 = 0$ to p_2. It is easy to verify that

$$f = \sigma_2' \circ \sigma_1' \circ \sigma_0' \circ \sigma_2^{-1} \circ \sigma_1^{-1} \circ \sigma_0^{-1},$$

where $\sigma_0, \sigma_1, \sigma_2$ are the σ-processes at p_0, p_1, p_2. In other words, a factorization of f whose existence was established in Theorem 3 has the form: $\varphi = \sigma_2 \circ \sigma_1 \circ \sigma_0$ and $\psi = \sigma_2' \circ \sigma_1' \circ \sigma_0'$, where $\sigma_0, \sigma_1, \sigma_2$ are the σ-processes at p_0, p_1, p_2, and $\sigma_0', \sigma_1', \sigma_2'$ are the σ-processes that contract proper transforms of the lines $x_0 = 0$, $x_1 = 0$, $x_2 = 0$ (see Fig. 12).

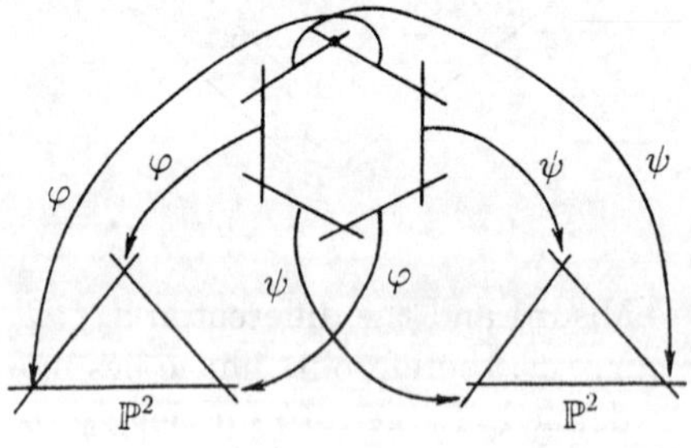

Fig. 12

The quadratic transformations are of interest because they provide examples of birational non-regular automorphisms of $\mathbb{P}^2$ (one can easily prove that projective transformations are the only automorphisms of the plane). They are important in view of the following result.

Theorem 4 (Noether's theorem). *Any birational automorphism of the plane can be factored into a product of a projective transformation and quadratic transformations* (*corresponding to different choices of the points* p_0, p_1, p_2).

Noether's theorem describes generators of the group of birational transformations of the plane. One can also describe the relations between the generators (Gizatulin (1984), Iskovskikh (1985)).

6.3. Contraction. In the above discussion, the σ-process has played the fundamental role, so it is natural to try to find an intrinsic characterization of it. The following *contraction criterion of Castelnuovo-Enriques* provides such a characterization (Hartshorne (1977)).

Theorem 5. *A smooth curve C on a smooth surface Y can be contracted to a nonsingular point $p \in X$ (i. e. there is a birational morphism $f\colon Y \to X$ with X smooth and $f(C) = p$ which is an isomorphism of $Y \backslash C$ and $X \backslash p$) if and only if $C \cong \mathbb{P}^1$ and $(C^2) = -1$.*

The morphism f is constructed in the form φ_D, where D is a suitable divisor (compare Sect. 3.1). It will contract C to a point if $(C, D) = 0$, because D is equivalent to a hyperplane section of the surface $\varphi_D(Y)$ missing the point p if $\varphi_D(C) = p$. Therefore we choose a suitable very ample divisor H, and let $D = H + mC$, where $m = (H, C)$.

First, we would like the map φ_D to be a morphism, i. e. the stalk of the sheaf $\mathcal{O}(D)$ has to be generated by its sections. This is clear at the points $x \notin C$. To prove this for $x \in C$, we establish that the restriction

$$H^0(Y, \mathcal{O}_Y(D)) \to H^0(C, \mathcal{O}_Y(D)|_C)$$

is an epimorphism: the sheaf $\mathcal{O}(D)|_C$ corresponds to a divisor of degree 0 on $\mathbb{P}^1$, so it is isomorphic to $\mathcal{O}_C$. To establish the epimorphism, we have to choose H such that $H^1(Y, \mathcal{O}(H)) = 0$; according to general properties of sheaf cohomology (CAV or Hartshorne (1977)) that is always possible by replacing H by nH, $n \gg 0$. Then the required assertion easily follows from the exact sequences.

$$0 \to \mathcal{O}_Y(H+(k-1)C) \to \mathcal{O}_Y(H+kC) \to \mathcal{O}_C(H+kC) \to 0 \ (k = 0, \ldots, m),$$

since we know the cohomology of sheaves on $C \cong \mathbb{P}^1$ corresponding to divisors.

Clearly the morphism φ_D contracts C, and is an isomorphism away from C since there $\mathcal{O}(D) \cong \mathcal{O}(H)$ and H is very ample.

However, the proof is not complete yet, because we have not established that $p = \varphi_D(C)$ is a simple point on the surface $X = \varphi_D(Y)$. In the above discussion, we have not even used that $(C^2) = -1$ – the condition $(C^2) < 0$ was suffice. In fact, one can easily describe the point p for arbitrary $(C^2) = -n$ (p is singular in general). Its complete local ring $\widehat{\mathcal{O}}_p$ is isomorphic to the ring of formal power series of the form

$$\sum_{i+j \equiv 0(n)} a_{ij} t_1^i t_2^j .$$

So it is regular only if $(C^2) = -1$.

We know that the contraction $f\colon Y \to X$ coincides with a σ-process (Theorem 3).

The curves C that satisfy Theorem 5, i. e. $C \cong \mathbb{P}^1$ and $(C^2) = -1$, are called (-1)-*curves* or *exceptional curves of the first kind.*

Example 3. Let $f\colon X \to \mathbb{P}^2$ be a blowing-up of two points in $\mathbb{P}^2$. If l is a line in $\mathbb{P}^2$ through the blown up points and l' its proper transform, then $(l'^2) = -1$. By Theorem 5, there exists a contraction $g\colon X \to Y$ such that $g(l') = p$. We have already encounted the resulting surface, namely Y is a quadric (Example 1 of Sect. 6.2).

Example 4. Let C be a curve of genus 2 with an equation $y = f(x)$ where $\deg f = 5$. Let $X = C \times C$ and $Y = X/G$, where $G = \{1, g\}$ and $g(c_1, c_2) = (c_2, c_1)$. Let Γ be the curve $\{(c, i(c))\}$, where i is the automorphism $(x, y) \mapsto (x, -y)$, and L its image in Y. According to Examples 5 of Sect. 4.2, $L \cong \mathbb{P}^1$ and $(L^2) = -1$ (formula (13) of Sect. 4.2). Therefore L can be contracted by a σ-process. We obtain a surface whose points parametrize the elements of $\mathrm{Cl}^0(C) = \mathrm{Pic}^0(C)$ (compare Example 5 of Sect. 2). It coincides with the surface $J(C)$, the Jacobian of C.

Example 5. Let $f\colon X \to B$ be a ruled surface (Example 4 of Sect. 2), $F_b = f^{-1}(b)$ one of its fibers, and $x \in F_b$ a point. Let $\sigma\colon X' \to X$ be the σ-process at x. The morphism

$$f' = f \circ \sigma : X' \to B$$

has fibers $(f')^{-1}(b') = f^{-1}(b')$ for $b' \neq b$, and $(f')^{-1}(b) = \sigma'(F_b) + L$ where $L = \sigma^{-1}(x)$. Since $(F_b^2) = 0$, we get $(\sigma'(F_b)^2) = -1$ by (8) of Sect. 4.2. Further, since $\sigma'(F_b) \cong F_b \cong \mathbb{P}^1$, we can contract the curve $\sigma(F_b)$ to a point by a σ-process $\sigma'\colon X' \to \overline{X}$ (Castelnuovo-Enriques criterion). Set $\overline{f} = f' \circ (\sigma')^{-1}$. Clearly $\overline{f}\colon \overline{X} \to B$ is a morphism, and its fiber $(\overline{f})^{-1}(b)$ coincides with $\sigma'(L) \cong \mathbb{P}^1$. Thus $\overline{X}$ is again a ruled surface, and

$$\varphi = \sigma' \circ \sigma^{-1} : X \to \overline{X}$$

is a birational transformation which is not an isomorphism. This birational transformation is said to be *elementary* and is denoted by elm_x.

Theorem 5 brings us to an important contraction problem: What configurations of curves $E \subset Y$ can be contracted to a point by a morphism $f\colon Y \to X$ that is an isomorphism away from E? Theorem 5 provides the answer in case the point $p = f(X)$ is simple. The situation is, however, more delicate for singular points. An important necessary condition is that if $C_1, \ldots, C_r$ is a collection of contractible curves, then the matrix (C_i, C_j) is negative defined.

The proof is very simple. One can easily construct a divisor $D = \sum_{i=1}^r m_i C_i$, $m_i > 0$, such that $(D, C_i) < 0$ for $i = 1, \ldots, r$. (Take a curve $H \subset Y$ distinct from each C_i and intersecting each C_i; consider a function g regular at the point p that is the image of each C_i, and equal 0 on the curve $f(H)$. Then $f^*(g) = \sum_{i=1}^r m_i C_i + F$, where $(C_i, F) > 0$ for $i = 1, \ldots, r$, and we may take $D = \sum_{i=1}^r m_i C_i$.) The required result follows from the following assertion from linear algebra.

Lemma. *Let $A = (\alpha_{ij})$ be a matrix with $\alpha_{ij} > 0$ for $i \neq j$. Let v be a vector with positive coordinates such that Av has negative coordinates. Then A is negative defined.*

For $k = \mathbb{C}$, this necessary condition is also sufficient, however, X will be, in general, a complex space which is not necessary an algebraic variety. For an arbitrary field k, this condition is also sufficient, though in the category of algebraic spaces only.

Singular points of surfaces are outside the scope of the present survey. However, we would like to touch upon one type of singularities since they naturally arise in the study of smooth surfaces.

A point $x \in X$ of a normal surface is said to be a *Du Val singularity*, if there is a smooth surface Y and a morphism $f\colon Y \to X$ that contracts curves $C_1, \ldots, C_r$ (and only these curves) to x, and $(C_i, K_Y) = 0$ for $i = 1, \ldots, r$.

(Such points are also called the *Klein singularities, rational double points, simple* or *canonical singularities.*)

The surface Y (as well as the morphism $f\colon Y \to X$) is called the *minimal resolution* of the singular point x. If X has several Du Val singularities, than there is a minimal resolution $Y \to X$ of all the Du Val singularities.

One can show that the normality of the singular points imply that the corresponding set $\cup C_i$ is connected, i. e., one cannot divide $\{C_i\}$ in two groups such that $(C_i, C_j) = 0$ if C_i and C_j belong to different groups.

As we have seen, the matrix (C_i, C_j) is negative defined; in particular $(C_i^2) < 0$. It follows at once from the condition $(C_i, K_Y) = 0$, together with (9) of Sect. 4.2, that the C_i's are smooth curves isomorphic to $\mathbb{P}^1$ and $(C_i^2) = -2$. Such curves are said to be (-2)-*curves.* Since $(C_i + C_j)^2 < 0$, we get $(C_i, C_j) \leq 1$ for $i \neq j$, i. e., C_i and C_j may intersect transversely only.

The matrices with these conditions are well known and are listed in the theory of roots of simple Lie algebras. The answer is given using graphs whose vertices correspond to the curves C_i, and two vertices are connected by a segment if the corresponding curves intersect. Only the following graphs are possible (see (Bourbaki (1968))):

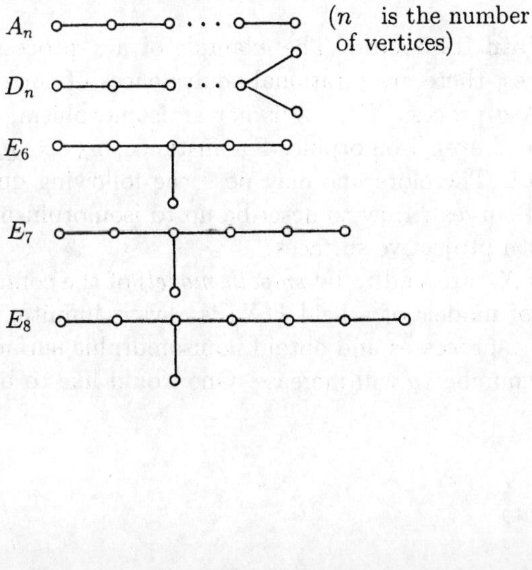

(For $k = \mathbb{C}$, each graph uniquely determines a sufficiently small neighborhood of the point x, as a complex space.) The singularities can be given by the following equations (Artin (1962)):

$$\begin{aligned} A_n&\colon\ x^2 + y^2 + z^{n+1} = 0\,, \\ D_n&\colon\ x^2 + y^2 z + z^{n-1} = 0\,, \quad n \geq 4\,, \\ E_6&\colon\ x^2 + y^3 + z^5 = 0\,, \\ E_7&\colon\ x^2 + y^3 + yz^3 = 0\,, \\ E_8&\colon\ x^2 + y^3 + z^7 = 0\,. \end{aligned}$$

The simplest Du Val singularity is A_1, which is a quadric cone.

The role of the Du Val singularities is that they "do not affect the canonical class". For instance, let $\{X_t\}$ be a flat family of surfaces whose fibers are smooth except for X_0 and X_0 has only Du Val singularities. Then there is a smooth surface Y and a birational morphism $Y \to X_0$ such that the main invariants, namely the geometric genus p_g, q, and (K^2), of Y and X_t ($t \neq 0$) are the same. Therefore, its is natural to include surfaces with Du Val singularities in families of surfaces, as "harmless" degenerations. Here we see a sharp contrast between surfaces and curves where the presence of singularities always decreases the genus of the normalization.

Furthermore, on surfaces with Du Val singularities, it is possible to develop the theory of divisors (in particular, to define canonical divisors) and intersection numbers, though with values in the field $\mathbb{Q}$ of rational numbers only. For example, let $Q^0 \subset \mathbb{P}^3$ be a quadric cone with vertex $0 \in Q^0$. Then 0 is the simplest Du Val singularity in the sense that the minimal resolution $Q' \to Q^0$ contains a unique (-2)-curve. If L is a ruling of Q^0, then (L^2) can be determined as follows. Clearly $2L$ is equivalent to a hyperplane section H of Q^0; since $(H^2) = 2$, we get $(2L, 2L) = 2$ hence $(L^2) = 1/2$.

§ 7. Minimal Models

7.1. The Main Theorem. The example of a σ-process shows that, in contrast to curves, there are birational equivalences of surfaces that are not isomorphisms. A σ-process $X' \to X$ is not an isomorphism, and the surfaces X' and X are not even isomorphic. For instance, $\varrho_{X'} = \varrho_X + 1$ according to (4) of Sect. 6.1. Therefore one may pose the following question (which is trivial in case of curves): How to describe up to isomorphism all birationally equivalent smooth projective surfaces?

Such surfaces X_α are said to be *smooth models* of the common field $k(X_\alpha)$. Clearly the set of models of a field $k(X)$ is always infinite, namely, we may perform various σ-processes and obtain nonisomorphic surfaces because each time the Picard number ϱ will increase. One would like to be able to choose (canonically if possible) certain special models, so we make the following definition.

Definition. A smooth projective surface X is called a *minimal model* if every birational morphism $f\colon X \to Y$ on a smooth surface Y is an isomorphism.

According to Theorem 3 of Sect. 6.2, the morphism f can be factored in a product of σ-processes, and by the Castelnuovo-Enriques criterion (Theorem 5 of Sect. 6.3), this is possible only if X contains an (-1)-curve. Therefore minimal models are precisely the surfaces without (-1)-curves.

If a surfaces X contains a (-1)-curve L, then by the same theorem, there is a σ-process $X \to X_1$ contracting L to a point. Next we can apply the same argument to X_1, and so on. This process will terminate, since the Picard number ϱ decreases by 1 on each step. As the result, we obtain a minimal model, i. e., any smooth projective surface is obtained from a minimal model by finitely many σ-processes.

Thus we have reduced our problem to a description of minimal models that are birationally equivalent. However, even such models are not unique. For example, we know that $\mathbb{P}^2$ and the quadric $\mathbb{P}^1 \times \mathbb{P}^1$ are birationally equivalent (Example 1 of Sect. 2). Both surfaces are minimal models, since they do not contain curves C with $(C^2) < 0$. In the case of $\mathbb{P}^2$, this follows because $N_{\mathbb{P}^2} = \mathbb{Z}h$, and for $C = nh$, we get $(C^2) = n^2$. For the quadric, this follows from Example 5 of Sect. 5.2 (Fig. 3).

Furthermore, all rational ruled surfaces, except $\mathbf{F}_1$, constructed in Example 4 of Sect. 2 are minimal models. Indeed, $\mathbf{F}_n$ contains a unique curve C with $(C^2) < 0$; it is the section S with $(S^2) = -n$ constructed in Example 4 of Sect. 2. We obtain infinitely many nonisomorphic minimal models. Utilizing the notion of elementary transformation (Example 5 of Sect. 6.3), one can show that this is the case for arbitrary ruled surfaces, namely, such a surface has infinitely many birationally equivalent nonisomorphic models. For a more detailed exposition of the theory of ruled surfaces, see Sect. 13.

So, the following two problems arise: 1) description of the set of minimal models, and 2) description of birational equivalences between the models. To begin with, we consider the first problem. We have presented, above, very special examples of minimal models, namely, $\mathbb{P}^2$ and ruled surfaces. Now, we will describe a much more general class ("general case"). This is related to the following useful notion.

Definition. A divisor D (and the corresponding class $d \in \operatorname{Pic} X$) is said to be *numerically effective* (or nef) if $(D, D') \geq 0$ for every $D' \geq 0$.

(Clearly it will suffice to verify this condition for curves D' only.) In other words, the image of the divisor D in the group N_X must belong to the convex cone $\overline{A}$ introduced in Sect. 5. As we have proved in Sect. 5, for such a divisor, we always get $(D^2) \geq 0$.

The connection with the notion of minimal model is based on the following trivial remark. If the class K_X is nef, then the surface X is a minimal model. Indeed, for a (-1)-curve L, we get $(L, K_X) = -1$ by the adjunction formula.

Surprisingly, the above examples exhaust all types of minimal models.

Theorem 1. *If X is a minimal model, then either its canonical class is* nef, *or it is isomorphic to a ruled surface or $\mathbb{P}^2$.*

A large portion of the present section is devoted to a description of ideas that form a basis of the proof of this fundamental theorem. The proof is based on a more detailed study of the convex cones $\overline{E}$ and Ω_+ described in Sect. 5.

Since every divisor is a linear combination of curves, the cone $\overline{E}$ coincides with the closed convex hull of the cone Ω_+ and the set of classes $c \in N_X$ that contain curves C with $(C^2) < 0$. Such curves, as well as the corresponding vectors in N_X and $N_X \otimes \mathbb{R}$ are called *exceptional.* They have an important property that reflects the rigidity of exceptional curves (compare Sect. 4).

Lemma 1. *Any exceptional vector in $N_X \otimes \mathbb{R}$ has a cone neighborhood that does not contain other exceptional vectors.*

(In other words, in the projective space $\mathbb{P}(N_X \otimes \mathbb{R})$, the points corresponding to exceptional vectors form a discrete set in the complement of the set $\mathbb{P}(\overline{\Omega}_+)$, where $\overline{\Omega}_+$ denotes the closure of Ω_+.)

Indeed, if C is a curve with $(C^2) < 0$, then the half-space $(C, D) < 0$ is such a cone neighborhood, since $(C, C') \geq 0$ for every curve $C' \neq C$ (Fig. 13).

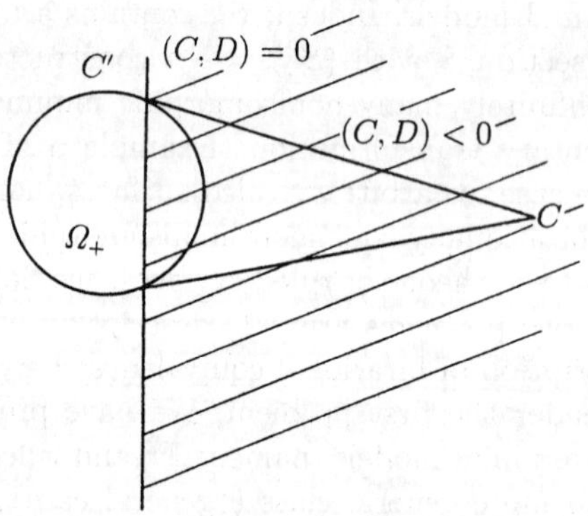

Fig. 13

However, as we have already mentioned before, we may have infinitely many exceptional curves, and then they accumulate near the boundary of $\overline{\Omega}_+$. In any case, a union of $\overline{\Omega}_+$ and all the exceptional vectors is closed. It follows that the cone $\overline{E}$ is just the convex hull of exceptional vectors and the cone $\overline{\Omega}_+$. Clearly any exceptional vector is *extremal* for the cone $\overline{E}$, i. e. cannot be written in the form $\alpha x + \beta y$, $\alpha > 0$, $\beta > 0$, $x, y \in \overline{E}$.

In the case of minimal models, we can make the picture more precise. This is based on the following simple observation.

Lemma 2. *An exceptional curve C with $(C, K_X) < 0$ is a (-1)-curve.*

The assertion follows at once from the adjunction formula. Since $(C^2) + (C, K_X) \geq -2$, we get $(C^2) = -1$ in our case, and in view of (10) of Sect. 4.2, $g(C) = 0$ and C is smooth, i. e. $C \cong \mathbb{P}^1$.

Corollary. *For a minimal model, the half-space $(K_X, x) < 0$ does not contain exceptional vectors.*

7.2. Proof of the Main Theorem. Now we may proceed to the proof of Theorem 1. We will assume that the canonical class K_X of a minimal model X is not nef, and consider two cases: I. $\varrho_X \geq 2$ (in this case, we will prove that X is a ruled surface), II. $\varrho_X = 1$ (we will prove that $X \cong \mathbb{P}^2$).

I. *The class K_X is not* nef *and $\varrho_X \geq 2$.*

To prove that X is a ruled surface, we have to find a pencil $\varphi\colon X \to B$ with a fiber $C_b = \varphi^{-1}(b)$, $b \in B$, such that $C_b \cong \mathbb{P}^1$. Clearly $(C_b^2) = (C_b, C_b') = 0$ (since all the C_b's are algebraically equivalent). By the adjunction formula (Sect. 4.2, (8)), $(C_b, K) = -2$. Conversely, if for an irreducible curve $C \subset X$, $(C^2) = 0$ and $(C, K) < 0$, then arguing as in the proof of Lemma 2, we get

$$(C, K) = -2\,, \quad g(C) = 0\,, \quad C \cong \mathbb{P}^1\,.$$

This explains why the key to a proof of the theorem in Case I is to find a curve C with $(C^2) = 0$ and $(C, K) < 0$. The idea of our proof will become more accessible if we divide the proof in two parts:

A. $\varrho_X = 2$. Then the cone $\overline{E}$ is just an angle (Fig. 14).

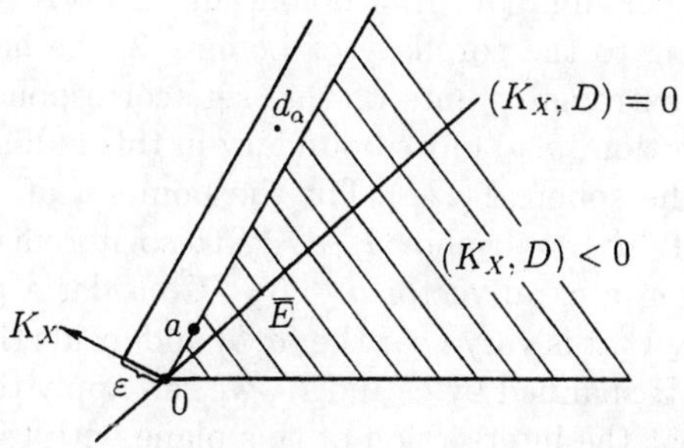

Fig. 14

Since the class K is not nef, a certain portion of the angle $\overline{E}$ will lie in the half-plane $(K, D) < 0$. As we have seen, this half-plane does not contain exceptional curves, so the extremal ray $0a$ lies in the angle $\overline{A}$, i. e. $(a^2) = 0$. Since $(a, K) < 0$, it remains only to establish that $a \in N_X \otimes \mathbb{Q}$ (we already know that $a \in N_X \otimes \mathbb{R}$!). Assuming to the contrary, we derive that for any $\varepsilon > 0$, there are vectors $d_\alpha = \alpha a + \varepsilon K \in N_X$ with arbitrary large α. Take ε with $0 < \varepsilon < 1/2$. Then for a sufficiently large α:

$$\frac{(d_\alpha^2) - (d_\alpha, K)}{2} + 1 > 0\,, \tag{1}$$

$$(d_\alpha, h) > 0\,, \tag{2}$$

$$(d_\alpha^2) < 0\,, \tag{3}$$

where h is an ample class. It follows from the Riemann-Roch inequality (Sect. 5.1 (3)), (1), and (2) that $l(d_\alpha) > 0$, contradicting the corollary of Lemma 2.

B. $\varrho_X > 2$. Although our argument works in general, we will draw, in Fig. 15, the case $\varrho_X = 3$, i. e. the projective plane $\mathbb{P}(N_X \otimes \mathbb{R})$.

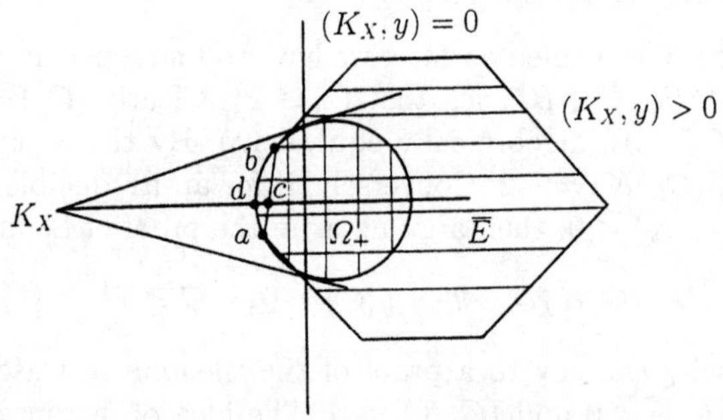

Fig. 15

By assumption, the line $(K, y) = 0$ contains interior points of the convex set $\mathbb{P}(\overline{E})$. According to the corollary of Lemma 2, the half-space $(K, y) < 0$ does not contain extremal points of this set, corresponding to exceptional curves. Hence a portion (a, b) of its boundary in this half-space coincides with the boundary of the sphere $\mathbb{P}(\Omega_+)$. For the points d of the portion (a, b) of the sphere, we get $(d^2) = 0$ and $(K, d) < 0$, so it remains to find a point corresponding to an integral vector $d \in N_X$. Consider a point corresponding to a vector $C \in N_X$ that is very near the (a, b) and inside the sphere $\mathbb{P}(\Omega_+)$. To the plane in $N_X \otimes \mathbb{R}$ spanned by C and K, we can apply the argument of Part **A**, so it follows that the intersection of this plane with the (a, b) corresponds to the class $d \in N_X$.

Thus, in both cases, we were able to find a divisor D corresponding to an extremal ray of the cone $\overline{E}$ and such that $(D^2) = 0$ and $(K, D) < 0$. Moreover, for an ample H, we get $(D, H) \geq 0$, and by the index theorem (Sect. 5.2, Theorem 1) $(D, H) > 0$. Again, by the Riemann-Roch inequality, $l(nD) \to \infty$ as $n \to \infty$. If a linear system $|nD|$, $n \gg 0$, has the mobile part D'

and $nD = D' + D^0$, where D^0 is the fixed part, then since D is an extremal ray, we get that D, D', and D^0 are proportional. We can replace D by D', and assume that $D^0 = 0$. Since $(D^2) = 0$, φ_{nD} is a morphism and its image is a curve B'.

Now, we apply the Stein factorization (Sect. 2) to this morphism. Then $\varphi_{nD} = \pi \cdot f$, where $\pi\colon B \to B'$ is a finite morphism and $f\colon X \to B$ has connected fibers. Since $(F^2) = 0$ and $(K, F) < 0$ for a fiber F, we get, as before, $p_a(F) = 0$. This implies that $F \cong \mathbb{P}^1$ provided F is irreducible. But if F is reducible, then all its components E_i are proportional in N_X, since the class of F defines an extremal ray. Therefore $(F_i, F_j) = 0$ for all i and j, contradicting the connectedness of fibers. Therefore all the fibers of $f\colon X \to B$ are irreducible and isomorphic to $\mathbb{P}^1$.

We observe that $\varrho_X = 2$ for any ruled surface, so after the discussion of Case **B**, we conclude that it cannot be realized.

II. *The class* K_X *is not* nef *and* $\varrho_X = 1$.

In this case, all our arguments and figures degenerate. Surprisingly, the argument in this simplest case is the least straightforward. We will present the main idea in the case $k = \mathbb{C}$ (i. e. in characteristic 0) only. By assumption $N_X = \mathbb{Z}h$, where we can assume that h is ample, and $K = -nh$, $n > 0$, hence $p_g = l(K) = 0$, and by the Lefschetz theorem (Sect. 3.3, Example 5) every cycle is algebraic, i. e. $b_2 = \varrho_X = 1$. By Kodaira's theorem (Sect. 5.1) $h^1(X, \mathcal{O}(K)) = 0$, hence $h^1(X, \mathcal{O}) = 0$ by duality, i. e. $q = 0$ and $\chi(\mathcal{O}_X) = 1$. By Noether's formula (Sect. 5.1, (2)) $(K^2) = 9$. If

$$S_X \cong H^2(X, \mathbb{Z}) \cong \mathbb{Z} \oplus P,$$

where P is a finite group, then we can choose a free generator in S_X such that its image in N_X, say l, is also a free generator and $(l^2) = 1$ by Poincaré duality, moreover $K_X = -3l$. We are in the same situation as on $\mathbb{P}^2$; it is natural to assume that the map φ_L defines an isomorphism $X \xrightarrow{\sim} \mathbb{P}^2$, where L corresponds to a line. The latter is in fact true, and is established by a series of simple standard arguments which we have already employed several times. This concludes the proof of Theorem 1.

Note that conversely if $X = \mathbb{P}^2$ or X is a ruled surface, then the class K_X is not nef. In the former case, this follows since $K_X = -3L$, where L is a line. In the latter case, if C is a ruling on a ruled surface, i. e. a fiber of a morphism $X \to B$ to a curve B, then $(C^2) = 0$, and since $C \cong \mathbb{P}^1$, we get $(K_X, C) = -2$ by the adjunction formula (Sect. 4.2, (8)).

7.3. Uniqueness of a Minimal Model. Now, we come to the second question posed at the end of the last section.

In Sect. 13, we will describe rational surfaces, i. e. birationally equivalent to $\mathbb{P}^2$, and ruled surfaces. We will prove that the rational ruled surfaces are exhausted by the surfaces $\mathbf{F}_n$ (Sect. 2, Example 4); all these surfaces are minimal, except $\mathbf{F}_1$, and are not isomorphic to each other.

A ruled surface $X \to B$ with $g(B) > 0$ is not rational since $q = g(B) > 0$. The ruling, i. e. the morphism $f\colon X \to B$, is uniquely determined by the surfaces X – one can show that f coincides with the *Albanese map* $\alpha\colon X \to \operatorname{Alb} X$ and $B = \alpha(X)$. Thus ruled surfaces with different bases B cannot be birationally equivalent. On the other hand, all surfaces with the same base B are birationally equivalent, and, in particular, they are birationally equivalent to $\mathbb{P}^1 \times B$. This follows from the local triviality of the fibration $X \to B$, which will be established in Sect. 13. In Sect. 13, we will also prove that all these surfaces are obtained from $\mathbb{P}^1 \times B$ by a sequence of elementary transformations (see Sect. 6.3, Example 5).

It remains to consider the largest class, namely the surfaces with K_X nef. It is a remarkable fact that such a surface cannot be birationally equivalent to a surface of the other two classes or another surface in this class (provided they are not isomorphic).

Theorem 2. *If the canonical class of X is* nef, *then any birational map $Y \dashrightarrow X$ is a birational morphism.*

If $\varphi\colon Y \dashrightarrow X$ is not a birational morphism, then by the theorem on the resolution of points of indeterminacy (see Sect. 6.2, Theorem 2), there is a sequence of σ-processes

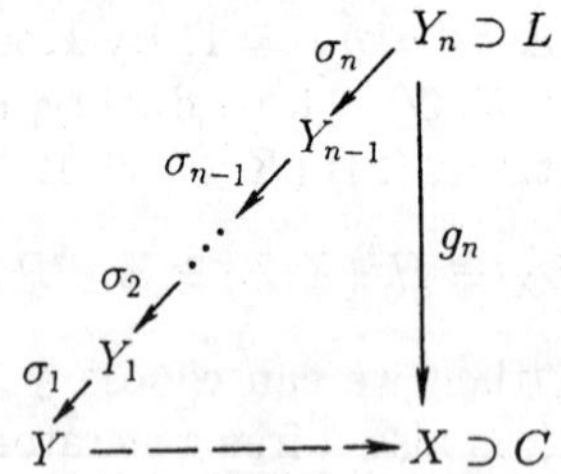

such that g_n is a morphism. Let $L \subset Y_n$ be a curve contracted by the σ-process σ_n. If $g_n(L)$ were a point, then clearly the σ-process σ_n would be unnecessary, i. e., g_{n-1} would already be a morphism. Thus we can assume that $g_n(L) = C$ is a curve. Factoring g_n in a product of σ-processes (Sect. 6.2, Theorem 3) and applying (9) of Sect. 6.1, we get $L = g_n'(C)$ and $(L, K_{Y_n}) \geq (C, K_X)$. Since K_X is nef, $(C, K_X) \geq 0$ hence $(L, K_{Y_n}) \geq 0$. However, L is a (-1)-curve, so $(L, K_{Y_n}) = -1$ (Sect. 4.2, adjunction formula).

Corollary 1. *If K_X and K_Y are* nef, *then any birational equivalence $Y \dashrightarrow X$ is an isomorphism.*

Indeed, by Theorem 2, the map $Y \dashrightarrow X$ and its inverse $X \dashrightarrow Y$ are birational morphism, i. e., $Y \cong X$ is an isomorphism.

Corollary 2. *A surface X with K_X* nef *is not birationally equivalent to $\mathbb{P}^2$ or any ruled surface.*

Indeed, let Y denote either $\mathbb{P}^2$ or any ruled surface. Suppose there is a birational equivalence φ: $Y \dashrightarrow X$. By Theorem 2, φ is a birational morphism. Since Y is a minimal surface (there are no (-1)-curves on Y, unless $Y = \mathbf{F}_1$ which is birationally equivalent to $\mathbb{P}^2$), φ: $Y \to X$ is an isomorphism. However, K_X is nef while K_Y is obviously not.

The notions and results discussed above have "relative" analogs for fibrations in curves f: $X \to B$. In this case, we consider only the morphisms φ: $Y \to X$ of fibrations f: $X \to B$ and g: $Y \to B$ that commute with the maps f and g, i. e. such that the following triangle is commutative

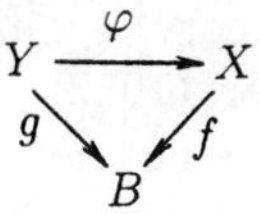

Let L be a (-1)-curve on Y. In our category, L may not be contractible, since the map $Y' \to B$ may fail to be a morphism after the contraction. (For example, if Y is a resolution of the points of indeterminacy of a map f': $Y' \dashrightarrow \mathbb{P}^1$.) However, if L lies in the fiber of g, then, after contracting L, we clearly get a morphism g': $Y' \to B$. Therefore a fibration $X \to B$ is said to be a *relative minimal model* if its fibers contain no (-1)-curves. Henceforth we will assume this.

A divisor D on X is said to be *relatively numerically effective* (or *relatively* nef) if $(D, C) \geq 0$ for every curve C that is a component of a fiber.

Lemma 3. *If f: $X \to B$ is a relative minimal model, and the genus of a general fiber is greater than* 0, *then the canonical class K_X is relatively* nef.

Since $F_b \approx F_{b'}$ for every $b, b' \in B$, we get $(F_b, C) = (F_{b'}, C)$ for every component C of F_b. Let

$$F_b = mC + \sum m_i C_i\,, \quad m > 0\,, \quad m_i \geq 0\,, \quad C \neq C_i\,.$$

Then $m(C^2) = -\sum m_i(C_i, C) \leq 0$, i. e., $(C^2) \leq 0$ and $(C^2) = 0$ if and only if $F_b = mC$ (we have used the connectedness of the fiber). By the adjunction formula (Sect. 4.2, (9)), we get $(F_b, K_X) \geq 0$ (since $(F_b^2) = 0$ and the genus of a general fiber > 0). Hence $F_b = mC$ implies $(C, K_X) \geq 0$. But if $F_b \neq mC$ then $(C^2) < 0$. So $(C, K_X) \geq 0$ by Lemma 2, since otherwise C would have been a (-1)-curve in the fiber.

Now it is easy to derive a "relative" analog of Theorem 2.

Theorem 3. *Let f:: $X \to B$ be a relative minimal model, and a genus of a general fiber* > 0. *Then any birational map φ: $Y \dashrightarrow X$ of a fibration g: $Y \to B$ that commutes with f and g is a birational morphism.*

Indeed, φ is a birational equivalence of the generic fibers G_ξ and F_ξ of the fibrations Y and X. However, a birational equivalence of smooth curves is an isomorphism. In particular, φ defines an isomorphism $\operatorname{Div} G_\xi \cong \operatorname{Div} F_\xi$, hence φ and φ^{-1} may contract curves on Y and X only if they belong to the fibers. However, according to Lemma 3, the canonical class K is numerically effective on such curves. So, we can repeat the proof of Theorem 2.

Clearly the relative case of Corollary 1 holds as well (the analog of Corollary 2 is trivial).

Remark. In the present section, we have followed an idea of Mori who has proposed to utilize the cone of effective 1-cycles $\overline{NE}(X)$ (see the remark at the end of Sect. 5) for building the theory of minimal models of algebraic varieties of arbitrary dimension $n \geq 2$ (Mori (1982)). If K_X is nef, then X is a minimal model by definition. In case K_X is not nef, the "negative part"

$$\{z \in \overline{NE}(X) | (z, K_X) < 0\}$$

of the cone $\overline{NE}(X)$ has a rather simple structure. It is a rational polyhedron, i. e., it is generated by a discrete set of extremal vectors (extremal rays). Given such an extremal vector R, we get a morphism $\varphi_R \colon X \to Y$ "contracting" R. In case of surfaces, it is either a contraction of a (-1)-curve, or a morphism defining a structure of a ruled surface $X \to B$, or $X = \mathbb{P}^2$ and $\overline{NE}(X) = \mathbb{R}_+$ is a positive half-line.

The picture is much more complicated for $n \geq 3$. Under the contraction $\varphi_R \colon X \to Y$, the variety Y may acquire singularities, though very mild. So, to continue the contraction process, one must have a similar theory for varieties with such singularities. It turned out that this is possible, although yet another difficulty emerged. On singular varieties, certain contractions of extremal vectors may take us outside the applicability of the theory. Here also a successful way out has been found (at least in dimension 3). One should point out that the minimal models constructed in this fashion have, as a rule, singularities if the dimension equals 3 or higher. As we have seen, in case of surfaces, the minimal models have no singularities, and the results coincide with the classical ones (all this is described, e. g., in (Mori (1982) and Kawamata-Matsuda-Matsuki (1987))).

§ 8. Birational Classification

8.1. Main Results. We proceed to the program outlined in Sect. 1, namely the classification of surfaces according to their canonical dimension $\kappa = 2, 1, 0, -\infty$, and numerical invariants (K^2), q, p, P_n. This classification, being the central result in the theory of surfaces, was one of the outstanding achievements of the Italian school of algebraic geometry. Throughout the rest of this survey (except Sect. 15), we always assume that the caracteristic of the field is 0.

We have already divided the surfaces in two types, namely: the surfaces with canonical class nef (they are always minimal models), and the surfaces whose minimal models have not nef canonical class. Since one can describe the second type explicitly – minimal models are ruled surfaces or are isomorphic to $\mathbb{P}^2$, we will restrict ourselves to the first type. So, henceforth, we assume that the canonical class K_X of a surface X is nef.

The classification of such surfaces is determined by the position of the canonical class (in the group N_X) relative to the positive half Ω_+ of the cone $(x^2) > 0$. Namely, the surface are divided in three classes (Fig. 16):

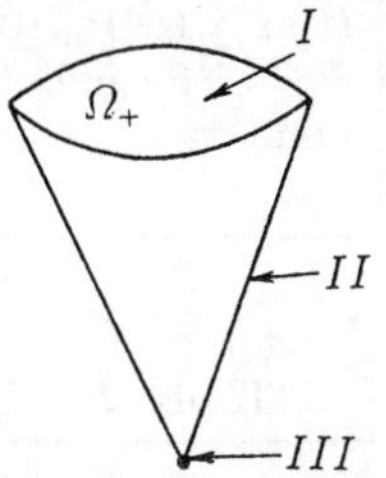

Fig. 16

I. K_X lies in the interior of the cone Ω_+, i.e. $(K^2) > 0$;
II. K_X lies on the boundary but $K_X \neq 0$ in N_X, i.e. $(K_X^2) = 0$ and $K_X \not\approx 0$;
III. K_X lies on the boundary and $K_X = 0$ in N_X, i.e. $K_X \approx 0$.

These three classes can also be characterized by the values of the canonical dimension κ, the growth of P_m as $m \to \infty$ (we write $P_m \sim m^r$ if P_m grows as a polynomial of degree r), a more precise description of the canonical class (it turns out that $K_X \approx 0$ only if $nK_X = 0$ in $\operatorname{Pic} K$ for a certain $n \in \{1, 2, 3, 4, 6\}$), and, in certain cases, by direct geometric constructions. The results of classification are collected in Table 1.

We may add the fourth class, namely the surfaces whose minimal models have not nef canonical class. They are characterized by the condition $\kappa = -\infty$. It follows from Table 1 that for these surfaces, $P_n = 0$ for $n \in \{1, 2, 3, 4, 6\}$. This can be replaced by a single condition: $P_{12} = 0$. Finally, by Theorem 1 of Sect. 7.1, their minimal models coincide with $\mathbb{P}^2$ or ruled surfaces.

The surfaces with K_X nef and $(K_X^2) > 0$ are said to be of *general type.* According to Table 1, $(K_X^2) = 0$ for the rest of the surfaces. We observe that all the surfaces in Class II are elliptic; however, certain surfaces in Class III can also be elliptic (but with $nK_X = 0$ for $n \in \{1, 2, 3, 4, 6\}$). Finally, the surfaces in Class III can be described more explicitly. The quotients of Abelian varieties by finite groups have the form $(E_1 \times E_2)/G$, where E_1 and E_2 are elliptic curves (as we mentioned in Sect. 2). Such surfaces are said to be *bi-elliptic.* We know the complete list of all possible curves E_1 and E_2 and groups G (Sect. 11). The quotients of surfaces of type $K3$ (by fixed-point-free involutions) are the *Enriques surfaces* (see Sect. 11).

To compare with Table 1, we have collected the data for curves in Table 2.

Table 1

Class	(K_X^2)	κ	P_m	Structure
I	$(K_X^2) > 0$	2	$P_m \sim m^2,\ P_2 \geq 2$	For $m \gg 0$, the map φ_{mK} is a birational morphism
II	$(K_X^2) = 0$, $nK_X \neq 0$ for $n \neq 0$	1	$P_m \sim m,\ P_m \geq 2$ if $m \in \{1, 2, 3, 4, 6\}$	Elliptic surfaces
III	$nK_X = 0$ $n \in \{1, 2, 3, 4, 6\}$	0	$(\forall m \geq 1)\ [P_m \leq 1]$, $P_n = 1$ for infinitely many values	Abelian varieties, $K3$ surfaces, or their quotients by by finite fixed-point-free groups of automorphisms

Table 2

Class	κ	g	$\deg K_X$	$\chi(\mathcal{O}_X)$	Structure
I	1	≥ 2	$2g - 2 > 0$	$1 - g < 0$	For $m \geq 3$, the map φ_{mK} is an isomorphic embedding
II	0	1	0	0	Elliptic curves
III	$-\infty$	0	$-2 < 0$	$1 > 0$	$\mathbb{P}^1$

The main purpose of this section is to describe proofs of the following theorems (henceforth, we always assume that K_X is nef).

Theorem 1. *If $(K_X^2) > 0$, then for sufficiently large m, the map φ_{mK} is a birational morphism contracting only (-2)-curves. Moreover, it contracts these curves to Du Val singularities, and maps X onto a normal surface. For these surfaces $\kappa(X) = 2$ and $P_2 \geq 2$.*

Theorem 2. *If $(K_X^2) = 0$ then $\kappa(X) \leq 1$; X is either an elliptic surface, an Abelian surface, a $K3$ surface, or an Enriques surface.*

We will discuss the surfaces of Class I in Sect. 9, and of Classes II and III in Sect. 10–12.

8.2. Discussion of Theorem 1. Since K_X is nef, we get $(K_X, H) > 0$ for a hyperplane section H. Indeed $(K_X, H) \geq 0$, and if $(K_X, H) = 0$, we derive a contradiction by the index theorem (Sect. 5.2, Theorem 1) because $(H^2) > 0$ and $(K_X^2) > 0$. Therefore $P_m \sim m^2$ by the Riemann-Roch inequality. In fact, by the vanishing theorem, we get a precise formula

$$P_m = \frac{m(m-1)}{2}(K_X^2) + \chi(\mathcal{O}_X)\,, \quad \text{for } m \geq 2\,, \tag{1}$$

which will be needed only later.

First of all, we will verify taht φ_{mK} is a birational equivalence for a sufficiently large m. This is quite easy. Indeed, $(K_X, H) > 0$ implies that $(mK_X - H, H) > 0$ and $(mK_X - H)^2 \to \infty$ as $m \to \infty$. As we have seen in Sect. 5, it follows that $l(mK_X - H) > 0$ for some $m > 0$, i. e. $mK_X = H + D$ with $D > 0$. Therefore, on $X \setminus \operatorname{Supp} D$, the map φ_{mK} coincides with φ_H, i. e. it is an inclusion.

The question when φ_{mK} is a morphism depends on the location of K_X in the cone $\overline{A}$. If K_X lies in the interior of $\overline{A}$, then our statement is a direct consequence of the Nakai-Moishezon ampleness criterion (Sect. 5.2). Assume that K_X lies on the boundary of $\overline{A}$, i. e., $(K_X, C) = 0$ for certain curves C. Then φ_{mK} contracts C for every m. We have seen in Sect. 6 that the curves C with $(K_X, C) = 0$ are (-2)-curves.

First, we will establish that the linear system $|mK_X|$ has no fixed components for sufficiently large m. Suppose E is such a component.

Consider an exact sequence

$$0 \to \mathcal{O}_X(mK_X - E) \to \mathcal{O}_X(mK_X) \to \mathcal{O}_E(mK_X|_E) \to 0\,.$$

Since K_X is nef, $(K_X, E) \geq 0$ and $h^0(E, \mathcal{O}_E(mK_X|_E)) \geq 1$, so it will suffice to show $h^1(X, mK_X - E) = 0$. We may apply the vanishing theorem (Sect. 5.1): $(mK_X - E)^2 > 0$ for $m \gg 0$, and $(mK_X - E, E) > 0$ for $m \gg 0$ provided $(K_X, E) > 0$. But if $(K_X, E) = 0$ and $(F, E) > 0$, then we cannot apply this argument. In this case $(E^2) = -r$, $E \cong \mathbb{P}^1$, and one can apply a stronger version of the vanishing theorem whose precise statement we omit (see, e. g. (Ramanujam (1978), Kawamata (1982), Vieweg (1982))). By increasing m, we can achieve that the system $|mK_X|$ has no fixed curves at all.

Similarly, one can establish the absence of base points for sufficiently large m. In other words, φ_{mK} is a morphism. We have already seen that it is a birational equivalence. Clearly only (-2)-curves are contracted. Moreover, their configurations are described by graphs of the table in Sect. 6.3, so φ_{mK} has only Du Val singularities. Now, we may use the fact that such singularities are normal, or apply the following general result that follows from elementary commutative algebra considerations. Let D be a divisor such that φ_D is a morphism. Then φ_{mD} is a normal variety for some $m > 0$. This is the so-called *theorem on projective normality* (Hartshorne (1977)). So the surface $\varphi_{mK}(X)$ is normal.

Another approach to the study of pluricanonical maps is based on the theory of vector bundles (Reider (1988)).

Remark 1. Given a surface X, one can show that for $m \gg 0$, all its images $\varphi_{mK}(X)$ are isomorphic and obtained from X by contracting to points all connected systems of (-2)-curves. This uniquely defined model of X is said to be *canonical* and is denoted by X_{can}.

Remark 2. Similarly to Theorem 1, one show that X_{can} can be constructed as follows. Let

$$R = \bigoplus_{m \geq 0} H^0(X, mK_X)$$

with a natural multiplication

$$H^0(X, mK_X) \otimes H^0(X, nK_X) \to H^0(X, (n+m)K_X)$$

(one may define R as the ring of regular differential 2-forms of all degrees on X). It is known that the graded ring R is finitely generated over $H^0(X, \mathcal{O}_X) \cong k$ and $X = \operatorname{Proj} R$ (see Mumford's appendix in (Zariski (1962))).

We have not discussed yet a proof of the assertion $P_2 \geq 2$ of Theorem 1. It requires new important ideas which we will now describe.

8.3. The Castelnuovo – de Franchis Inequality. Until now, our main and almost unique tool was the Riemann-Roch inequality. Magically, it allows us to derive that the space $L(D)$ has sufficiently many functions from the geometric properties of the divisor D, namely, (D^2) is sufficiently large. We applied this argument most often to $D = mK_X$ with $(K_X^2) > 0$ and m sufficiently large. However, the Riemann-Roch inequality is of no help if we are interested in a concrete value of m, and more so if $(K_X^2) = 0$. For instance, the formula (1) is of no help if $m = 2$ and we do not know the sign of the number $\chi(\mathcal{O}_X)$. Here we will employ a new kind of considerations.

Lemma 1 (Castelnuove – de Franchis inequality). *If the canonical class K_X is* nef, *then $e(X) \geq 0$.*

This result is based on the proposition of Sect. 2 and the proposition of Sect. 4.2. According to these propositions, if there exists a morphism $f\colon X \to C$ with connected fibers such that $g(C) \geq 0$ and $g(F) \geq 0$, where F is a general fiber of f, then $e(X) \geq e(C)e(F) \geq 0$. Moreover, it will suffice to verify that $g(C) > 0$, because $g(F) = 0$ implies $F \cong \mathbb{P}^1$ hence $(F, K_X) = -2$ by the adjunction formula, contrary to our assumption. In view of Corollary 2 of Theorem 2 in Sect. 6.2, it will suffice to construct f as a rational map because such a map will necessary be a morphism.

We will derive the lemma by contradiction. Suppose that $e(X) < 0$. Then one can even construct the required f with $g(C) \geq 2$. If there were such a morphism, then two linearly independent over k forms $\eta_1, \eta_2 \in H^0(C, \Omega_C^1)$ would produce the forms

$$\omega_i = f\eta_i \in H^0(X, \Omega_X^1), \qquad i = 1, 2$$

with $\omega_1 \wedge \omega_2 = 0$. The main idea of the proof is that the converse also holds, namely, if

$$\omega_1, \omega_2 \in H^0(X, \Omega_X^1), \qquad \omega_1 \notin k\omega_2, \quad \omega_1 \wedge \omega_2 = 0,$$

then there is a morphism $f\colon X \to C$ such that

$$\omega_i = f^*\eta_i\,, \qquad \eta_i \in H^0(C, \Omega_C^1)\,.$$

We will sketch the proof for $k = \mathbb{C}$.

Since $\omega_1 \wedge \omega_2 = 0$, we get $\omega_2 = \varphi\omega_1$, $\varphi \in k(X)$, $\varphi \notin \mathbb{C}$, i.e. φ defines a rational map $\varphi\colon X \to \mathbb{P}^1$.

Let $X' \to X \to \mathbb{P}^1$ be a resolution of points of indeterminacy, and

$$X' \xrightarrow{\psi} C \to \mathbb{P}^1$$

its Stein factorization. We consider the points that admit a system of coordinates (u, v) such that ψ can be written as $(u, v) \mapsto u$, i.e. $\psi = u$. Such points from an open set in X'. Next, we utilize that the forms ω_1 and ω_2 are closed (e.g., see Sect. 14). Then locally $\omega_1 = dh$, $\omega_2 = udh$ and since ω_2 is closed, $\partial h/\partial v = 0$. This shows that $\omega_1 = \psi^*\eta_1$ and $\omega_2 = \psi^*\eta_2$ in our open set, where η_1 and η_2 are forms holomorphic on a certain open subset of the curve C. One can easily prove that η_1 and η_2 can be extended holomorphically to the whole C hence $g(C) \geq 2$.

It remains to find $\omega_1, \omega_2 \in H^0(X, \Omega_X^1)$, $\omega_1 \notin k\omega_2$, $\omega_1 \wedge \omega_2 = 0$. We will apply the following elementary argument from linear algebra.

If E and F are finite dimensional spaces, $\dim F < 2(\dim E - 2)$, and ϱ is a homomorphism $\wedge^2 E \to F$, then there are $e_1, e_2 \in E$, $e_1 \notin ke_2$, such that $\varrho(e_1 \wedge e_2) = 0$.

Indeed, comparing the dimensions of $\operatorname{Ker} \varrho$ and the variety of decomposable bivectors, we conclude that they have a nontrivial intersection.

In our case, we have to verify the inequality $\varrho < 2(q-2)$. It may not hold for the surface X. However, we may assume that $b_1(X) > 0$, since $e(X) > 0$ otherwise. Therefore there are unramified coverings $X' \to X$ of arbitrary large degree N, and if $e(X) < 0$, then $e(X') = Ne(X)$ can be made arbitrary small. In view of (4) of Sect. 1, $e(X) \geq 2 - 4q + 2p$. So, if $e(X') < -6$, then we get the desired inequality for X' and derive a contradiction.

Corollary. *If the canonical class K_X is* nef, *then $\chi(\mathcal{O}_X) \geq 0$, and $\chi(\mathcal{O}_X) > 0$ for surfaces of general type.*

This follows at once from Lemma 1, Noether's formula, and $(K_X^2) \geq 0$; moreover, $(K_X^2) > 0$ provided X is a surface of general type.

The inequality $P_2 \geq 2$ (in Theorem 1) follows at once from (1) and the corollary since X is of general type.

8.4. Discussion of Theorem 2. Here $(K_X^2) = 0$. All the arguments are based on a construction of a pencil of elliptic curves on X, and an identification of X with an Abelian surface or a $K3$ surface when there are no such pencils.

First, assume $P_m(X) \geq 2$ for a suitable $m > 0$. If the linear system $|mK_X|$ has a fixed part $\overline{D}$, then $mK_X = \overline{D} + D$, where D runs over divisors of a mobile linear system Δ. Since $(K_X^2) = 0$, we get $(K_X, \overline{D}) + (K_X, D) = 0$. Hence $(K_X, \overline{D}) = (K_X, D) = 0$ because K_X is nef. This, in turn, implies that

$(\overline{D}, D) + (D^2) = 0$. Since Δ is a mobile system, $(\overline{D}, D) \geq 0$ and $(D^2) \geq 0$, so $(\overline{D}, D) = (D^2) = 0$.

It follows that Δ has no base points. Hence φ_Δ is a morphism. Since $\varphi_{mK} = \varphi_\Delta$, φ_{mK} is also a morphism. Further, since Δ consists of the inverse images of hyperplane sections of $\varphi_\Delta(X)$, and $(D^2) = 0$ for $D \in \Delta$, we deduce that $\varphi_\Delta(X) = \varphi_{mK}(X)$ is a curve, i.e. $\kappa(X) = 1$. Let

$$X \xrightarrow{f} B \to \varphi_{mK}(X)$$

be the Stein factorization of φ_{mK}. We get $(F_b^2) = 0$ where $F_b = f^{-1}(b)$. By Bertini's general theorem, only finitely many fibers of f have singularities. But for a smooth fiber C, we get $(C^2) = 0$ and $(C, K_X) = 0$, i.e. C is an elliptic curve. Thus X is an elliptic surface.

So, it remains to consider the case: $P_m \leq 1$ for all $m \geq 1$. Then $p = 0$ or $p = 1$, and $q \leq p + 1$ by Lemma 1 of Sect. 8.3. So, we have to consider several special cases when p and q are small and $P_m \leq 1$. We obtain several special types of surfaces. Therefore such a sorting out is not a trivial task; it is similar to the argument at the end of the proof of Theorem 1 in Sect. 7.1. We will describe only the key points.

Type A. $\chi(\mathcal{O}_X) > 0$. Then we have the following possibilities: 1) $p = 1$ and $q = 0$, 2) $p = q = 1$, 3) $p = q = 0$.

If $p = 1$ and $q = 0$, then by the Riemann-Roch theorem $\chi(\mathcal{O}_X(2K_X)) = \chi(\mathcal{O}_X) = 2$, i.e. $h^0(X, \mathcal{O}_X(2K_X)) + h^2(X, \mathcal{O}_X(2K_X)) \geq 2$. We can assume that $P_2 = h^0(X, \mathcal{O}_X(2K_X)) \leq 1$ hence $h^2(X, \mathcal{O}_X(2K_X)) \geq 1$. By duality $h^0(X, \mathcal{O}_X(-K_X)) \geq 1$. Since K_X is nef, this may happen only if $K_X = 0$, i.e. X is a $K3$ surface.

If $p = q = 1$, then $\dim \operatorname{Pic}^0 X = 1$ and there is a $\sigma \in \operatorname{Pic}^0 X$ such that $\sigma \neq 0$ and $2\sigma = 0$. By the Riemann-Roch inequality, $l(\sigma) + l(K - \sigma) \geq \chi(\mathcal{O}_X) = 1$ hence $l(K - \sigma) \geq 1$ because $l(\sigma) = 0$. If $D \in |K - \sigma|$ and $K_0 \in |K|$, then $2D \in |2K|$ and $2K_0 \in |2K|$. Since we can assume $P_2 \leq 1$, we get $2D = 2K_0$ (as divisors!) hence $D = K_0$, contradicting $\sigma \neq 0$. Thus this possibility cannot be realized.

If $p = q = 0$, then $\chi(\mathcal{O}_X(mK_X)) = 1$ for all $m \in \mathbb{Z}$. Since $l(-K_X) = 0$ (otherwise $K_X \sim 0$ and $p = 1$), we get $P_2 \geq 1$ (by the Riemann-Roch inequality). As we know, the surface with $p = q = 0$ and $2K_X \sim 0$ are the *Enriques surfaces.* For them $\kappa = 0$ (they are discussed in Sect. 11). These surfaces are quotients of $K3$ surfaces by a group of order 2. But if $2K_X \not\sim 0$, then $l(-2K_X) = 0$ and by duality $P_3 \geq 1$. Choosing $D_2 \in |2K_X|$ and $D_3 \in |3K_X|$, we conclude that $|6K_X|$ contains two distinct divisors, $3D_2 \neq 2D_3$, i.e. $P_6 \geq 2$.

Type B. $\chi(\mathcal{O}_X) = 0$. Then we have two possibilities: 1) $p = 1$ and $q = 2$, or 2) $p = 0$ and $q = 1$.

If $p = 1$ and $q = 2$, we employ the Albanese map α: $X \to A$, $\dim A = 2$. Then $\alpha(X)$ generates A as a group provided we assume that $0 \in \alpha(X)$. It follows that $\dim \alpha(X) > 0$. If $\alpha(X)$ is a curve B, then $g(B) > 1$ by elementary properties of Abelian varieties (2-dimensional Abelian varieties do not contain curves with $g(B) = 0$, and are not generated by a curve with $g(B) = 1$). We get a fibration $X \to B$. Now we apply formulas (5) of Sect. 2 for $e(X)$. If F is a general fiber, then $g(F) > 0$. If $g(F) > 1$, then $e(X) > 0$ by the propositions of Sect. 2 and 4, which contradicts $e(X) = 12\chi(\mathcal{O}_X) = 0$. But if $g(F) = 1$, we get a pencil of elliptic curves.

If we assume $\dim \alpha(X) = 2$, i. e. $\alpha(X) = A$, then X is an Abelian variety. Indeed, consider a form $\alpha^*(\omega)$, where $\omega \in H^0(A, \Omega_A^2)$, $\omega \neq 0$. Clearly $(\alpha^*(\omega)) \in K_X$. Let $(\alpha^*(\omega)) = \sum n_i C_i \neq 0$, where C_i are curves. Since $(\alpha^*(\omega), K_X) = 0$ and K_X is nef, we get $(C_i, K_X) = 0$, hence $(C_i^2) = -2$ or $(C_i^2) = 0$. Therefore the C_i's are rational or elliptic curves. For a rational C_i, $\alpha(C_i)$ is a point on A. By the contraction criterion of Sect. 6.3, all the $\alpha(C_i)$'s could not be points, since then the intersection matrix (C_i, C_j) would be negative definite, a contradiction.

Now, if $\dim \alpha(C_i) = 1$, then C_i and $C = \alpha(C_i)$ are elliptic curves. By elementary properties of Abelian varieties, we can assume that C is a subgroup of A, i. e. $C = \pi^{-1}(0)$, where $0 \in B = A/C$ and $\pi\colon A \to B$ is a homomorphism. Therefore the system $|m_0 K|$ contains the curve $(\pi\alpha)^*(0)$ for a suitable m_0. Since $l(mC) \to \infty$ as $m \to \infty$, we get $l(mK) \to \infty$.

It remains to consider the case $(\alpha^*(\omega)) = 0$. In this case the map $\alpha\colon X \to A$ is an unramified covering, and unramified coverings of Abelian varieties are Abelian. Indeed, for $k = \mathbb{C}$, $A = \mathbb{C}^2/\Omega$ where Ω is a lattice in $\mathbb{C}^2$, and an unramified covering $A' \to A$ has the form $A' = \mathbb{C}^2/\Omega'$, where $\Omega' \subset \Omega$.

In the case $p = 0$ and $q = 1$, we again consider the Albanese map $\alpha\colon X \to A$, where $\dim A = 1$. Its fibers are connected, because $g(A) \leq g(A') \leq q = 1$ in the Stein factorization

$$X \xrightarrow{\alpha'} A' \to \mathbb{A}\,.$$

Hence A' is an elliptic curve. So, by the universal property of the Albanese varieties, we get $A' = A$. Furthermore, α is a smooth morphism. This follows from the propositions of Sect. 2 and Sect. 4.2, since $e(X) = 0$ by the Noether's formula. A general fiber F is nonrational, so $g(F) \geq 1$. If $g(F) = 1$, we get a pencil of elliptic curves. We assume that $g(F) > 1$. We will use the following important nontrivial assertion.

Lemma 2. *Let $\varphi\colon X \to B$ be a smooth morphism with connected fibers and $g(B) \leq 1$. Then all the fibers are isomorphic. Moreover, if $g(B) = 0$ then $X \cong B \times F$, and if $g(B) = 1$ then there is an unramified covering $B' \to B$ such that*

$$X \underset{B}{\times} B' \cong F \times B', \tag{2}$$

where F is a fiber of φ.

One may explain this as follows. The set of all curves of genus $g > 1$ up to isomorphism have a structure of an algebraic variety M_g, called the *moduli variety.* It is known (see, e. g. (Ahlfors-Bers (1960), Mumford-Fogarty (1982))) that M_g can be represented as a quotient of a bounded domain $T_g \subset \mathbb{C}^{3g-3}$, called the *Teichmüller space*, by a certain discrete group Γ_g. The group Γ_g contains a subgroup Γ' of finite index, which acts fixed-point-free on T_g. The quotient $M'_g = T_g/\Gamma_g$ is a finite covering of M_g.

The morphism φ: $X \to B$ defines the map ψ: $B \to M_g$, sending a point $b \in B$ to $\psi(b) \in M_g$ corresponding to the fiber $F_b = \varphi^{-1}(b)$. The latter map is also a morphism – this condition is a part of the definition of the moduli variety. One can obtain a map ψ': $B' \to M'_g$ of an unramified covering $B' \to B$ by utilizing the construction of the moduli variety (consider families of curves with a rigidity condition). We get $g(B') \leq 1$.

The map ψ' can be lifted to a map of the universal coverings. If $g(B) = 0$, then $B' = B \cong \mathbb{P}^1$, and we get a map $\mathbb{P}^1 \to T_g$, which is a map to a point by Liouville's theorem. If $g(B) = 1$, then also $g(B') = 1$, and the map of the universal coverings, $\mathbb{C} \to T_g$, is also constant since T_g is a bounded domain.

The argument with Teichmüller's space can be replaced by a more "elementary" argument. First, we map a point $b \in B$ to the Jacobian $J(F_b)$ which, in turn, corresponds to a point of the Siegel upper half-plane. Clearly the latter is isomorphic to a bounded domain. We then apply the Torelli theorem (Shokurov (1988)). Since $\psi(B)$ is a point of M_g, the generic fiber of $X \to B$ is isomorphic to a curve F, perhaps over an extension of the field $k(B)$. This means that

$$X \underset{B}{\times} B' \cong F \times B'$$

for a suitable covering $B' \to B$. It is easy to verify that if $B' \to B$ has ramification points, then the family has degenerate fibers. So $B' \to B$ is unramified, and the lemma follows.

Going back to the casse $p = 0$ and $q = 1$, we observe that the projection π: $F \times B' = X \times_B B' \to X$ is an unramified covering by construction. Hence $\pi^*(K_X) = K_{F\times B'}$. Since $\pi^*(K_{F\times B'}) = nK_X$ for a suitable $n > 0$, we get $\pi_*(mK_{F\times B'}) = mnK_X$. By (4) of Sect. 2, $l(mK_{F\times B'}) \to \infty$ as $m \to \infty$, so $l(mnK_X) \to \infty$. As we know, the latter implies the existence of elliptic pencils on X.

§ 9. Surfaces of General Type

9.1. Moduli. The surfaces of general type are analogous to curves of genus $g > 1$. For them one would like to obtain a picture similar to the one we get for curves. Namely, one would like to be able to choose certain discrete invariants similar to the genus, and parametrize the surfaces with given discrete invariants by points of a certain finite dimensional variety. In the present section, we will discuss these questions.

We have encountered the following integral invariants:

$$e(X), \quad (K_X^2) \quad p, \quad q, \quad b_1, \quad b_2.$$

Further, we have the following relations:

$$e(X) = 2 - 2b_1 + b_2\,, \quad b_1 = 2q\,, \quad (K_X^2) + e(X) = 12(1 - q + p)\,.$$

So, we might assume that $e(X)$, (K_X^2), and p are independent. However, by (Sect. 8.2, (1)), given (K_X^2) and $\chi(\mathcal{O}_X)$, then P_m ($m \geq 2$) may take only finitely many values. Furthermore, since $p \leq P_2$ and $1 - q + p \geq 0$ by Lemma 1 of Sect. 8.3, the same holds for p and q. Therefore, in the sequel, we take (K_X^2) and $e(X)$ as the basic invariants.

A parametrization of surfaces with given (K_X^2) and $e(X)$ by points of a finite dimensional variety is based on the same ideas as the ones employed for curves. One constructs a certain projective embedding $\varphi\colon X \hookrightarrow \mathbb{P}^N$, which is uniquely determined by X up to projective transformations. Then one proves that for the surfaces X with given invariants (K_X^2) and $e(X)$, the number N and the degree of the surface $\varphi(X)$ in $\mathbb{P}^N$ can take only finitely many values. Thus the problem is reduced to a projective classification of surfaces of a given degree in a given projective space $\mathbb{P}^N$ up to projective transformations (or a version of this problem).

The embedding $\varphi\colon X \hookrightarrow \mathbb{P}^N$ is constructed as in the case of curves, namely, we consider pluricanonical embeddings φ_{mK}. Given a surface of general type, for sufficiently large m, the map φ_{mK} corresponding to the m-th power of the canonical class is a birational morphism, which contracts (-2)-curves only; moreover, we obtain only Du Val singularities. In order to be able to consider all the surfaces together, we have to verify the existence of φ_{mK} with the same m for all the surfaces. All curves of genus $g > 1$ can be embedded by φ_{3K}. A similar assertion holds for surfaces.

Theorem 1. *For any minimal surface X of general type, the map φ_{5K} is a birational morphism to $\mathbb{P}^n$, where $N = l(5K) - 1$, and the image $\varphi_{5K}(X)$ is normal with Du Val singularities only* (see (Barth-Peters-Van de Ven (1984), Bombieri (1973), Reider (1988))).

According to (Sect. 8.2, (1)), $\varphi_{5K}(X) \subset \mathbb{P}^N$, where $N = 10(K_X^2) + \chi(\mathcal{O}_X) - 1$, and it is a surfaces of degree $((5K_X)^2) = 25(K_X^2)$. Moreover, its Hilbert polynomial equals

$$(25/2)(K_X^2)T^2 - (5/2)(K_X^2)T + \chi(\mathcal{O}_X)$$

(see CAV). The surfaces of a given degree in a given projective space $\mathbb{P}^N$ are parametrized by the points of a finite dimensional variety – moduli variety.

One may also use that the 2-dimensional subschemes $Y \subset \mathbb{P}^N$ with a given Hilbert polynomial $P(T)$ are parametrized by the points of a finite dimensional scheme $\mathrm{Hilb}^P(\mathbb{P}^N)$ (Grothendieck (1962), Exp. 221). Its open subset U parametrizes the images $\varphi_{5K}(X)$ of smooth surfaces with given invariants (K_X^2) and $e(X)$. Moreover, the surfaces X and X' are isomorphic if and only if their images $\varphi_{5K}(X)$ and $\varphi_{5K}(X')$ are projectively equivalent. In other words, the group $PGL(N+1, k)$ acts on U, and the surfaces X are in a one-to-one correspondence with the orbits of this group on U.

Thus we obtain a one-to-one correspondence between the surfaces of a given type and the elements of the set $U/PGL(N+1,k)$. As we know, the question when a quotient of an algebraic variety by an algebraic group can be given a structure of an algebraic variety is rather delicate. Such a quotient set may even be "wild" as a topological space – if there are nonclosed orbits then it is non-Hausdorff. However, in our case, one is able to establish (by a highly nontrivial argument) the existence of an algebraic structure on the quotient. We get the following theorem (Gieseker (1977), Mumford (1977), Mumford-Fogarty (1982)).

Theorem 2. *The minimal surfaces X with the given invariants $n = (K_X^2)$ and $m = e(X)$ are classified by a quasiprojective moduli variety $M_{m,n}$.*

We conclude our discussion of this topic by observing that φ_{mK} may not be birational for $m \leq 4$. However, the map φ_{3K} is a birational equivalence except for a few special cases.

Theorem 3. *Let X be a surface of general type. Then the map φ_{3K} is a birational morphism except when $(K_X^2) = 2$ and $p_g(X) = 3$, or $(K_X^2) = 1$ and $p_g(X) = 2$.*

In these exceptional cases, φ_{3K} is not a birational morphism.

9.2. Geography of Surfaces (Barth-Peters-Van de Ven (1984), Chen (1987)). We now proceed to the second question necessary for the classification of surfaces of general type: Which values of (K_X^2) and $e(X)$ are realized? For minimal surfaces of general type, the description of all possible pairs of numbers $n = (K_X^2)$ and $m = e(X)$ is called the *geography of surfaces.*

First, we gather all the necessary conditions we already know:

$$n > 0\,, \quad m > 0\,, \quad n + m \equiv 0 (\mathrm{mod}\, 12)$$

(by the Riemann-Roch theorem). We also have two important inequalities. The first one is *Noether's inequality*

$$p \leq \tfrac{1}{2}(K_X^2) + 2\,. \tag{1}$$

One can easily derive it by applying Clifford's theorem on algebraic curves (Shokurov (1988)) to curves of the linear system $|K|$. It follows from Noether's inequality that

$$5(K_X^2) - e(X) + 36 \geq 0\,, \quad \text{if } (K_X^2) \text{ even}\,; \tag{2}$$

$$5(K_X^2) - e(X) + 30 \geq 0\,, \quad \text{if } (K_X^2) \text{ odd}\,. \tag{3}$$

One has to take into account that $\chi(\mathcal{O}_X) \leq p + 1$, and Noether's formula (Sect. 5.1, (2)) which gives an expression for χ in terms of (K_X^2) and $e(X)$.

The second one is the *Bogomolov-Miyaoka-Yau inequality* (Algebraic Surfaces (1981), Miyaoka (1977)):

$$(K_X^2) \leq 3e(X)\,. \tag{4}$$

The proof of (4) is much more delicate. There are two methods.

One method is based on the same idea as the proof of Lemma 1 in Sect. 8.3. However, those ideas produce a much weaker inequality: $(K_X^2) \leq 8e(X)$. An essentially new ingredient is to employ, in addition to Ω_X^1, all symmetric powers $S^n\Omega_X^1$ (see (Van de Ven (1978)) and Bogomolov's report in (Algebraic Surfaces (1981))).

Another method works only for $k = \mathbb{C}$. It is based on a construction of a particular Riemann metric (Calabi-Yau metric). The expression $3e(X) - (K_X^2)$ can be written as an integral over X of a certain nonnegative density, hence it is nonnegative. Moreover, this approach allows to obtain the following important addition to the inequality (4). For a surface X of general type, (4) is an equality if and only if X is isomorphic to the quotient B/G, where $B \subset \mathbb{C}^2$ is a ball $|z_1|^2 + |z_1|^2 < 1$ and G a discrete group of its automorphisms (Yau (1977)).

We will draw, in the (m, n)-plane, the domain D corresponding to the above necessary conditions. It is bounded from below by the lines $m + n = 12$ and $m = 5n + 36$, and from above by the line $n = 3m$ (see Fig. 17).

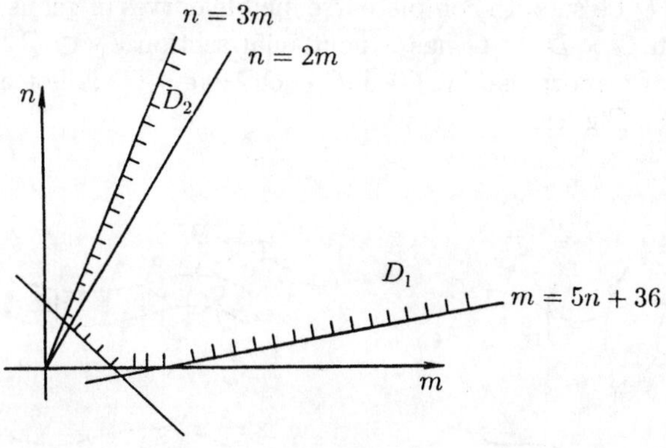

Fig. 17

Let the pairing on $H^2(X, \mathbb{R})$ has type (b^+, b^-). The *signature of a surface*, $\tau = b^+ - b^-$, plays an essential role in the "geography". By the Hodge theory

$$b^+ = 2p + 1\,, \quad b^- = b_2 - 2p - 1\,.$$

So $\tau = 4p + 2 - b_2$. Now, by the Riemann-Roch theorem $\tau = (K^2 - 2e)/3$. The line $\tau = 0$ (i.e. $n = 2m$) divides the domain D in two parts: the lower part D_1 (corresponding to $\tau < 0$) and the upper part D_2 (corresponding to $\tau > 0$). The vast majority of surfaces lie in D_1. It is easy to construct the corresponding examples and prove that almost all points of D_1 are realized.

Theorem 4. *Given a pair* $(n,m) \in D_1$ *with* $n+m \equiv 0 (\mathrm{mod}\, 12)$, *there exists a minimal surface with* $(K_X^2) = n$ *and* $e(X) = m$, *except perhaps in the following cases*:

$$n - 2m + 3k = 0\,, \qquad k = 1, 2, 3, 5, 7\,.$$

One can obtain boundary points of D_2 (i. e. points on the lines $n = 3m$ and $n = 2m$) in the form U/G, where U is a bounded homogeneous domain in $\mathbb{C}^2$, and G a discrete group of automorphisms acting freely on U. It was established by E. Cartan that such domains are isomorphic either to a ball B or a polydisk $D \times D$, where D is the disk $|z| < 1$. In both cases, by *Hirzebruch's proportionality theorem*, $(K_X^2)/e(X)$ depends only on U, where $X = U/G$, and it equals $(K^2)/e$ for the symmetric space dual to U, which is $\mathbb{P}^2$ for B and $\mathbb{P}^1 \times \mathbb{P}^1$ for $D \times D$, hence it equals 3 or 2, respectively (Hirzebruch (1987)).

An important class of surfaces corresponding to the points of D_2 are the so-called *Kodaira surfaces.*

Example. A Kodaira surface gives an example of a smooth fibration $X \to B$ in curves that is not locally trivial. For such a fibration, we get $g(B) > 1$. (The case $p = 0$ and $q = 1$ is discussed in Theorem 2 of Sect. 8.1.)

Let C and D be smooth complete irreducible curves of genus > 1 such that the projection $C \times D \to C$ has a nontrivial section $\Gamma_0 \subset C \times D$. Thus Γ_0 is the graph of a morphism γ: $C \to D$ such that $\gamma(C)$ is not a point (hence $\gamma(C) = D$); see Fig. 18.

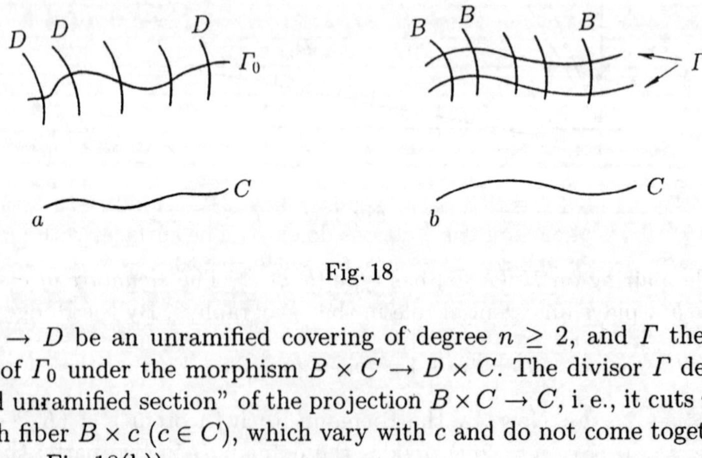

Fig. 18

Let $B \to D$ be an unramified covering of degree $n \geq 2$, and Γ the inverse image of Γ_0 under the morphism $B \times C \to D \times C$. The divisor Γ defines an "n-fold unramified section" of the projection $B \times C \to C$, i. e., it cuts n points on each fiber $B \times c$ ($c \in C$), which vary with c and do not come together (for $n = 2$, see Fig. 18(b)).

We will assume the existence of a cyclic covering $X \to B \times C$ of degree $r \geq 2$ that is branched along Γ. Then the fibers of the projection $X \to C$, denoted by F_c, are r-fold cyclic coverings of the fiber of $B \times C$ with n distinct branch points $(B \times c) \cap \Gamma$. Therefore, by the formula for the genus of a covering (Shokurov (1988)), all this curves have the same genus:

$$rg + (r-1)(n-2)/2\,,$$

where g is the genus of B, i. e., the fibration has no degenerate fibers. Since the section Γ_0 is nontrivial, the branch points of $\Gamma_0 \to B$ vary with c, hence the fibration is not locally trivial. Its is easy to verify that one can indeed realize the above construction. The surfaces obtain this way are called *Kodaira surfaces.* If the genus of the curve B equals g, then by a simple calculation:

$$\frac{(K_X^2)}{e(X)} = 2 + \frac{1 - \dfrac{1}{r^2}}{(2g-1) - \dfrac{1}{r}}\,. \tag{5}$$

It follows that all these surfaces correspond to points of D_2 (the right-hand side of (5) takes values between 2 and 7/3); see, e. g. (Barth-Peters-Van de Ven (1984)).

Although surfaces corresponding to the boundary points of D_2 as well as Kodaira examples have infinite fundamental groups $\pi_1(X)$, there are many examples of simply connected surfaces corresponding to points of D_2.

It is known that the "slopes" of points of D are distributed rather evently – the numbers $(K_X^2)/e(X)$ are everywhere dense on the segment $[1/5, 3]$.

9.3. Almost Rational Surfaces. We will conclude with a discussion of surfaces of general type with $p = q = 0$. These surfaces are interesting, because they are "similar to rational" and related to a general problem of investigating conditions that characterize rational surfaces.

Let X be such a surface. Then $\chi(\mathcal{O}_X) = 1$ hence

$$(K_X^2) + e(X) = 12\,, \quad b_1 = 0\,, \quad e(X) = 2 + b_2\,, \quad (K_X^2) + b_2 = 10\,.$$

Since $(K_X^2) > 0$ and $b_2 > 0$, these surfaces are divided in 9 classes according to the value of $(K_X^2) = 9, 8, \ldots, 1$ (and $b_2 = 1, 2, \ldots, 9$). It was established that the surfaces of each of the 9 classes do exist. The surfaces with $(K_X^2) = 9$ and $b_2 = 1$ are of interest, because for such surfaces, p and q as well as (K_X^2) and b_2 – i. e., all the encounted numerical invariants – are the same as for the plane $\mathbb{P}^2$. Therefore they are sometime called *fake planes.* Since they satisfy the equality $(K_X^2) = 3e(X)$, these surfaces can be represented, as we mentioned above, in the form B/G, where $B \subset \mathbb{C}^2$ is a ball and G its discrete group of automorphisms. It seems natural to construct these surfaces as such quotients, however, no one was able, until now, to construct corresponding groups G. A surface of general type with $p = q = 0$ and $(K_X^2) = 9$ was, in fact, constructed utilizing a p-adic uniformization (Mumford (1970b)).

Similarly, the surface of general type with $p = q = 0$, $(K_X^2) = 8$, and $b_2 = 2$ are called *fake quadrics.* One can construct an example of such a surface in the form $X = (C_1 \times C_2)/G$, where C_1 and C_2 are curves of genus g_1 and g_2, and G a finite group acting on C_1 and C_2 and diagonally on $C_1 \times C_2$. If $|G| = (g_1 - 1)(g_2 - 1)$ then $(K_X^2) = 8$, and if $C_1/G \cong C_2/G \cong \mathbb{P}^1$ then $p = q = 0$ for X. A group with these properties acting freely can be constructed as follows. Let $C_1 \cong C_2$ with an equation $x_0^5 + x_1^5 + x_2^5 = 0$, and $G = \mathbb{Z}/5 \times \mathbb{Z}/5$. The group G acts on C_1 by the formula

$$g(x_0 : x_1 : x_2) = (x_0 : \varepsilon_1 x_1 : \varepsilon_2 x_2)\,, \qquad \varepsilon_1^5 = \varepsilon_2^5 = 1\,,$$

and on C_2 by φ_g, where $\varphi \in \mathrm{Aut}(\mathbb{Z}/5 \times \mathbb{Z}/5)$ is given by the matrix $\begin{pmatrix} 1 & 2 \\ 3 & 5 \end{pmatrix} \in GL(2, \mathbb{F}_5) \cong \mathrm{Aut}(\mathbb{Z}/5 \times \mathbb{Z}/5)$ (Beauville (1978)).

Finally, another extremal case, namely:

$$p = q = 0\,, \quad (K_X^2) = 1\,, \quad b_2 = 9$$

is interesting, because there are simply connected surfaces of general type with such invariants. This shows that the conditions $p = q = 0$ and $\pi_1(X) = 0$ are not sufficient to characterize the plane $\mathbb{P}^2$ (Barlow (1982)). In Sect. 10, we will explain how to construct a similar example in the class of elliptic surfaces.

Finally, we will mention how the above discussion leads to the following characterization of $\mathbb{P}^2$ over $k = \mathbb{C}$: if a surface X is homeomorphic to $\mathbb{P}^2$, then it is isomorphic to $\mathbb{P}^2$.

Indeed, by the assumption, $b_1 = q = 0$ and $b_2 = 1$, hence all cycles are algebraic and $p = 0$ (according to the trivial part of Lefschetz's theorem from Sect. 3.3). If K_X is not nef, then X is isomorphic to $\mathbb{P}^2$ by Theorem 1 of Sect. 7.1 (for ruled surfaces $b_2 \neq 1$). But if K_X is nef, then $(K_X^2) = 9$ by the Riemann-Roch theorem (Sect. 5.1, (2)). Hence X is a surface of general type. Since $e(X) = 3$, we get $(K_X^2) = 3e(X)$, so $X \cong B/G$, where B is a ball and $G \cong \pi_1(X)$. However, $\mathbb{P}^2$ and X are simply connected, a contradiction.

It is unknown (at the time of writing) whether there are differential manifolds homeomorphic but not diffeomorphic to the complex plane $\mathbb{P}^2(\mathbb{C})$, i. e. whether $\mathbb{P}^2$ has a smooth structure different from the canonical one (an "exotic" smooth structure).

§ 10. Elliptic Surfaces

10.1. Families of Groups. Recall that X is an elliptic surface if there exists a morphism $f \colon X \to B$ onto a smooth curve whose fibers are connected and the genus of the generic fiber is 1 (Sect. 2, Example 5). The generic fiber F_ξ of f is an elliptic curve over the field $k(B)$ (it may not posses a rational point). Since such fields have a lot in common with algebraic number fields (both types of fields are "1-dimensional"), the theory of elliptic surfaces is analogues to the arithmetic of elliptic curves.

It follows from the theory of elliptic curves (the Riemann-Roch theorem for curves) that if F_ξ has a rational point $\mathcal{O}$ over $k(B)$ (which means that the fibration $f: X \to B$ has a section $\varphi: B \to X$), then this curve is isomorphic over $k(B)$ to a curve given by an equation $y^2 = x^3 + ax + b$ (in affine coordinates). In the sequel, to simplify the formulas, we assume that $\operatorname{char} k \neq 2, 3$. Thus an elliptic surface is birationally equivalent to a surface in $\mathbb{P}^2 \times B$ given by an equation

$$y^2 = x^3 + \alpha x + \beta \qquad (\alpha, \beta \in k(B)). \tag{1}$$

This equation is called the *Weierstrass normal form.* The surface (1) is not necessary smooth, although its general fiber is smooth. It can have only isolated singular points, which belong to finitely many fibers.

Again, by the Riemann-Roch theorem for curves, all the elements of the group $\operatorname{Pic}^0 F_\xi$, in the exact sequence (3) of Sect. 3.5, have representatives of the form $\alpha - 0$, where α is a point of F_ξ rational over $k(B)$, i.e., $\alpha \in F_\xi(k(B))$ and the group law in $\operatorname{Pic}^0 F_\xi$ makes $F_\xi(k(B))$ into a group. It is easy to verify that the projection $f: X \to B$ defines an isomorphism $\operatorname{Pic}^0 B \cong \operatorname{Pic}^0 X$, except for a trivial case when $X \cong B \times E$ with E an elliptic curve over k (in other words, $q(X) = g(B)$ where $g(B)$ is the genus of B). Therefore, if $X \not\cong B \times E$, then the group $\operatorname{Pic}^0 X$ goes to zero when we restrict $\operatorname{Pic} X$ to F_ξ. Since the image of the restriction coincides with the whole group $\operatorname{Pic} F_\xi$, the latter is a discrete group. Furthermore, it is finitely generated since $\operatorname{Pic} X / \operatorname{Pic}^0 X \cong S_X$ is finitely generated. So, the same holds for the group $\operatorname{Pic}^0 F_\xi$, which is the group $F_\xi(k(B))$ of points of the generic fiber, i.e. the points of the curve F_ξ given by (1) with coordinates in $k(B)$. This is the geometric analog of the Mordell-Weil theorem for elliptic curves over an algebraic number field.

We may assume that an elliptic surface $f: X \to B$ is a relatively minimal model (by contracting (-1)-curves of the fibers); see the end of Sect. 7. Henceforth, we will always assume this. Of course, we do not necessarily get a minimal model in general, i.e., X may contain (-1)-curves that cannot be contracted without giving up the condition that f is a morphism.

Example 1. Let C_1 and C_2 be two elliptic curves in $\mathbb{P}^2$ with equations $F_1 = 0$ and $F_2 = 0$ that intersect in 9 points $(x_1, \ldots, x_9)$. Let $\sigma: X \to \mathbb{P}^2$ be the blowing-up of these 9 points. Then a rational map $f: X \to \mathbb{P}^1$ given by $(F_1 : F_2)$ is a morphism, and X is an elliptic surface. Its fibers correspond to the curves $\lambda_1 F_1 + \lambda_2 F_2 = 0$. By properties of σ-processes, the curves $L_i \subset X$, $L_i = \sigma^{-1}(x_i)$, are sections of our elliptic surface. Moreover, they are (-1)-curves. However, after contracting any one of those curves the map to B fails to be a morphism (Fig. 19).

We will state an important corollary of Theorem 3 from Sect. 7.3.

Theorem 1. *A birational equivalence $X \dashrightarrow X$ that commute with f: $X \to B$ is an automorphism* (*we assume that X is a relatively minimal model*).

Example 2. We assume that $f: X \to B$ has a section $\varphi: B \to X$, $\varphi(B) = S_0$. Then any nonsingular fiber F has a structure of an Abelian group with $o = S_0 \cdot F_\xi$ (intersection of S_0 with F_ξ) as its zero element. Assume there is another section S. It defines a point $\alpha = S \cdot F$ on each nonsingular fiber as

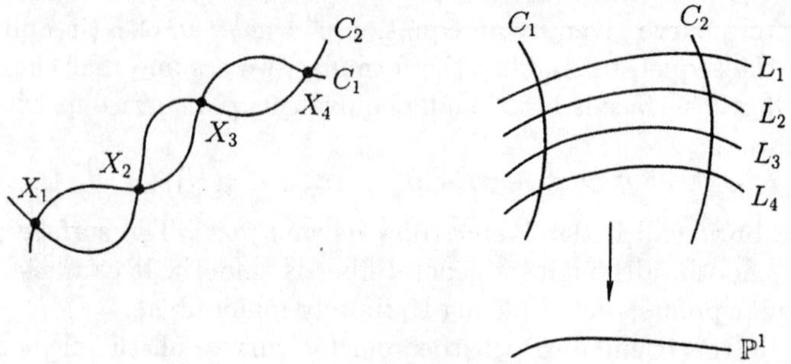

Fig. 19

well as the translation t_S by the point. The translation t_S, by the section S, is defined on an open subset of X obtained by removing singular fibers. It defines a rational map of X (this is clear if we consider it on the generic fiber F_ξ which is nonsingular). By Theorem 1, the rational map t_S is a morphism, even an automorphism, of X.

Example 3. With the notation of Example 1, we choose a pencil of cubics $\lambda F + \mu G = 0$ such that its singular fibers are irreducible curves (with a unique singular double point); in fact, this is the "general case". A surface obained by blowing up 9 points of $\{F = 0, G = 0\}$ defines an elliptic family $f\colon X \to \mathbb{P}^1$ with rational curves L_i $(i = 1, \ldots, 9)$ as sections. Set $S_0 = L_1$, and denote the image of L_i in N_X by α_i. Then in the sequence (3) of Sect. 3.5:

$$\alpha_i - \alpha_1 \in (\operatorname{Pic} X)_F, \quad \Phi = \mathbb{Z}F.$$

On the other hand, $\alpha_1, \ldots, \alpha_g$ are independent in N_X (Sect. 6.1, (2)), hence $\alpha_i - \alpha_1$ $(i = 2, \ldots, 9)$ are independent in $(\operatorname{Pic} X)_F$, and t_{α_i} $(i = 2, \ldots, 9)$ define a group of automorphisms of X isomorphic to $\mathbb{Z}^8$. In particular, given any section L_i, we obtain, with a help of this group, infinitely many sections L such that $(L^2) = -1$ and $L \cong \mathbb{P}^1$, i. e., X has *infinitely many* (-1)*-curves.*

Another application of the above construction is that $\alpha_i - \alpha_1$ define 8 independent rational points in the group $F_\xi(k(\mathbb{P}^1)) = F_\xi(k(t))$, $t = \lambda/\mu$. The above construction also makes sense for $k = \mathbb{Q}$. Letting $t = c \in \mathbb{Q}$, we obtain from F_ξ a cubic curve F_c: $\lambda F + \mu G = 0$ over $\mathbb{Q}$, and $\alpha_i - \alpha_1$ will produce 8 independent points on F_c. According to the Hilbert "irreducibility theorem", these points will be independent on F_c provided c is sufficiently general, i. e. $\operatorname{rk} F_c(\mathbb{Q}) \geq 8$. This gives one of methods for constructing cubic curves of large rank over $\mathbb{Q}$.

An elliptic curve over a field k is an algebraic group. It is natural to assume that an elliptic fibration $X \to B$ is, in a sense, a "family of algebraic groups", i. e., one can introduce an algebraic structure on its fibers F_b which depends "rationally" on $b \in B$. Now we will examine in what sense this is possible.

To introduce a structure of an algebraic group on an elliptic curve E, it is necessary to choose a point o – a zero element. Similarly, to make an elliptic fibration $X \to B$ into a family of groups, one has to choose a zero point o_b in each fiber F_b, which depend ratonally on b, i. e., a section φ: $B \to X$. We assume the existence of such a section. It gives a rational point (over $k(B)$) of the generic fiber F_ξ. This allows us to introduce a structure of an algebraic group over $k(B)$ in F_ξ, i. e., we can define a morphism μ: $F_\xi \times F_\xi \to F_\xi$ of curves over $k(B)$. Geometrically, this gives a rational map

$$\mu : X \underset{B}{\times} X \to X ,$$

where $X \times_B X$ is the fiber product over B. Then $\mu(x, x') = x \oplus x'$ if $x, x' \in F_b$, where F_b is a smooth fiber, and $x \oplus x'$ is the sum with respect to the group low on F_b such that $\varphi(b)$ is a zero element. Our first aim is to describe the domain where the map μ is regular.

Theorem 2. *The map μ is regular at the points $(x, x') \in F_b \times F_b$ provided x and x' are simple points of F_b.*

The theorem follows at once from Theorem 1.

If S is any section defining a point $\eta \in F_\xi$ of the generic fiber, then the corresponding translation t_S defines an automorphism t_S: $X \to X$ which commutes with the projection $X \to B$ (see Example 2). Clearly, for two sections S and S':

$$\mu(t_S(x), t_{S'}(x')) = t_{S+S'}\mu(x, x') .$$

Since one can find a local section through any nonsingular point of a fiber, we can extend μ from the points where it is regular to all the points. (We considered a sufficiently small neighborhood of the fiber F_b but the question of regularity of a map is a local one.)

Thus each singular fiber F_b is also a group provided we remove its singular points. In particular, one has to remove all multiple components. We denote this group by $F_b^{\#}$. In general this group is not connected. Let S_0 be a section that defines the group law. Then its connected component of zero is precisely the component intersecting S_0. If we denote it by F_b^0, then the group $F_b^{\#}/F_b^0$ is finite.

A union of all the sets $F_b^{\#}$, $b \in B$ (note that $F_b^{\#} = F_b$ provided F_b is smooth) is an open subset $X^{\#} \subset X$, and, by Theorem 2, it defines a family of groups over B (in general non-proper since some of the fibers are not projective). In other words, we get a group scheme over B, called the *Neron model* of the surface X. Similarly, the fibers F_b^0 define a family $X^0 \subset X^{\#}$. Clearly X^0, $X^{\#}$, and X are different only at the points of singular fibers.

Example 4. Let $X \to B$ be a fibration, given by (1), whose fibers are smooth elliptic curves, i. e., we are given a family of elliptic curves parametrized by the curve B. Since an elliptic curve is uniquely determined up to isomorphism by the value of the absolute invariant

$$j = \frac{a^3}{4\alpha^3 + 27\beta^2},$$

we get that j is regular on the projective base B, hence it is a constant. Therefore all the fibers are isomorphic to a unique elliptic curve E. In particular, the generic fiber F_ξ is isomorphic to E, though over a finite extension $k(B')$ of $k(B)$ only. The covering $B' \to B$ has to be unramified, since otherwise the family $X \to B$ would have had singular fibers. One can choose $B' \to B$ to be a Galois covering. Clearly we get an isomorphism

$$X \underset{B}{\times} B' \cong E \times B'$$

preserving the group law (i. e., $X \to B$ is a twisted form of a constant group scheme $E \times B$). We have a simple procedure for constructing such fibrations. One has to take a fixed-point-free group Γ of automorphisms of B' such that $B'/\Gamma = B$, and a monomorphism $\varphi\colon \Gamma \to \operatorname{Aut}^0 C$, where $\operatorname{Aut}^0 C$ is the automorphism group preserving the group structure on C. Let

$$G \subset \operatorname{Aut}(B' \times C), \qquad G = \{(g, \varphi(g)) | g \in \Gamma\}.$$

Then $X = (B' \times C)/G$. Since $|\operatorname{Aut}^0 C| \in \{1, 2, 3, 4, 6\}$, $|G|$ takes similar values. In particular, if B is an elliptic curve, then its unramified covering B' is also an elliptic curve, and Γ consists of its translations. This way we obtain a series of examples of bi-elliptic surfaces (Sect. 2, Example 8). Clearly for such surfaces

$$nK_X = 0, \qquad n \in \{1, 2, 3, 4, 6\}.$$

10.2. Singular Fibers. We will describe singular fibers of elliptic fibrations.

Example 5. Let $\pi\colon B' \to B$ be a cyclic covering of degree n of the base B that has a ramification of order n at a point $b_0' \in B'$. Let s be an automorphism of the covering. Let E be an elliptic curve, $a \in E$ its point of order n, and $\gamma\colon E \to E$ the automorphism $x \to x + a$ of order n. We consider the product $X' = G' \times E$, its automorphism $g = (s, \gamma)$, and the quotient $X = X'/\{g\}$. Clearly we get a morphism

$$f : X \to B = B'/\{s\}.$$

If $b \in B$ is not a branch point of π and $\pi^{-1}(b) = (b_1', \dots, b_n')$, then the inverse image of b in X' consists of the curves $b_1' \times E, \dots, b_n' \times E$. Clearly the fiber $f^{-1}(b)$ is nonsingular and isomorphic to E. But if $b_0 = \pi(b_0')$, then the curve $b_0' \times E$ is invariant with respect to g, which acts on it as s. If t is a local parameter at b_0, we can take t as a local equation of the fiber $f^{-1}(b_0)$. It follows that F_{b_0} equals nE', where $E' = E/\{\gamma\}$. For b near b_0, the fiber F_b winds up onto F_{b_0}, and covers it n times when $b \to b_0$. We get an example of a multiple fiber.

One encounters similar constructions in 3-dimensional topology (with a circle in place of the torus E). In topology, the fibrations in circles with certain fibers multiple are called *Seifert fibrations.*

A fiber F_b of an elliptic fibration is said to be *multiple* if the multiplicity of each component of F_b is greater than 1. Clearly, if a surfaces has at least one multiple fiber F_{b_0}, then it has no sections, because we get $(C, F_{b_0}) > 1$ for every curve C that intersects fibers. The converse holds locally, namely, in a neighborhood (for $k = \mathbb{C}$ in complex topology, and in general in etale topology) of a non-multiple (i. e. ordinary) fiber, there is a section. We just take a component that is a part of a singular fiber with multiplicity 1, and consider a curve passing transversely through a simple point of this component.

We will describe the structure of singular ordinary fibers. To begin with, we consider an arbitrary fibration in curves.

Proposition. *Let* $F = \sum_{i=1}^{r} n_i C_i$ *be a singular fiber of a family* $f\colon X \to B$. *Then for* $x_1, \dots x_r \in \mathbb{R}$,

$$\sum (C_i, C_j) x_i x_j \le 0$$

with equality if and only if the vector $(x_i, \dots x_r)$ *is a scalar multiple of* $(n_1, \dots, n_r)$.

Indeed, $(C_i, C_j) \ge 0$ for $i \ne j$, $(C_i, F) = 0$, and the fiber is connected. The result now follows by a simple "linear algebra" argument.

The above proposition is similar to the Hodge index theorem (Sect. 5.2, Theorem 1) and the contractibiltiy criterion of Sect. 6.3.

Going back to elliptic surfaces, we observe that for a nondegenerate fiber F, we get $(F^2) = 0$, and $(K_X, F) = 0$ by the adjunction formula (Sect. 4.2, (8)). Hence $(K_X, F) = 0$ for any fiber. In particular, if a fiber C is irreducible, then it is smooth or $g(C) = 0$ and $\delta = 1$ by (Sect. 4.2, (9)). In the latter case, it follows at once that C has either a double point with distinct tangents or a double point of type $y^2 = x^3$.

But if the fiber is reducible, then each component C_i satisfies the condition $(C_i^2) < 0$ hence $(C_i, K_X) \ge 0$ (we assume that the surface X is relatively minimal).

Now, $(K_X, F) = 0$ implies that $(C_i, K_X) = 0$, so $(C_i^2) = -2$ and $g(C_i) = 0$ by (Sect. 4.2, (9)). By the proposition, if the fiber has components C_i and C_j, then $(C_i, C_j) \le 2$ with $(C_i, C_j) \le 1$ provided the fiber has at least three components. All the matrices (a_{ij}) that satisfy the proposition, i. e. negative semidefined with $a_{ij} \in \{-2, 0, 1\}$, are described in the theory of roots of simple algebras, see (Bourbaki (1968)). They are described by graphs, as in Sect. 6.3 (see Table 1). Note that in a few simplest cases, the graph does not determine the configuration of components (e. g., $(C_i, C_j) = 2$ may correspond either to two points of intersection or a tangency at one point). In Table 1, those simplest cases are described in the last two rows; here the components correspond to lines of the graph instead of points.

Table 1

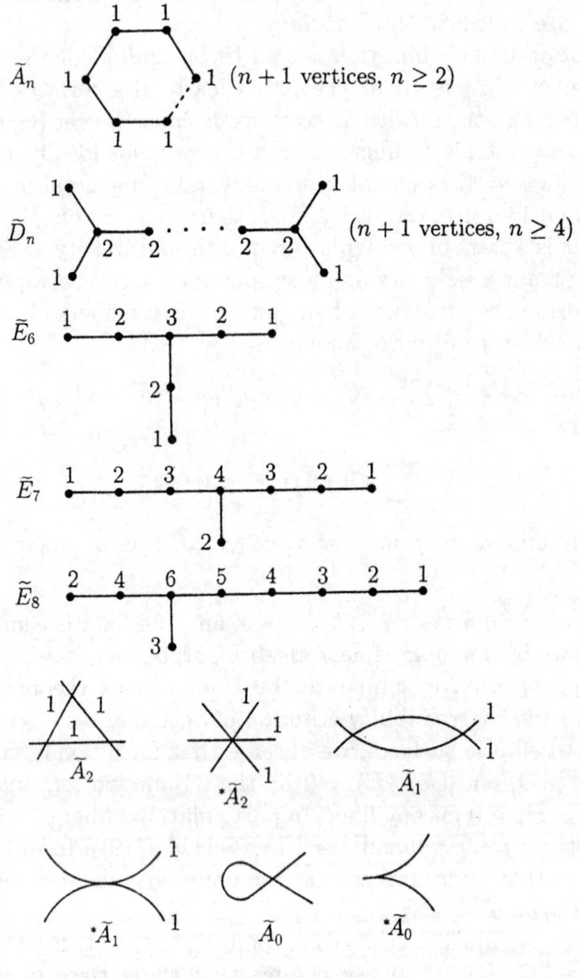

(the numbers denote the multiplicities of curves in the fiber)

In addition to the notation from the theory of roots, another notation, introduced by Kodaira, is often used. The correspondence is described in Table 2.

Table 2

$\widetilde{A}_n$	$\widetilde{D}_n$	$\widetilde{E}_6$	$\widetilde{E}_7$	$\widetilde{E}_8$	$^*\widetilde{A}_2$	$^*\widetilde{A}_1$	$^*\widetilde{A}_0$
I_{n+1}	I^*_{n-4}	IV^*	III^*	II^*	IV	III	II

Table 3

$r = s = 0$ $d > 2$ $\widetilde{A}_n$ $n = d - 1$	$r = 2,\ s = 3$ $d > 6$ $\widetilde{D}_n,\ n > 4$ $n = d - 2$	$r = s = 1$ ${}^*\widetilde{A}_0$ $d = 2$	$r = 1,\ s \geq 2$ ${}^*\widetilde{A}_1$ $d = 3$	$r \geq 2,\ s = 2$ ${}^*\widetilde{A}_2$ $d = 4$
$r = 2,\ s = 3$ $\widetilde{D}_4$ $d = 6$	$r \geq s,\ s = 4$ $\widetilde{E}_6$ $d = 8$	$r = 3,\ s \geq 5$ $\widetilde{E}_7$ $d = 9$	$r \geq 4,\ s = 5$ $\widetilde{E}_8$ $d = 10$	

Table 1 is very similar to the list of Du Val singularities described in Sect. 6.3. They are related as follows. If we remove a component of multiplicity 1 in any fiber of Table 1, we obtain a graph of a Du Val singularity, which is denoted by the same symbol without "$\sim$". Therefore all the components, except one, can be contracted to a Du Val singularity, and the remaining component will give an irreducible fiber of the new (singular) elliptic surface. It turns out that all such surfaces are described by equations of type (1) (at least locally in a neighborhood of the fiber), i. e., they have the Weierstrass normal form. Conversely, the original surface is the minimal resolution of its singularities.

Table 3 shows how to recover the original fiber by its Weierstrass form. We denote the orders of zeros of the functions α and β at $b \in B$ by $v_b(\alpha) = r$ and $v_b(\beta) = s$. If $r \geq 4$ and $s \geq 6$, then we apply the transformation $x = t^2u$, $y = t^3v$ (t is a local parameter at b) and obtain an equation of type (1) where r has decreased by 4 and s by 6. Therefore we will assume that either $r < 4$ or $s < 6$. Furthermore, in Table 3, $d = v_b(D)$ where $D = 4\alpha^3 + 27\beta^2$.

We observe that in all the cases, $d = v_b(D)$ coincides with the Euler characteristic of the fiber – a fact that may be explained and proved in a more general context.

The type of the group $F_b^\# / F_b^0$ is given in Table 4.

Table 4

F_b	$\widetilde{A}_n,\ {}^*\widetilde{A}_n$	$\widetilde{D}_n$ $n \equiv 1 (\mathrm{mod}\, 2)$	$\widetilde{D}_n$ $n \equiv 0 (\mathrm{mod}\, 2)$	$\widetilde{E}_6$	$\widetilde{E}_7$	$\widetilde{E}_8$
$F_b^\# / F_b^0$	$\mathbb{Z}/(n+1)\mathbb{Z}$	$\mathbb{Z}/4\mathbb{Z}$	$\mathbb{Z}/2\mathbb{Z} \oplus \mathbb{Z}/2\mathbb{Z}$	$\mathbb{Z}/3\mathbb{Z}$	$\mathbb{Z}/2\mathbb{Z}$	0

The 1-dimensional connected group F_b^0 is an elliptic curve provided F_b is nondegenerate; it is isomorphic to the multiplicative group for fibers of type A_n, and the additive group in all other cases.

The root theory is employed not only as an axillary device to obtain the list of types of singular fibers (Table 1). It also reflects all the properties of those fibers, described above. We will describe the connection between the root theory and the theory of lattices associated with singular fibers.

In the lattice N_X, the sublattice

$$F^{\perp} = \{x \in N_X \,|\, (x, F) = 0\}$$

is nonpositive definite. The elements $x \in F^{\perp}$ such that $(x^2) = -2$ are said to be its *roots*. They correspond to the classes that contain linear combinations of components of fibers D with $(D^2) = -2$. The components are simple roots. Clearly the scalar product (x, y) induces a scalar product in the lattice $F^{\perp}/F\mathbb{Z}$. Its sublattice R generated by the roots is negative definite, and the roots form the root system R_F. Its simple roots correspond to the components of fibers that do not intersect the zero section S_0. Let $R_F = \oplus R_i$ be an orthogonal decomposition into irreducible systems. The irreducible factors R_i correspond to various reducible fibers of $f\colon X \to B$. If $Q(R_i)$ is a sublattice generated by the root system R_i, and $Q(R_i)^*$ the dual lattice, then

$$Q(R_i)^*/Q(R_i) \cong F_b^{\#}/F_b^0\,,$$

where F_b is the reducible fiber corresponding to the system R_i.

Finally, the multiplicities of components of a fiber have the following meaning. Assume that F_b corresponds to the root system R_i. It contains a basis $r_1, \dots, r_m$ of roots such that any root can be written as $\pm \sum a_j r_j$, $a_j \geq 0$. There is a maximal positive root $\varrho = \sum b_j r_j$ such that if $r = \sum a_j r_j$ is a root then $b_j \geq a_j$. If C_j denotes the component corresponding to a simple root r_j, and C_0 is the component intersecting S_0, then $F_b = C_0 + \sum b_j r_j$.

There is an important procedure for simplifying a degenerate fiber by passing to a covering of the base. If $B' \to B$ is such a covering, then the surface $X' = X \times_B B'$ is not necessary smooth in general. The minimal desingularization of X' gives an elliptic surface $\overline{X} \to B'$ with simpler fibers as a rule. For a suitable choice of a covering, one can obtain a surface without multiple fibers and with singular fibers of type $\widetilde{A}_n$ only. Such singular fibers are said to be *stable*. Moreover, the fibers of type $\widetilde{D}_n$ become of type A_{2n+3}, and the rest become nonsingular fibers (one can easily verify this with a help of Table 3). Such a simplification is a rather special case of the general semi-stable reduction theorem for curves (Artin-Winters (1971)).

The same procedure implies that multiple fibers can be always obtained by a procedure similar to the one described in Example 2. One has only to consider an appropriate family $X' \to B'$ without multiple fibers in place of $B' \times E$. It follows that a multiple fiber F_b is always of the form $m\overline{F}$, where $\overline{F}$ is an ordinary fiber such that the group $\overline{F}^0$ is either an elliptic curve or a multiplicative group (it has to contain points of finite order!), i. e., $\overline{F}$ is either a smooth curve or a fiber of type $\widetilde{A}_n$.

As we have seen in our argument for the adjunction formula (Sect. 5.2, (9)), the restriction of the sheaf $\mathcal{O}_X(C+K_X)$ to a smooth curve C gives the sheaf $\mathcal{O}_C(K_C)$. Given an elliptic surface X, it follows that the restriction of $\mathcal{O}_X(K_X)$ to any nondegenerate fiber F_b is the sheaf $\mathcal{O}_{F_b}$. Hence a suitable multiple of K_X contains a divisor consisting of components of fibers, i. e., $rK_X \sim \sum n_i C_i$ where C_i are components of fibers. However, we have seen that $(C, K_X) = 0$ for any component C of a singular fiber (provided X is a relatively minimal model). Now, by the last proposition, K_X is proportional to an integral combination of fibers, i. e. $rK_X = \sum n_i F_{b_i}$. If $f\colon X \to B$ has a section, then K_X equals such a combination. So, for an elliptic surface with a section, we get $K_X = f^*(M)$, where M is a suitable divisor class on B. One can easily show that $\mathcal{O}_B(M) = \mathcal{O}_B(K_B) \otimes R^1 f_* \mathcal{O}_X$. It is also easy to show that $\deg R^1 f_*\mathcal{O}_X = \chi(\mathcal{O}_X)$ (Barth-Peters-Van de Ven (1984), Bombieri-Husemöller (1975)). Hence

$$K_X = f^*(M), \quad \deg M = 2g - 2 + \chi(\mathcal{O}_X). \tag{2}$$

On the other hand, $\deg M = (K_X, S)$ for any section S of the family $X \to B$. Hence

$$(S^2) = -\chi(\mathcal{O}_X). \tag{3}$$

In particular, if X is a rational surface then $(S^2) = -1$, i. e., every section is a (-1)-curve.

One can also easily calculate other invariants of the surface. If $X \cong B \times E$ where E is an elliptic curve, then $q(X) = g(B) + 1$. In all other cases $q(X) = g(B)$. By the Riemann-Roch theorem, $e(X) = 12\chi(\mathcal{O}_X)$, and by (Sect. 2, (5)), $e(X) = \sum e(F_b)$. Moreover, one can read off the values of $e(F_b)$ from the list of singular fibers in Table 1. Clearly $e(X) \geq 0$ and $\chi(\mathcal{O}_X) = e(X)/12 \geq 0$.

In particular, if $B \cong \mathbb{P}^1$ and $X \ncong \mathbb{P}^1 \times E$ (we will not consider the trivial case when $X \cong \mathbb{P}^1 \times E$), then $K_X \sim (r-2)E$ for a suitable integer $r \geq 0$, $\chi(\mathcal{O}_X) = r$, $q = 0$, and $p_g = r - 1$. One can easily prove that Weierstrass normal form can be taken as follows:

$$y^2 = x^3 + a(t)x + b(t), \tag{4}$$

where $a(t)$ and $b(t)$ are polynomials of degree $4r$ and $6r$. This gives a complete picture of the elliptic surfaces that have sections. For a given integer r, they form a finite dimensional family parametrized by the coefficients of the polynomials a and b in (4). The polynomials a and b satisfy the following conditions. They do not have a common root of multiplicity ≥ 4 in a and of multiplicity ≥ 6 in b, and $4a^3 + 27b^2 \not\equiv 0$. Furthermore,

$$a \to \alpha^2 a, \quad b \to \alpha^3 b, \quad \alpha \in k, \quad \alpha \neq 0,$$

is the only transformation that maps the surface (4) into an isomorphic one.

10.3. Jacobian Fibration (Barth-Peters-Van de Ven (1984), Ogg (1962), Shafarevich (1961)). Now, we will consider an elliptic fibration $X \to B$ without sections. Then it is impossible to introduce a group structure on fibers that rationally depends on a point b, since it is even impossible to choose rationally a zero point. An example is provided by an Abelian surface that contains an elliptic curve E as a subgroup. The homomorphism $A \to A/E = B$ defines an elliptic fibration whose fibers are isomorphic to E. If there were a section φ: $B \to A$, then by elementary properties of Abelian varieties, we could choose φ to be a homomorphism, and then $A \cong E \times B$ (as groups).

It is easy to construct examples of Abelian varieties over $\mathbb{C}$ that are not products utilizing a representation $A = \mathbb{C}^2/\Omega$. Even if there are no sections, we still have an addition operation $A * E \to A$, i. e., an action of E on each fiber of the fibration $A \to B$. One may interpret this operation as a fiberwise action of the fibration $J = E \times B$ on X, i. e., as a morphism $J \times_B X \to X$. Furthermore, J is already a fibration with a section. We have a similar situation in general.

To an elliptic fibration $X \to B$, we associate an elliptic fibration $J \to B$ which has a section, and a rational map ψ: $J \times_B X \to X$ that commutes with projections to B and has the following properties.

1) ψ is regular on the set of nonsingular points of fibers of J and X;

2) if F_b is an ordinary fiber of the family X, then the restriction of ψ to $G_b^\# \times F_b^\#$ (where G_b is a fiber of the family J), denoted by ψ_b, defines a fixed-point-free and transitive action of the group $G_b^\#$ (recall that $F_b^\#$ and $G_b^\#$ denote the subsets of simple points of F_b and G_b).

The family J is uniquely determined by those properties, and is called the *Jacobian family* of the family X.

So, $F_b^\#$ is a homogeneous space of the group $G_b^\#$ whose elements act without fixed points. Therefore, given a point $x_0 \in F_b^\#$, the map $g \to \psi(y, x_0)$ ($y \in G_b^\#$) defines an isomorphism between $F_b^\#$ and $G_b^\#$, which however depends on the choice of x_0. If one were able to "algebraically" choose a point in each fiber, then we would get an isomorphism between J and X. But this amounts to the existence of a section of X, contrary to our assumption.

The relationship between the group $G_b^\#$ and the curve $F_b^\#$ is analogous to the relationship between an affine space $\mathbb{A}$ and the corresponding vector space $\mathbb{V}$. Both are special cases of the general notion of *principal homogeneous space*. Thus an elliptic fibration $X^\# = \cup G_b^\#$ is a fibration of principal homogeneous spaces or, in other words, a principal homogeneous space of its Jacobian fibration. Henceforth, the fiberwise operation $J^\# \times_B X^\# \to X^\#$ is regarded as addition.

It is helpful to compare this situation with its simple model, namely, the theory of 1-dimensional vector bundles with a base B. Given such a bundle $L \to B$, the space $L \backslash S_0$, where S_0 is a zero section, is a principal homogeneous space over the multiplicative group $\mathbb{G}_m$, i. e. over the fibration $\mathbb{G}_m \times B$ of groups.

The elliptic fibrations are classified by their Jacobian fibrations, which are assumed to be known. One can introduce a group structure in the set $I(J)$ of all elliptic fibrations with a given Jacobian fibration. A sum of fibrations X' and X'' is a fibration X such that there is a rational map η: $X' \times_B X'' \to X$, commuting with the projections to the base, which is regular on the set of simple points of fibers and related to the action of J by the condition that fiberwise

$$\eta(x' + y', x'' + y'') = y' + y'' + \eta(x', x''), \quad x' \in F'_b, \quad x'' \in F''_b, \quad y', y'' \in G_b,$$

where F' and F'' are fibers of X' and X''. Such a principal homogeneous space X exists and is unique. The zero element is the Jacobian fibration J itself. This group is similar to the group $H^1(X, \mathcal{O}_X^*)$ of line bundles. The group $I(J)$ is torsion. If $C \subset X$ is a curve, not contained in the fibers of the projection $X \to B$, and the map $C \to B$ has degree n, then the element of the group $I(J)$ corresponding to X is annihilated by n.

The vast majority of elements of $I(J)$ arise from multiple fibers. If F_b is a multiple fiber of $X \to B$, then $F_b = m\overline{F}_b$, where F_b is a smooth curve or a fiber of type $\widetilde{A}_n$. It turns out that the fiberwise action $J \times_B X \to X$ can be extended, as a morphism, to the fibers $G_b \subset J$ and F_b, however, the action of $G_b^{\#}$ on $F_b^{\#}$ will now have a stabilizer cyclic subgroup of order m, which is determined by a point of order m in the group G_b^0. One can see it in Example 5 of Sect. 10.2, where $J = B \times E$, and the multiple fiber is isomorphic to $E/\{a\}$. Thus, to each multiple fiber of X, we associate an invariant: a point of finite order of the corresponding fiber $F_b \subset J$. We get a homomorphism

$$\tau \colon I(J) \to \bigoplus_{b \in B} (G_b^0)_t\,, \tag{5}$$

where $(G_b^0)_t$ is the subgroup of elements of finite order in G_b^0.

This homomorphism is an epimorphism provided J has at least one singular fiber. The existence of a multiple fiber is even an obstruction to the existence of a differentiable section (for $k = \mathbb{C}$), i.e., if $\tau \neq 0$ then X is not even isomorphic to J as a differentiable fibration. Thus the map τ is similar to the map that associates to a line bundle L its characteristic class $c(L) \in H^2(X, \mathbb{Z})$ (if we regard L as a principal homogeneous space over $\mathbb{C}^* \times B$).

It remains to describe the kernel $I_0(J)$ of τ. Assume $k = \mathbb{C}$. If J has at least one singular fiber, then $I_0(J)$ consists of fibratons having differentiable sections and isomorphic to J as differentiable fibrations. Therefore $I_0(J)$ is analogous to the Picard variety. We study it with a help of an exact sequence similar to the exponential sequence $1 \to \mathbb{Z} \to \mathcal{O} \to \mathcal{O}^* \to 1$ (Sect. 3.3, (2)):

$$0 \to \Gamma \to \mathcal{F} \to \mathcal{G}^0 \to 0\,, \tag{6}$$

where $\mathcal{G}^0$ is the sheaf of local sections of the family $J^0 = \cup_{b \in B} G_b^0$, $\mathcal{F}$ is the sheaf of sections of a 1-dimensional bundle whose fiber over $b \in B$ is the tangent space of G_b^0 at its zero point, and Γ is the sheaf of discrete groups

$R^1 f_* \mathbb{Z}$ whose fiber over $b \in B$ is $H_1(G_b^0, \mathbb{Z})$. The exact sequence (6) extends the representation $E = \mathbb{C}/\Omega$, $\Omega = H_1(E, \mathbb{Z})$, of an individual elliptic curve to the family J.

The group $I_0(J)$ is isomorphic to the *Brauer group* of the surface J, and to $(\mathbb{Q}/\mathbb{Z})^r$ up to a finite group, where $r = b_2 - \varrho$ is the "number of transcendental cycles" (Kodaira (1864–1969), Chap. V, Sect. 3).

Example 6. Let A be a 2-dimensional Abelian variety containing a 1-dimensional Abelian variety (elliptic curve) E. Let $A/E = B$ be an elliptic curve as well.

The homomorphism $A \to B$ gives A a structure of a fibration over B whose fibers are isomorphic to E (compare with the example at the beginning of this section). The translations by elements $a \in E$ gives it a structure of a principal homogeneous space over $E \times B$; moreover, it has no sections provided $A \not\cong E \times B$. One can easily show that this way we obtain all the elements of the group $I_0(E \times B)$. In other words, the group $I_0(E \times B)$ is isomorphic to the group $\mathrm{Ext}(B, E)$ of extensions of B by E.

10.4. Classification (Barth-Peters-Van de Ven (1984), Bombieri-Husemöller (1975)). The canonical class of an arbitrary elliptic surface can be calculated by a formula similar to (2), provided we take into account multiple fibers (Algebraic Surfaces (1965)). Let $n_1 E_1, \ldots, n_r E_r$ be the multiple fibers, where the E_i's are not divisible in N_X. Then

$$K_X = f^*(M) + \sum (n_i - 1) E_i\,, \quad M \in \mathrm{Cl}(B)\,, \qquad \deg M = 2g - 2 + \chi(\mathcal{O}_X)\,. \tag{7}$$

Since $n_i E_i = F$, it follows from (7) that in $N_X \otimes \mathbb{Q}$:

$$K_X = \gamma(X) F \qquad \gamma(X) = 2g - 2 + \chi(\mathcal{O}_X) + \sum \left(1 - \frac{1}{n_i}\right)\,. \tag{8}$$

The elliptic surfaces are divided in 3 classes according to the value of $\gamma(X)$: $\gamma(X) > 0$, $\gamma(X) = 0$, or $\gamma(X) < 0$.

In the first case, mK_X is numerically equivalent to an effective divisor for a suitable $m > 0$; in the second case, $K_X \approxeq 0$; and in the third case, $mK_X \approxeq D > 0$ for a suitable $m < 0$. Clearly, in the first case, $mK_X \approxeq m'F$ hence $\kappa(X) = 1$; in the third case, $\kappa(X) = -\infty$; and in the second case, $\kappa(X) = 0$ if $m_1 K_X \sim 0$ for a suitable $m_1 \neq 0$, and $\kappa(X) = -\infty$ otherwise. We will see below that if $\gamma(X) = 0$ then always $m_1 K_X \sim 0$ for a suitable $m_1 \neq 0$, so $\kappa(X) = 0$.

A surface in the first or third class has a unique elliptic pencil – its irreducible members are characterized by the condition that they are irreducible components of curves of the linear system $|mK_X|$ for sufficiently large $m > 0$ if $\gamma(X) > 0$, and for sufficiently small $m < 0$ if $\gamma(X) < 0$.

A surface in the second class can have several distinct pencils.

Example 7. An Abelian variety $X = E_1 \times E_2$, where E_1 and E_2 are elliptic curves, has two pencils: $X \to E_1$ and $X \to E_2$.

Example 8. If a surface $X_4 \subset \mathbb{P}^3$ contains a line, then a plane H through L cuts X_4 in L and an additional cubic curve, and the pencil of planes through L gives a pencil of cubic curves. If X_4 contains two lines, then it contains two distinct elliptic pencils.

The case $\gamma(X) > 0$ (i. e. $\kappa(X) = 1$) is in fact the "general" one. By the Riemann-Roch theorem, $\chi(\mathcal{O}_X) = e(X)/12$, and it follows from the inequality $e(X) \geq 0$ and the proposition of Sect. 2 that it occurs if either $g \geq 2$, or $g = 1$ and $\chi(\mathcal{O}_X) \geq 1$, or $g = 0$ and $\chi(\mathcal{O}_X) \geq 2$.

One can easily describe the remaining cases. If $g = 1$ we always get $\gamma(X) \geq 0$. If $\gamma(X) = 0$ then $\chi(\mathcal{O}_X) = 0$, and consequently $e(X) = 0$ by the Riemann-Roch theorem (Sect. 5.1, (2)). It follows from the propositions of Sect. 2 and Sect. 4.2 that there are no singular fibers. Moreover, by Lemma 2 of Sect. 8.4, for a suitable unramified Galois covering $B' \to B$, we get $g(B') = 1$ and $X \times_B B' \cong E \times B$, where E is an elliptic curve. Hence $X = (E \times B')/G$, where G is a finite automorphism group acting freely on $E \times B'$, i. e., X is a bi-elliptic surface.

Consider the Jacobian fibration $J = J(X)$. By Example 4 of Sect. 10.1, there exists an elliptic curve E and an unramified covering $B' \to B$ such that

$$g(B') = 1\,, \quad J \underset{B}{\times} B' \cong E \times B'\,, \quad J \cong (E \times B')/G\,,$$

where $|G| \in \{1, 2, 3, 4, 6\}$. For $X' = X \times_B B'$, we get $J(X') \cong E \times B'$, and X' is an Abelian variety: $X' \cong A \supset E$ and $A/E \cong B'$, by Example 6 of Sect. 10.3. Since $A = X' \to X$ is an unramified covering of degree $|G|$ and $K_A = 0$, we get $nK_X = 0$ for $n \in \{2, 3, 4, 6\}$. Thus always $12K_X = 0$.

If $g = 0$, we have to find all solutions of the relations

$$\gamma(X) = 0; \quad \sum_1^r \left(1 - \frac{1}{n_i}\right) = 2 - \chi(\mathcal{O}_X)\,, \quad \chi(\mathcal{O}_X) = 0, 1, 2\,, \tag{9}$$

$$\gamma(X) < 0; \quad \sum_1^r \left(1 - \frac{1}{n_i}\right) < 2 - \chi(\mathcal{O}_X)\,, \quad \chi(\mathcal{O}_X) = 0, 1, 2\,. \tag{10}$$

If $\gamma(X) = 0$, i. e. for (9), the answer is given in Table 5.

Clearly we get $q = 0$ and $p = 1$ in Case 1, and $q = p = 0$ in Case 2. In Cases 3–6, we get $q = 1$ and $p = 0$, and no new surfaces arise. Indeed, in those case, the Albanese variety is an elliptic curve B, and the Albanese map $\alpha\colon X \to B$ brings us back to the previously considered case: $\gamma(X) = 0$ and $g(B) = 1$.

So, if $\gamma(X) = 0$ then always $12K_X = 0$; in particular $\kappa(X) = 0$.

But if $\gamma(X) < 0$ then $-K_X$ is nef. Therefore, if X is a minimal model, then it is ruled by Theorem 1 of Sect. 7.1. But if X is not a minimal model,

Table 5

N	$\chi(\mathcal{O}_X)$	r	Multiplicity of fibers	K_X	Name of surface	$\kappa(X)$
1	2	0		0	$K3$	0
2	1	2	(2, 2)	$2K_X \sim 0$	Enriques	0
3	0	4	(2, 2, 2, 2)	$2K_X \sim 0$	bi-elliptic	0
4	0	3	(3, 3, 3)	$3K_X \sim 0$	bi-elliptic	0
5	0	3	(2, 4, 4)	$4K_X \sim 0$	bi-elliptic	0
6	0	3	(2, 3, 6)	$6K_X \sim 0$	bi-elliptic	0

we consider a morphism $X \to Y$ onto its minimal model. It follows from (Sect. 6.1, (2)) that $(K_Y^2) > 0$ and $(D, K_Y) = (\sigma^*(D), K_X) \leq 0$ for $D > 0$. Therefore Y is again a rational or ruled surface (one can easily verify that Y is in fact a rational surface).

10.5. Applications. Elliptic surfaces have many applications. We will describe several of them.

1. In the theory of elliptic curves, it is known that if $E \subset \mathbb{P}^2$ is a smooth cubic curve with a given zero point o and an addition law, then its points $p_1, \dots, p_9$ are cut out by a curve G of degree 3 if and only if $\sum p_i = 0$ (in the sense of that addition law). Then $\lambda E + \mu G$ is a pencil of cubics. After blowing up the points p_i, we obtain a surface X with a pencil of elliptic curves (Sect. 10.1, Example 1). Similarly, if $\sum p_i = \varepsilon_n$ is a point of order n on E, then the cycle $n \sum p_i$ is cut out by a curve G of degree $3n$ that has singular points of multiplicity n at all the points p_i. By (Sect. 6.2, (12)), the normalization of G has genus 1. The curves $\lambda E^n + \mu G$ form a pencil of curves of genus 1 in $\mathbb{P}^2$.

After blowing up the points p_i, we again obtain a surface Y with a pencil of elliptic curves and a multiple fiber of multiplicity n corresponding to the curve E^n. The surface constructed above is its Jacobian fibration. Such pencils of elliptic curves are called *Halphen pencils*. One can show that any rational elliptic surface is obtained from a Halphen pencil (one has to establish only that the points p_i merge, as well as the degenerations of curves into curves of type $\widetilde{A}_n$ and $^*\widetilde{A}_n$, $n = 0, 1, 2$). This means that any pencil of curves of birational genus 1 (i. e. the normalizations have genus 1) in $\mathbb{P}^2$ is obtained from a suitable Halphen pencil by a birational automorphism of the plane (Dolgachev (1966a)).

2. Again let X be an elliptic surface obtained by blowing up 9 points of intersection of two cubics in $\mathbb{P}^2$. Let $I(X)$ be the group of principally homogeneous spaces with the Jacobian fibration X. Let p_1 and p_2, $p_1 \neq p_2$, be two prime numbers. Let $Y_{p_1 p_2} \in I(X)$ be a surface with two multiple fibers of multiplicities p_1 and p_2. It is not hard to show that $q = p_g = 0$ for every surface $Y \in I(X)$. Furthermore, one can prove that $\pi_1(Y_{p_1 p_2}) = 0$.

Indeed, $\pi_1(F) \to \pi_1(Y)$ is an epimorphism, where F is a fiber of an elliptic family Y. Hence $\pi_1(Y)$ is an Abelian group, and since $q = 0$, it is finite.

By duality, the torsion in $\pi_1(Y) = H_1(Y_{p_1,p_2},\mathbb{Z})$ coincides with the torsion in $H_2(Y_{p_1p_2},\mathbb{Z}) \cong \operatorname{Pic} Y_{p_1p_2}$. Suppose $D \in \operatorname{Div} Y_{p_1p_2}$ and $nD \sim 0$ for $n > 1$, i. e. $D \approx 0$. Then it follows at once from the Riemann-Roch theorem that $l(K - D) \geq 1$. Hence D is equivalent to a linear combination of components of fibers. If the degenerate fibers of X are irreducible (i. e., they are of type $\widetilde{A}_0$ or ${}^*\widetilde{A}_0$), then $D \sim mF + k_1E_1 + k_2E_2$, where p_iE_i are multiple fibers. We can assume that $0 \leq k_i < m$. This together with the relations $p_iE_i \sim F$ and $nD \sim nmF + nk_1E_1 + nk_2E_2$ imply that $p_1p_2m + k_1p_2 + k_2p_1 = 0$, and we derive a contradiction.

On the other hand, the surface $Y_{p_1p_2}$ is not rational, because the elliptic family on $Y_{p_1p_2}$ has two multiple fibers, while on rational surfaces, an elliptic family comes from a Halphen pencil and has a single multiple fiber. So, we have constructed an example of a nonrational surface with $q = p = \pi_1(X) = 0$ (Dolgachev (1966b)).

An analogous example in the class of surfaces of general type was mentioned at the end of Sect. 9.

3. We get $p_g = 0$ for X and $Y_{p_1p_2}$ (see above). Hence $H_2(\mathbb{Z}) \cong \operatorname{Pic} X \cong N_X$ in both cases. For X, the quadratic form on N_X has the form $x_1^2 - x_2^2 - \ldots - x_{10}^2$ by (Sect. 6.1, (2)). By standard formulas, for $Y_{p_1p_2}$, we get $b_2 = 0$ and the index (i. e. the difference between the number of positive and negative squares in the form on $N_Y = H_2(Y,\mathbb{Z})$) equals -8. Acording to a theorem of Rokhlin, if for a 4-dimensional simply connected differentiable manifold X, the form on $H_2(X)$ is even then the index is divisible by 16. Therefore the above form is odd for $Y_{p_1p_2}$. However, according to a general theorem on indefinite unimodular quadratic forms, they are determined by their type (the number of positive and negative squares) and parity. So the forms for X and $Y_{p_1p_2}$ are equivalent.

Finally, simply connected 4-manifolds are determined by the quadratic form up to homotopy type, and according to Freedman's result up to homeomorphism. Therefore X and $Y_{p_1p_2}$ have the same homotopy type. At the same time, Donaldson constructed an invariant that allows to distinguis such non-diffeomorphic manifolds. In particular, Y_{2p_1} and Y_{2p_2} have distinct Donaldson invariant provided $p_1 \neq p_2$. Thus there are infinitely many 4-manifolds of the same homotopy type (and even topological type) that are not diffeomorphic to each other (Van de Ven (1987)).

Recently, by developing the same methods, several new results were obtained on the topology of algebraic varieties.

Examples 2 and 3 show the existence of homeomorphic simply connected algebraic surfaces having different canonical dimension κ: $\kappa(X) = -\infty$ but $\kappa(Y_{p_1p_2}) = 1$. The situation changes if we replace homeomorphisms by diffeomorphisms. It was recently proved that simply connected diffeomorphic algebraic surfaces have the same canonical dimension (Podstrigach-Tjurin (1992), Podstrigach (1994)).

It was established that on any simply connected algebraic (projective) surface X, with a possible exception of a plane with at most 7 points blown up and a quadric, there exists an "exotic" smooth structures, i. e., there exists a 4-dimensional smooth manifold homeomorphic but not diffeomorphic to $X(\mathbb{C})$.

It was also established the existence of infinitely many distinct "exotic" smooth structures on a plane with 9 points blown up and on a $K3$ surface. In the later case, any such smooth structure admits a quasicomplex structure, which is however nonintegrable. This examples show that 4-dimensional simply connected smooth compact manifolds differ from simply connected manifolds of dimension ≥ 5, which admit only finitely many smooth structures according to S. P. Novikov's results.

Given a fixed integer $d \geq 8$, it was established that all algebraic surfaces obtained by blowing up the plane in d points are homeomorphic (but not diffeomorphic in general).

§ 11. Surfaces of Canonical Dimension 0

As before, we will consider only the surfaces that are minimal models.

By Theorem 2 of Sect. 8.1, a surface with $\kappa = 0$ is isomorphic either to an Abelian surface or a $K3$ surface, or it has an elliptic pencil. Moreover, in Theorem 2 of Sect. 8.1, we have described surfaces with an elliptic pencil and $\kappa = 0$. Thus we get the following 4 types of surfaces with $\kappa = 0$:

I. $K3$ surfaces,
II. Enriques surfaces,
III. Abelian surfaces,
IV. Bi-elliptic surfaces.

The $K3$ surfaces are discussed in the next section. Here we will consider the remaining cases in detail.

11.1. Enriques Surfaces. Let X be an Enriques surfaces. By definition, $2K_X \sim 0$ and $p(X) = q(X) = 0$.

Proposition. *Let X be an Enriques surfaces. Then*

$$\chi(\mathcal{O}_X) = 1\,, \quad l(X) = 12\,, \quad b_2 = 10\,, \quad H_2(X, \mathbb{Z}) \cong \mathbb{Z}^{10} \oplus \mathbb{Z}/2\mathbb{Z}\,,$$
$$\pi_1(X) \cong H_1(X, \mathbb{Z}) \cong \mathbb{Z}/2\mathbb{Z}\,.$$

The universal covering of X is a $K3$ surface. The Picard variety equals 0 and all 2-dimensional cycles are algebraic, i. e. $S_X \cong H_2(X, \mathbb{Z})$.

One can calculate $\chi(\mathcal{O}_X)$, $l(X) = 12$, and b_2 with a help of Noether's formula (Sect. 5.1, (2)). Since the dimension of the Picard variety equals $q(X) = 0$, it is trivial. Since $p(X) = 0$, by the Lefschetz theorem (Sect. 3.3), all cycles are algebraic, i. e., $S_X \cong H_2(X, \mathbb{Z})$. Since $K_X \neq 0$ in S_X and $2K_X = 0$, there is an element of order 2 in $S_X \cong H_2(X, \mathbb{Z})$. It follows from the universal coefficient formula that $H_1(X, \mathbb{Z})$ has an element of order 2 as well, so there is an unramified covering $Y \to X$ of degree 2. Then $K_Y = 0$ and $e(Y) = 24$, hence $q(Y) = 0$, i.e., Y is a $K3$ surface. In the next section, we will explain why all $K3$ surfaces are simply connected. So, Y is the universal covering of X, hence $\pi_1(X) = \mathbb{Z}/2\mathbb{Z}$.

In (Sect. 2, Example 9), we have constructed an Enriques surface as a quotient of a $K3$ surface given by the equations

$$f_1 + h_1 = 0\,, \quad f_2 + h_2 = 0\,, \quad f_3 + h_3 = 0\,, \tag{1}$$

where $f_i(x_0, x_1, x_2)$ and $h_i(x_3, x_4, x_5)$ are quadratic forms, by the involution

$$(x_0 : x_1 : x_2 : x_3 : x_4 : x_5) \mapsto (x_0 : x_1 : x_2 : -x_3 : -x_4 : -x_5)\,.$$

One can show that any Enriques surface can be obtained this way (Verra (1983)). This presentation allows us to determine the number of parameters the Enriques surfaces depend on. Since each of the quadratic forms f_1, f_2, f_3, h_1, h_2, h_3 has 6 coefficients, 36 coefficients appear in the equations. Such a surface is transformed into an isomorphic one under linear transformations of 3 variables x_0, x_1, x_2, 3 variables x_3, x_4, x_5 and 3 equations (1). All together $3^2 \cdot 3 = 27$ coefficients take part in those transformations. However, the multiplication of all the variables x_0, x_1, x_2, x_3, x_4, x_5 by $\alpha \neq 0$ coincides with the multiplication of all equations (1) by α^2. So, it remains 26 independent coefficients. Therefore the set of nonequivalent Enriques surfaces depends on $36 - 26 = 10$ parameters and form a connected 10-dimensional variety.

One can justify the above heuristic calculations. In fact, by the general deformation theory of analytic varieties, the modul variety of deformations of a variety X exists (at least locally) and is smooth provided $h^2(X, \mathcal{T}_X) = 0$ ($\mathcal{T}_X$ is the sheaf of sections of the tangent bundle); moreover, its dimension equals $h^1(X, \mathcal{T}_X)$. One can easily show that $h^2(X, \mathcal{T}_X) = 0$ and $h^1(X, \mathcal{T}_X) = 10$ for Enriques surfaces.

We will present two more constructions of Enriques surfaces.

A. Let $Q = \mathbb{P}^1 \times \mathbb{P}^1$ be a quadric, and B a smooth curve in the class $4H$, where H is a plane section. One can easily establish the existence of a double covering $Y \to Q$ branched along a given divisor D, provided $D \equiv 0 (\bmod 2)$. In particular, there is such a covering $Y \to Q$ with B as its branch locus. Since $K_Q = -2H$, straightforward calculations show that $K_Y = 0$ and Y is a $K3$ surface.

Consider an automorphism τ of Q acting as involution $(x_0 : x_1) \mapsto (x_0 : -x_1)$ on each $\mathbb{P}^1$. It has 4 fixed points on Q: p_1, p_2, p_3, p_4. Clearly one can choose B to be invariant with respect to τ and missing the points p_1, p_2, p_3, p_4. Then τ can be lifted to Y, and we get an involution τ' of Y. If ν is an automorphism of the covering $Y \to Q$, then $g = \tau\nu$ clearly has no fixed points, and $X = Y/G$ is an Enriques surface, where $G = \{1, g\}$.

An advantage of this presentation is that it makes evident the existence of two pencils of elliptic curves on X. Indeed, the projection

$$Y \to \mathbb{P}^1 \times \mathbb{P}^1 \xrightarrow{\pi_i} \mathbb{P}^1$$

(π_i is the projection on the i-th factor, $i = 1, 2$) defines a pencil of curves, each curve being a double covering of the corresponding line $\pi_i^{-1}(c)$, $c \in \mathbb{P}^1$, branched in 4 points of intersection $\pi_i^{-1}(c) \subset B$, i. e. an elliptic curve. One can descent both pencils to X. Each pencil gives precisely the pencil used in the construction of Enriques surface in (Sect. 8.1, Theorem 2).

The above construction can be generalized to the case when B has singularities such that the corresponding singularities of Y are Du Val; we then resolve those singularities. However, this construction does not give all the Enriques surfaces too, only the "general" ones: with two distinct elliptic pencils. One can consider a similar construction with the quadric Q replaced by a quadric cone – this way we obtain "special" Enriques surfaces that have a single elliptic pencil (Verra (1983)).

B. The classical construction of an Enriques surface (due to Enriques himself) defines it as the normalization of a surface of degree 6 in $\mathbb{P}^3$ passing twice through the edges of a tetrahedron. Take the coordinate system defined by our tetrahedron. Then the equation of the surface of degree 6 takes the form

$$x^2y^2z^2 + x^2y^2 + x^2z^2 + y^2z^2 + xyzf_2(x, y, z) = 0\,, \qquad (2)$$

where f_2 is a polynomial of degree 2.

The formula $K_X = (n-4)H$ for the canonical class of a smooth surface X of degree n in $\mathbb{P}^3$, where H is a plane section (Sect. 2, Example 1), can be generalized to the case when X is the normalization of a surface $X' \subset \mathbb{P}^3$ of degree n passing twice through curves $C_1, \ldots, C_r$ (and without more complicated singularities):

$$K_X = (n-4)H - \sum C_i\,.$$

In our case, the system $|K_X|$ consists of quadrics passing through the edges of the tetrahedron. Since there are no such quadrics, the system $|K_X|$ is empty, i. e. $p_g(X) = 0$. The system $|2K_X|$ consists of surfaces of degree 4 passing through the edges of the tetrahedron. There exists a (reducible) surface with this property, namely a sum of the faces of our tetrahedron. Therefore $P_2(X) > 0$. As before, it is not hard to show that $2K_X = 0$.

We observe that not all Enriques surfaces can be represented in the form (2). It can only be done for the surfaces we called "general" in the discussion of the previous construction (e. g., see Algebraic Surfaces (1965)).

11.2. Abelian Surfaces (Mumford (1970a)). By definition, an *Abelian surface* is a projective variety that is an algebraic group (which is necessary Abelian). For such a surface, $K_X = 0$ since there is a differential form $\omega \in H^0(X, \Omega_X^2)$ invariant with respect to all translations, which obviously satisfies $(\omega) = 0$. Furthermore, since the tangent bundle (of any algebraic group) is trivial, $e(X) = 0$ hence $\chi(\mathcal{O}_X) = 0$, so $q = 2$. In Sect. 8, we have seen that the conditions $K_X = 0$ and $q = 2$ characterize Abelian surfaces. Throughout the rest of Sect. 11.2, we denote by X an Abelian surface.

By adjunction, we get $(C^2) \equiv 0 (\mathrm{mod}\, 2)$ for any curve $C \subset X$, hence $(D^2) \equiv 0 (\mathrm{mod}\, 2)$ for any $D \in S_X$. By a suitable translation $t_a\colon x \mapsto x + a$, we can transform C into a curve $t_a(C)$ that is in a general position with respect to C. On the other hand, the curves $t_a(C)$, $a \in X$, form an algebraic family, hence they are algebraically equivalent. Therefore $(C^2) = (C, t_a(C)) \geq 0$, i. e., there are no exceptional curves on X (as we mentioned before, there are no nontrivial maps of rational curves to Abelian varieties).

The simplest invariant of the lattice N_X is the number

$$\pi(X) = \min \left\{ \tfrac{1}{2}(C^2) | (C^2) > 0,\ C \in N_X \right\} .$$

For example, if $X = E_1 \times E_2$, where E_1 and E_2 are elliptic curves, then for $C = (E_1 \times e_2) + (e_1 \times E_2)$, we get $(C^2) = 2$ and $\pi(X) = 1$. If $X = J(C)$ is the Jacobian of a curve C of genus 2 (Sect. 2, Example 6), then C can be embedded in X: $c \mapsto (c - c_0) \in \mathrm{Pic}^0 X$ (for a fixed point c_0). We then get $(C^2) = 2$ by adjunction, hence $\pi(X) = 1$, as before. However, there are Abelian surfaces with arbitrary $\pi(X)$. For example, if E_1 and E_2 are sufficiently general elliptic curves, $e_1 \in E_1$ and $e_2 \in E_2$ points of order n, $a = (e_1, e_2) \in E_1 \times E_2$, and $X = (E_1 \times E_2)/\{a\}$, then it is not hard to verify that $\pi(X) = n$.

An Abelian surface X with an ample divisor class h is called a *polarized* (Abelian) *surface*, and the number $(h^2)/2$ is called the *degree of polarization*. So, $\pi(X)$ is the smallest degree of polarization of X. If $\pi(X) = 1$, then the Abelian surface with the corresponding divisor class is said to be *principally polarized*.

One can show that an Abelian surface with $\pi(X) = 1$ is either the Jacobian of a curve of genus 2 or has the form $E_1 \times E_2$ (these two cases intersect). Any Abelian variety can be represented, as in the above example, in the form X/Γ, where X is an Abelian variety with $\pi(X) = 1$ and $\Gamma \subset X$ a finite subgroup.

One can show that all Abelian surfaces X with a given invariant $\pi(X)$ can be embedded into the same projective space as surfaces of the same degree. Their images form an irreducible family of surfaces. From this, one can deduce that they are parameterized up to isomorphism by points of an irreducible variety. The dimension of this "moduli variety" equals 3 (for every value of π).

One can describe the picture more explicitly for $k = \mathbb{C}$. Any complex torus of dimension 2 has the form $\mathbb{C}^2/\Omega$, where $\Omega \subset \mathbb{C}^2$ is a 4-dimensional lattice. Let e_1, e_2, e_3, e_4 be its basis. We can take $e_1 = (1, 0)$ and $e_2 = (0, 1)$ in $\mathbb{C}^2$. Then the lattice is given by the vectors $e_3 = (\alpha, \beta)$ and $e_4 = (\gamma, \delta)$. The independence of e_1, e_2, e_3, e_4 over $\mathbb{R}$ is expressed by

$$\begin{vmatrix} \mathrm{Im}\,\alpha & \mathrm{Im}\,\beta \\ \mathrm{Im}\,\gamma & \mathrm{Im}\,\delta \end{vmatrix} \neq 0 . \tag{3}$$

So, any torus is determined by four parameters α, β, γ, δ satisfying (3).

We have already mentioned (Sect. 2, Example 6) that an arbitrary torus $X = \mathbb{C}^2/\Omega$ is not necessary isomorphic (as a complex analytic variety) to an algebraic surface. In fact, if the cycles σ_{ij} form a natural basis in $H_2(X, \mathbb{Z})$, i. e., they are images of $e_i \wedge e_j$, then any algebraic cycle $C = \sum a_{ij}\sigma_{ij}$ gives a relation

$$\sum a_{ij} \begin{vmatrix} \xi_i & \xi_j \\ \eta_i & \eta_j \end{vmatrix} = 0, \tag{4}$$

where (ξ_i, η_i) are the coordinates of the basic vectors e_i. One can rewrite (4) in the form

$$CAC^T = 0 \tag{5}$$

where $A = (a_{ij})$ is a skew-symmetric matrix of type $(4,4)$, and

$$C = \begin{pmatrix} \xi_1 & \dots & \xi_4 \\ \eta_1 & \dots & \eta_4 \end{pmatrix},$$

is the "period matrix" of type $(2,4)$. One can strengthen a bit the relations (4) and (5) as follows. Consider the number

$$\varphi(\lambda_1, \lambda_2) = \int_C \sum_{i,j=1,2} \lambda_i \overline{\lambda}_j dz_i \wedge \overline{dz_j}.$$

Simple calculations show that $\varphi(\lambda_1, \lambda_2) \geq 0$, with equality if and only if the inverse image of C in $\mathbb{C}^2$ is a line parallel to the line $\overline{\lambda}_1 z_1 + \overline{\lambda}_2 z_2 = 0$. In the latter case, C is a translation of an elliptic curve and $(C^2) = 0$. Therefore, if C is assumed to be ample, then $\varphi(\lambda_1, \lambda_2)$ is a positive definite Hermitian form . We can rewrite this as

$$CAC^* > 0 \tag{6}$$

where ">" means positive definite.

Theorem 1. *A torus $\mathbb{C}^2/\Omega$ with a period matrix C is isomorphic to an algebraic surface if and only if the conditions* (5) *and* (6) *hold with A a suitable skew-symmetric integer matrix.*

We have already established the necessity. The sufficiency is established by a direct construction of an embedding into a projective space with a help of θ-functions.

To find a more explicit form of (5) and (6), one has to choose appropriate bases in $\mathbb{C}^2$ and Ω.

A choice of basis in Ω yields a substitution $A \mapsto M^T AM$, where M is a unimodular integer matrix. It is well known that A can be reduced to the form

$$A = \begin{pmatrix} 0 & D \\ -D & 0 \end{pmatrix},$$

where $D = \begin{pmatrix} \delta_1 & 0 \\ 0 & \delta_2 \end{pmatrix}$, $\delta_1 \,|\, \delta_2$. Since D enter everywhere up to a scalar multiple, we can even assume that $D = \begin{pmatrix} 1 & 0 \\ 0 & d \end{pmatrix}$. Considering, as before, the basis e_1 and e_2 in $\mathbb{C}^2$, we will reduce C to the form $C = (E, U)$, where E is the identity and U an arbitrary matrix of order 2. Then one can easily verify that (5) and (6) take the form

$$(UD)^T = UD\,, \tag{7}$$

$$\operatorname{Im}(UD) > 0\,, \tag{8}$$

where $\operatorname{Im}(UD)$ is the matrix consisting of imaginary parts, and the sign "$>$" means positive definite.

Recall that the matrix A came from the coefficients of the algebraic cycle $s = \sum a_{ij}\sigma_{ij} \in H_2(X, \mathbb{Z})$. After identifying $H_2(X, \mathbb{Z})$ with $H^2(X, \mathbb{Z}) = \wedge^2 H^1(X, \mathbb{Z})$, we may view A as a bivector and calculate (s, s) as a square of that bivector. One can easily derive that $(s, s) = 2d$. Thus we see that d coincides with the invariant $\pi(X)$ introduced above.

Fixing d (i. e. $\pi(X)$) and D, we set $Z = UD$. Then (8) describes an open set in the 3-dimensional space $\mathbb{H}_2$ of symmetric (by (7)) matrix Z.

If $\varrho(X) > 0$, then, in addition to the cycle C with $(C^2) = d$ used in the derivation of (5) and (6), there exist a cycle C' that is not a scalar multiple of C. It gives a relation similar to (7), i. e. a quadratic relation on the elements of the matrix $Z \in \mathbb{H}_2$. Thus we get countably many surfaces in the domain $\mathbb{H}_2$, and if a point Z does not belong to any of those surfaces, then $\varrho(X) = 1$ for the corresponding surface X. But if Z belongs to one of the surfaces only, then $\varrho(X) = 2$; if Z belongs to the intersection of two surfaces, then $\varrho(X) = 3$, and if Z belongs to the intersection of three surfaces, then $\varrho(X) = 4$ (Fig. 20).

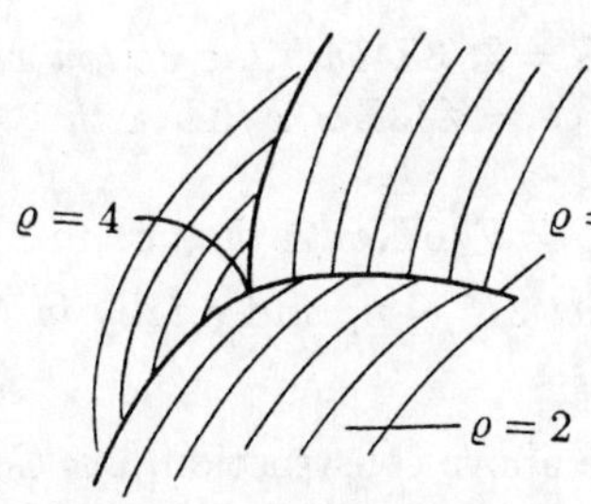

Fig. 20

11.3. Bi-elliptic Surfaces. By definition, these are the surfaces $X = (E \times B)/G$, where E and B are elliptic curves, and G is a finite subgroup of the translation group of E, which acts on B not only by translations but by nontrivial group automorphisms as well. It turns out that there are only a few possibilities for such constructions, and they all can be described explicitly.

Since G is a finite subgroup of the automorphism group of the curve B, it is a semidirect product $T \rtimes H$, where $T \subset B$ is a subgroup of translations, and H a nontrivial subgroup of the automorphism group of B preserving the group structure. By the theory of elliptic curves, the group of such automorphisms is isomorphic to $\mathbb{Z}/n\mathbb{Z}$ with $n \in \{2,3,4,6\}$. Consequently H has the same form as well. On the other hand, G is a subgroup of the translation group of the curve E, which is clearly Abelian. Therefore the above semidirect product is in fact a direct product $T \times H$, which means that the elements of T are invariant with respect to H. One can easily list all the fixed points of the action of H, namely:

for the reflection $x \mapsto -x \Rightarrow$ the points of order 2;

for the curve $\mathbb{C}/\mathbb{Z} + i\mathbb{Z}$, $i = \sqrt{-1}$, and the automorphism $x \mapsto ix \Rightarrow$ the points 0 and $(1+i)/2$;

for the curve $\mathbb{C}/\mathbb{Z} + \varrho\mathbb{Z}$, $\varrho = \exp(2\pi i/3)$, and the automorphism $x \mapsto \varrho x \Rightarrow$ the points 0 and $\pm(1-\varrho)/3$;

for the same curve and the automorphism $x \mapsto -\varrho x \Rightarrow$ only the point 0.

Furthermore, since $G = T \times H$ is a subgroup of translations of E, it is generated by 2 elements, except when $G = B_2 \times \mathbb{Z}/2\mathbb{Z}$ where $B_2 \subset B$ is a subgroup of points of order 2.

From the above discussion, one can easily obtain a complete list of bi-elliptic surfaces (Beauville (1978)).

Theorem 2. *The following is the list of all bi-elliptic surfaces:*

1) $G = \mathbb{Z}/2\mathbb{Z}$ *acts on* B *as the reflection* $x \mapsto -x$*;*

2) $G = \mathbb{Z}/2\mathbb{Z} \times \mathbb{Z}/2\mathbb{Z}$ *acts on* B *via* $x \mapsto -x$ *and* $x \mapsto x + \varepsilon$*, where* $\varepsilon \in B_2$*;*

3) $B = \mathbb{C}/\mathbb{Z} + i\mathbb{Z}$ *and* $G = \mathbb{Z}/4\mathbb{Z}$ *with the action* $x \mapsto ix$*;*

4) $B = \mathbb{C}/\mathbb{Z} + i\mathbb{Z}$ *and* $G = \mathbb{Z}/4\mathbb{Z} \times \mathbb{Z}/2\mathbb{Z}$ *with the action* $x \mapsto ix$, $x \mapsto x + (1+i)/2$*;*

5) $B = \mathbb{C}/\mathbb{Z} + \varrho\mathbb{Z}$ *and* $G = \mathbb{Z}/3\mathbb{Z}$ *with the action* $x \mapsto \varrho x$*;*

6) $B = \mathbb{C}/\mathbb{Z} + \varrho\mathbb{Z}$ *and* $G = \mathbb{Z}/3\mathbb{Z} \times \mathbb{Z}/3\mathbb{Z}$ *with the action* $x \mapsto \varrho x$, $x \mapsto x + (1-\varrho)/3$*;*

7) $B = \mathbb{C}/\mathbb{Z} + \varrho\mathbb{Z}$ *and* $G = \mathbb{Z}/6\mathbb{Z}$ *with the action* $x \mapsto -\varrho x$.

In Cases 1 *and* 6, *we get* $3K \sim 0$*; and finally in Case* 7, *we get* $6K \sim 0$ *thus* $12K \sim 0$ *in all the cases.*

Remark. If $T = 0$ in the above construction, i. e. G consists of group automorphisms of the group B, then we get surfaces constructed in Example 4 of Sect. 10.1. They correspond to Cases 1, 3, 5, and 7.

§ 12. $K3$ Surfaces

12.1. Main Invariants. Recall (see Sect. 2, Example 7) that an algebraic surface X is a $K3$ *surface* if $K_X = 0$ and $H^1(X, \mathcal{O}_X) = 0$.

Then it follows from Noether's formula (Sect. 5.1, (2)) that

$$b_2 = 22\,, \quad p_g = 1\,, \quad h^{1,1} = 20\,.$$

By adjunction (Sect. 4.2, (9)), (C^2) is even and $(C^2) \geq -2$ for an irreducible curve C. Moreover, $(C^2) = -2$ if and only if C is a smooth rational curve. Such curves are called (-2)-*curves.* If C is a smooth curve of genus 1 (elliptic) then $(C^2) = 0$. If $(C^2) = 0$, then C is either elliptic, or rational with one double point that is a node or a cusp (the latter follows at once from (Sect. 4.2, (9))).

The principal invariant of a $K3$ surface is its Severi group (since $H^1(X, \mathcal{O}_X) = 0$, we get $S_X = \operatorname{Pic} X$). It follows at once from the Riemann-Roch theorem that S_X has no torsion, i. e. $S_X = N_X$. The lattice S_X is even, i. e., x^2 is even for $x \in N_X$. This follows by linearity, since (C^2) is even provided C is an irreducible curve. Since the ground field has characteristic 0 (as we always assume), the rank ϱ of S_X is at most $h^{1,1} = 20$. The simplest invariant of X and S_X is the minimal value of $(x^2)/2$, where $x \in S_X$ and $(x^2) > 0$. We call it the *class of the surface* X, and denote by π_X.

For example, if $X \to \mathbb{P}^2$ is a double covering branched along a curve of degree 6, then $\pi_X = 1$. For a smooth surface of degree 4 in $\mathbb{P}^3$, we get $\pi_X = 2$. For a complete intersection of type $(2, 3)$ in $\mathbb{P}^4$, we get $\pi_X = 3$; and for a complete intersection of type $(2, 2, 2)$ in $\mathbb{P}^5$, we get $\pi_X = 4$.

In Sect. 5.2, we have introduced two dual cones, E and $\overline{A}$. Here we will use the cone $\overline{A}$, the closure of the cone of ample divisors. We know that $\overline{A} \subset \Omega_+$, where Ω_+ is the positive half of the cone Ω: $(x^2) \geq 0$ subject to the condition $(x, h) > 0$ for an ample divisor class h. By the Riemann-Roch theorem, given a divisor $D \in \Omega$ on a $K3$ surface, then either $D > 0$ or $-D > 0$; moreover, $D > 0$ provided $D \in \Omega_+$. The cone $\overline{A}$ is defined by the condition $(x, c) \geq 0$, where c is the divisor class of a curve C.

A simple argument shows that given a divisor $D > 0$ on a $K3$ surface, then $(D^2) > 0$ implies $(D, C) > 0$ for any irreducible curve C, with possible exceptions (-2)-curves.

Therefore $x \in \overline{A}$ if $(x, c) \geq 0$ for all classes c that contain (-2)-curves.

In the projective space $\mathbb{P}(S_X \otimes \mathbb{R})$, the interior of the cone Ω_+ defines an open subset U (U is the so-called *Cayley-Klein model* of the Lobachevski geometry). The intersection of the half-spaces $(x, c) \geq 0$ for all $c \in S_X$ containing (-2)-curves is a convex polyhedron (possibly with infinitely many vertices tending to infinity). By the above discussion, this polyhedron coincides with the image of $\overline{A}$ in $\mathbb{P}(S_X \otimes \mathbb{R})$, which we also denote by $\overline{A}$ (Fig. 21).

The polyhedron $\overline{A}$ has the following very important interpretation. It is easy to verify that any element $c \in S_X$ such that $c^2 = -2$ defines an automorphism $x \mapsto x + (x, c)c$ preserving the scalar product in S_X given by the intersection pairing. Such automorphisms are called *reflections*, and they

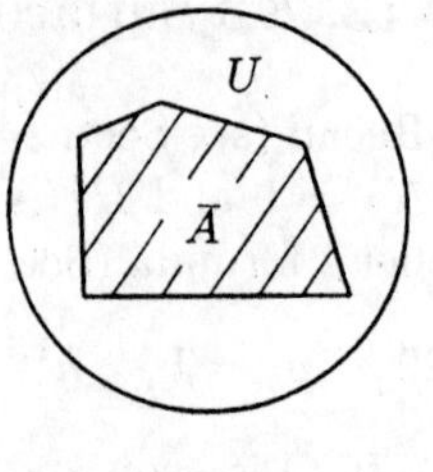

Fig. 21

generate a subgroup $W(S_X)$ of the group $\operatorname{Aut} S_X$ of all automorphisms of S preserving the scalar product. The reflections (as any automorphisms of $\operatorname{Aut} S_X$) define motions of the Lobachevski space, and $W(S_X)$ is a discrete group of motions. It follows from the theory of reflection groups that $\bar{A}$ is the fundamental domain of the group $W(S_X)$.

12.2. Projective Geometry. Let $D \subset X$ be a very ample divisor, and $\varphi_D\colon X \to \mathbb{P}^N$ the embedding corresponding to the linear system $|D|$. Then $h^1(X, \mathcal{O}(D)) = 0$, and by the Riemann-Roch theorem, $\varphi_D(X)$ is a surface of degree $2g-2$ in a g-dimensional space (where $g = N$); moreover, a general hyperplane section of this surface is a canonical curve of genus g.

We will consider a more general situation, namely: C is an irreducible curve on a $K3$ surface X, $(C^2) > 0$, and φ_C is a rational map corresponding to the linear system $|C|$.

Theorem 1. *The system $|C|$ has no base points, i. e. φ_C is a morphism. Its image $\varphi_C(X)$ is a normal surface, and general curves of $|C|$ are smooth. The degree of $\varphi_C\colon X \to \varphi_C(X)$ equals* 1 *or* 2. *If* $\deg \varphi_C = 1$, *then no smooth curve of $|C|$ is hyperelliptic. But if* $\deg \varphi_C = 2$, *then all the smooth curves of $|C|$ are hyperelliptic. In the letter case $|2C|$ defines a birational morphism.*

Now, we will describe both possibilities, namely: $\deg \varphi_C = 1$ or 2. It turns out that the case $\deg \varphi_C = 2$ is an exceptional one (we may view this case as an analog of hyperelliptic curves in the theory of $K3$ surfaces).

Theorem 2. *If* $\deg \varphi_C = 1$, *then the surface $\varphi_C(X)$ has only Du Val singularities of type A_n, D_n, E_6, E_7, E_8 (compare Set.* 6.3). *The morphism φ_C: $X \to \varphi_C(X)$ is a minimal resolution of those singularities, and it contracts the configuration of rational curves described at the end of Sect.* 6. *We always get* $\deg \varphi_C = 1$ *provided $\pi_X > 6$. Thus every $K3$ surface with $\pi > 6$ is isomorphic to a surface of degree 2π in $\mathbb{P}^{\pi+1}$.*

Theorem 3. *If* $\deg \varphi_C = 2$, *then $\varphi_C(X)$ is isomorphic either to $\mathbb{P}^2$, or a ruled surface $\mathbf{F}_n$, or a cone over a rational curve Γ_n of degree n in $\mathbb{P}^n$ (Γ_n is the image of $\mathbb{P}^1$ under the Veronese map v_n). In all the cases, we get $n \in \{0, 1, 2, 3, 4\}$. In particular, we can choose C with $(C^2) = 2\pi_X$. Then $\varphi_C(X) = \mathbb{P}^2$ if $\pi = 1$. But if $\pi \in \{2, 3, 4, 5, 6\}$, then $\varphi_C(X) = \mathbf{F}_{\pi-2}$, or $\varphi_C(X)$ is a cone over $\Gamma_{\pi-2}$.*

We can explicitly describe the branch curve of the double covering φ_C: $X \to \varphi_C(X)$. If $\varphi_C(X) = \mathbb{P}^2$, then it is a curve of degree 6, which is either smooth or has singularities such that the corresponding double covering has Du Val singularities. If $\varphi_C(X) = \mathbf{F}_{\pi-2}$, then the branch curve lies in the class $-2K_{\mathbf{F}_{\pi-2}} = 4S + 2\pi f$, where f is the class of a fiber of the ruled surface $\mathbf{F}_{\pi-2}$, and S is a section with $(S^2) = -\pi + 2$ (compare Sect. 2, Example 4) and similar restrictions on singularities. One has a similar description in case of the cone cover $\Gamma_{\pi-2}$ (Seminar Palaiseau (1985), Exp. IV).

12.3. Topology. Let X be a $K3$ surface over $\mathbb{C}$. For the corresponding 4-dimensional topological manifold $X(\mathbb{C})$, we get $b_1 = 0$ and $b_2 = 22$. Furthermore, $X(\mathbb{C})$ is simply connected (later we will explain how to derive it). Therefore $H^1(X,\mathbb{Z}) = 0$ and $H^2(X,\mathbb{Z})$ has no torsion. The lattice $H^2(X,\mathbb{Z})$ is the main topological invariant (henceforth, we denote $X(\mathbb{C})$ simply by X). This lattice is even; this is established similarly to the assertion that S_X is even utilizing the topological analog of the adjunction formula:

$$(x,x) \equiv (w_2, x)(\operatorname{mod} 2)\,, \qquad x \in H^2(X,\mathbb{Z})\,,$$

where w_2 is the second Stiefel-Whitney class such that $w_2 \equiv K_X(\operatorname{mod} 2)$.

By Poincaré duality, the lattice $H^2(X,\mathbb{Z})$ is unimodular, i. e., it has a basis $\{e_i\}$ with the Gram determinant $|(e_i, e_j)| = \pm 1$. Finally, by the Hodge theory (e. g., see Sect. 14), the space $H^2(X,\mathbb{R}) = H^2(X,\mathbb{Z}) \otimes \mathbb{R}$ has type $(3,19)$.

By a general theorem in the theory of quadratic forms (Serre (1970)), an indefinite unimodular lattice E is uniquely determined by its parity and the type of the space $E \otimes \mathbb{R}$. This allows to describe the lattice $H^2(X,\mathbb{Z})$ explicitly. It is isomorphic to a direct sum of three 2-dimensional lattices U_2 with the Gram matrix $\begin{pmatrix} 0 & 1 \\ 1 & 0 \end{pmatrix}$ and two 8-dimensional lattices E_8, where E_8 is an even unimodular positive definite lattice. The latter can be described as follows:

$$E_8 = E' + e\mathbb{Z}\,, \quad E' \subset e_1\mathbb{Z} + \ldots + e\mathbb{Z}\,, \quad (e_i, e_j) = \delta_{ij}\,;$$

E' consists of $\sum_{i=1}^{8} a_i e_i$, $a_i \in \mathbb{Z}$, and $\sum a_i \equiv 0(\operatorname{mod} 2)$; $e = (\sum_{i=1}^{8} e_i)/2$. Since the lattices $H_2(X,\mathbb{Z})$ and $H^2(X,\mathbb{Z})$ are isomorphic, the same holds for $H_2(X,\mathbb{Z})$.

According to a well-known theorem in topology, a simply connected compact 4-dimensional manifold M is uniquely determined up to homotopy type by the lattice $H^2(M,\mathbb{Z})$. Therefore all $K3$ surface have the same homotopy type. In fact, they are even diffeomorphic but this requires a more delicate argument (see below).

An important property of $K3$ surfaces is their indecomposibility, namely, they cannot be represented as a connected sum of two smooth manifolds. It is quite possible that $K3$ surfaces belong to a short list of "elementary" simply connected manifolds, so that an arbitrary simply connected manifold is a connected sum of those "elementary" manifolds.

12.4. Analytic Geometry. The main analytic invariants of a $K3$ surface X defined over comples numbers are the integrals

$$\int_{\sigma_1} \omega, \ldots, \int_{\sigma_{22}} \omega$$

of a uniquely defined (up to a scalar multiple) 2-dimensional differential form $\omega \in H^0(\Omega_X^2)$, $\omega \neq 0$, along 22 basic cycles $\sigma_1, \ldots, \sigma_{22}$ of $H_2(X, \mathbb{Z})$.

Those numbers uniquely determine X. Precisely, we have the following theorem (Barth – Peters – Van de Ven (1984), Piatetski-Shapiro – Shafarevich (1971)).

Theorem 4 (an analog of Torelli's theorem for $K3$ surfaces). *Let X and X' be $K3$ surfaces, h and h' ample divisor classes, and ω and ω' holomorphic forms on X and X'. Let φ: $H_2(X, \mathbb{Z}) \to H_2(X', \mathbb{Z})$ be an isomorphism of lattices mapping $[h]$ to $[h']$. Let $\sigma_1 \ldots, \sigma_{22}$ be a basis of $H_2(X, \mathbb{Z})$ and $\sigma_1' = \varphi_*(\sigma_1), \ldots, \sigma_{22}' = \varphi_*(\sigma_{22})$ a basis of $H_2(X', \mathbb{Z})$ such that*

$$\int_{\sigma_i} \omega = \lambda \int_{\sigma_i'} \omega', \qquad i = 1, \ldots, 22, \quad \lambda \in \mathbb{C}.$$

Then the isomorphism φ is induced by a unique isomorphism f: $X \to X'$ of surfaces.

The point

$$p_X = \left(\int_{\sigma_1} \omega, \ldots, \int_{\sigma_{22}} \omega \right)$$

is called the *period of the surface* X. We will describe the points that arise this way. Let $e_1, \ldots, e_{22}$ be a basis of $H^2(X, \mathbb{Z})$ dual to the given basis $\sigma_1, \ldots, \sigma_{22}$ of $H_2(X, \mathbb{Z})$, and for $x = (x_1, \ldots, x_{22})$ and $y = (y_1, \ldots, y_22)$, set

$$\Phi(x, y) = \sum_{i,j=1}^{22} (e_1, e_j) x_i y_j .$$

Then the obvious relation $\omega \wedge \omega = 0$ can be written in the form

$$\Phi(x, x) = 0 , \tag{1}$$

where $x = p_X$, and since $\int_X \omega \wedge \overline{\omega} > 0$, we get

$$\Phi(x, \overline{x}) > 0 , \tag{2}$$

Finally, since h is an algebraic class, we get $\int_{[h]} \omega = 0$, hence a linear relation

$$h(x) = \sum h_i x_i = 0 , \tag{3}$$

where $[h] = \sum h_i \sigma_i$.

The relation (3) defines a linear subspace E of dimension 21. Since the vector $p_X \in E$ is defined only up to a scalar multiple, it is natural to consider it as a point in the projective space $\mathbb{P}(E)$ of dimension 20. The equation (1) defines a quadric of dimension 19 in $\mathbb{P}(E)$, and the inequality (2) defines an open subset of the quadric. We denote the latter by $\Omega(h)$.

Now, we consider an arbitrary lattice $\mathbf{L}$ isomorphic to the lattices $H_2(X, \mathbb{Z})$ for $K3$ surfaces, a vector $h \in \mathbf{L}$ with $h^2 > 0$, and a basis in $\mathbf{L}$. The relations (1), (2), and (3) define a domain $\Omega(h)$.

As we mentioned before, the quadratic form $\Phi(x,x)$ on $H^2(X,\mathbb{R}) = H^2(X,\mathbb{Z}) \otimes \mathbb{R}$ has type $(3,19)$. Since $(h^2) > 0$, its restriction to the subspace $h(x) = 0$ has type $(2,19)$. Hence, in a suitable homogeneous coordinate system, the domain $\Omega(h)$ is given by the relations

$$\begin{gathered} z_1^2 + z_2^2 - z_3^2 - \ldots - z_{21}^2 = 0\,, \\ |z_1|^2 + |z_2|^2 - |z_3|^2 - \ldots - |z_{21}|^2 > 0\,. \end{gathered} \tag{4}$$

Fortunately for the theory of $K3$ surfaces, this domain has appeared and was thoroughly investigated in other branches of mathematics. In the theory of automorphic functions, it is called the "Cartan domain of type IV" (Siegel (1949)), and in special relativity theory, it is called a "future tube" (Vladimirov-Sergeev (1985)). Of course, in the definition, one may replace 21 by any number $n > 2$.

This manifold is not connected. It is not hard to show that $z_1 \neq 0$ and $\operatorname{Im}(z_2/z_1) \neq 0$ on $\Omega(h)$, and it consists of two connected components: $\operatorname{Im}(z_2/z_1) > 0$ and $\operatorname{Im}(z_2/z_1) < 0$.

A large Lie group $G_h(\mathbb{R})$ acts on this manifold. It is the subgroup of the real orthogonal group of the quadratic form $\Phi(x,x)$ that leaves the vector h fixed. The manifold $\Omega(h)$ is homogeneous with respect to $G_h(\mathbb{R})$. Moreover, it is a symmetric homogeneous space, and can be mapped isomorphically onto a bounded domain in $\mathbb{C}^{19}$ (Siegel (1949)).

Theorem 5. *Any point $x \in \Omega(h)$ can be obtained in the form $x = p_X$ (i. e., as the period) of a certain $K3$ surface with an ample class $h \in H_2(X,\mathbb{Z})$.*

(Concerning the proofs of Theorems 4 and 5, see (Kulikov (1977), Piatetski-Shapiro – Shafarevich (1971), Seminar Palaiseau (1985)).)

Remark. If the lattice $\mathbf{L}$ contains a vector a with $(a^2) = -2$ and $a(x) = 0$, then the point $x \in \Omega(h)$ corresponds to a $K3$ surface with a non-ample class h. The map defined by h or $2h$ contracts all the vectors $a \in H_2(X,\mathbb{Z})$ with $(a^2) = -2$ and $a(X) = 0$ to Du Val singular points.

Let $\check{\Omega}(h)$ be the set of $x \in \Omega(h)$ without such vectors a. One can easily prove that it is open. There is a fibration $\mathcal{X}_h \to \check{\Omega}(h)$ over $\check{\Omega}(h)$, which is topologically trivial and whose fiber over $x \in \check{\Omega}(h)$ is the $K3$ surface X such that $p_X = x$.

We have seen that in order to associate to X a point p_X, one has to choose a basis $\{\sigma_i\}$ of $H_2(X, \mathbb{Z})$. We can always take a basis with a fixed Gram matrix (σ_i, σ_j). The fibers of $\mathcal{X}_h \to \check{\Omega}(h)$ are surfaces with a fixed basis of this type. Such objects are called *marked K3 surfaces*. The same surface has many structures of a marked surface. All those surfaces are transformed one into another by the group $G_h(\mathbb{Z})$ of integral automorphisms of the form $\Phi(x, x)$ that fix the vector h.

The quotient space $\check{\Omega}(h)/G_h(\mathbb{Z})$ describes $K3$ surfaces up to isomorphism.

One can also easily describe the dependence on a choice of an ample class h. We get the same surface by passing to another vector h' obtained from h by an automorphism of the lattice $\mathbf{L}$. According to the theory of quadratic forms, two vectors $h, h' \in \mathbf{L}$ are equivalent with respect to the group $\mathrm{Aut}(\mathbf{L})$ if $(h^2) = (h'^2)$ and h and h' are *primitive* (i. e. not divisible in $\mathbf{L}$ by an integer > 1). Obviously, we may assume that h is primitive. Then the variety $\check{\Omega}(h)$ is determined by the number (h^2) only. So, the family $\mathcal{X}_h$ describes all the surfaces with a gien $\pi = (h^2)/2$.

12.5. Applications. Theorems 4 and 5 allow us to describe the structure of $K3$ surfaces in detail.

Corollary 1. *All $K3$ surfaces with a given class are parametrized by points of a 19-dimensional irreducible analytic space $\check{\Omega}(h)/G_h(\mathbb{Z})$. One can prove that this space is an algebraic variety.*

On the level of marked surfaces where the situation is somewhat simpler, the complements $\Omega(h)\backslash\check{\Omega}(h)$ also correspond to $K3$ surfaces, whose (-2)-curves in the class a with $(a^2) = -2$ and $h(a) = 0$ are contracted to Du Val singular points. All the spaces $\Omega(h)$ are contained in the domain Ω defined by the conditions (1) and (2) only. The points $x \in \Omega$ that does not belong to any $\Omega(h)$ also correspond to $K3$ surfaces, though non-algebraic ones (see Sect. 14).

Corollary 2. *For every integer $n \geq 1$, there exists a $K3$ surface X with $S_X = \mathbb{Z}h$, $(h^2) = 2n$.*

Thus all values of the class π can be realized. They fill up "most" of the points of $\check{\Omega}(h)$ with $(h^2) = 2\pi$. The points corresponding to surfaces with one additional algebraic class a form a hyperplane $a(x) = 0$ by the Lefschetz theorem (Sect. 3.3). So we get the following assertion.

Corollary 3. *All the values of the Picard number ϱ, $1 \leq \varrho \leq 20$, are realized by $K3$ surfaces with a given invariant π.*

In fact, we have the following more precise result.

Corollary 4. *A lattice H is realized in the form $H \cong S_X$ for a certain $K3$ surface if and only if $H \otimes \mathbb{R}$ has type $(1, r-1)$, $r \leq 20$, and there is a primitive embedding $\varphi\colon H \hookrightarrow \mathbf{L}$ (i. e., if $nl \in \varphi(H)$ with $l \in \mathbf{L}$ and $n \in \mathbb{Z}$,* *then* $l \in \varphi(H)$). *The set of surfaces* X *such that* $S_X \supseteq H$ *for a suitable lattice* H *is parametrized by the points of an irreducible variety of dimension* $20 - r$, $r = \operatorname{rk} H$; *moreover,* $S_X = H$ *for "most" points.*

It follows from the existence of the fibration $\mathcal{X}_h \to \check{\Omega}(h)$ that all $K3$ surfaces with a given class π are diffeomorphic. By analyzing intersections of the subspaces $\Omega(h)$ in Ω, we can deduce a stronger assertion.

Corollary 5. *All* $K3$ *surfaces are diffeomorphic.*

Corollary 6. *All* $K3$ *surfaces are simply connected.*

In view of Corollary 5, it will suffice to verify the latter corollary for a single $K3$ surface, e. g., a Kummer surface, where the statement is rather elementary.

In particular, Theorem 4 can be applied to the study of automorphisms of $K3$ surfaces.

By the theorem, the group $\operatorname{Aut} X$ of automorphisms of a $K3$ surface X coincides with the group of those automorphisms of the lattice $H_2(X, \mathbb{Z})$ that transform the vector p_X to a scalar multiple, and a certain ample class h to an ample class. The latter is equivalent to the condition that an automorphism transforms effective cycles to effective ones, i. e. preserves the polyhedra $\overline{E}$ and $\overline{A}$.

On can show that the restriction of $\operatorname{Aut} X \subset \operatorname{Aut} H_2(X, \mathbb{Z})$ from $H_2(X, \mathbb{Z})$ to the subgroup of algebraic cycles S_X defines a homomorphism with finite kernel. The image of that homomorphism is a subgroup of finite index in the group of all automorphisms of the lattice S_X that preserve the polyhedron $\overline{A}$. The latter group is isomorphic to $\operatorname{Aut} S_X / W(S_X)$, where $W(S_X)$ is the group generated by reflections, defined in Sect. 1. So, the group $\operatorname{Aut} X$ is isomorphic to $\operatorname{Aut} S_X / W(S_X)$ "up to finite groups". In particular, we get the following corollary.

Corollary 7. *The group* $\operatorname{Aut} X$ *is finite if and only if the index* $(\operatorname{Aut} S_X : W(S_X))$ *is finite.*

This statement is equivalent to the condition that the polyhedron $\overline{A}$ has a finite volume in the Lobachevski space.

Thus the question concerning the finiteness of $\operatorname{Aut} X$ is completely determined by the lattice S_X. Precisely, this group is finite if $\varrho = 1$ ($\varrho = rkS_X$). For $\varrho = 2$, it is finite if S_X contains a vector x with $(x^2) = 0$ or $(x^2) = -2$. For $\varrho \geq 3$, there are only finitely many non-isomorphic lattices S_X that give a finite group $\operatorname{Aut} X$. For each ϱ, their number is given in the table (Nikulin (1981, 1984)):

$\varrho = \operatorname{rk} S_X$	3	4	5	6	7	8	9	10	11	12	13	14	15	16	17	18	19	20
Number of lattices S_X	27	14	10	10	9	12	10	9	4	4	3	3	1	1	1	1	0	0

In order to show how extensive is our knowledge of automorphisms of $K3$ surfaces, we will present results of the theory of symplectic automorphisms. An automorphism g of a $K3$ surface X multiplies a differential form $\omega \in H^0(\Omega_X^2)$, $\omega \neq 0$, by a scalar $\chi(g)$. It is said to be *symplectic* if $\chi(g) = 1$. There is a description of all finite groups of symplectic automorphisms. They are connected to a well-known group M_{23}, the simple Mathieu group, which is realized as a 4-transitive permutation group of the set Ω_{23} of 23 elements.

It was established that a finite group G can be realized as a group of symplectic automorphisms of a $K3$ surface if it can be embedded in a Mathieu group M_{23} so that its action on the set Ω_{23} divides the set into at least 5 orbits. We get 11 maximal subgroups of M_{23} with this property. For each of the subgroups, one can explicitly describe the $K3$ surfaces on which it acts. The group of symplectic automorphisms of the surface in $\mathbb{P}^3$ given by the equation

$$X^4 + Y^4 + Z^4 + T^4 + 12XYZ = 0$$

has the maximal order, namely 960. This is the only surface of degree 4 with such a group of linear automorphisms (Mukai (1988)).

12.6. Generalizations. The theory of $K3$ surfaces has so many beautiful features that a question arises whether it is only the simplest case of a more general theory. It is quite possible that there are several such theories as in case of elliptic curves: the theory of elliptic curves is a special case of the theory of curves of an arbitrary genus and the theory of Abelian varieties of an arbitrary dimension.

Presently, at least one such generalization of the theory of $K3$ surfaces is known, namely, the theory of *symplectic algebraic varieties.* By definition, a symplectic algebraic variety is a simply connected variety of even dimension $2r$ that has a differential form $\omega \in H^0(\Omega^2)$ such that ω^r does not vanish anywhere. Over $\mathbb{C}$, such varieties have beautiful differential-geometric properties, and they were extensively investigated with a help of integrals of the from ω along cycles $\sigma \in H_2(X, \mathbb{Z})$ (Beauville (1983), Seminar Palaiseau (1985)).

§ 13. Ruled and Rational Surfaces

13.1. Ruled Surfaces. We will consider the last class of surfaces, namely the surfaces of canonical dimension $-\infty$. They are characterized by the condition that the canonical class of their minimal models is not numerically effective. According to Theorem 1 of Sect. 7.1, they are either ruled (i. e. fibrations over a curve B with fiber $\mathbb{P}^1$) or the plane $\mathbb{P}^2$.

According to the classification presented in Sect. 8 and Theorem 2 of Sect. 8.1, for surfaces whose canonical class is nef, we get $P_m \neq 0$ for $m \in \{1, 2, 3, 4, 6\}$ hence $P_{12} \neq 0$. Thus we have the following theorem.

Theorem 1. *A surface is birationally equivalent to a ruled surface or the plane* $\mathbb{P}^2$ *if and only if* $P_{12} = 0$.

Let $X \to B$ be a ruled surface. We apply the general method for studying fibrations, i. e., consider its generic fiber as a genus 0 curve over the function field $R = k(B)$. We are now dealing with curves of genus 0 over a non-algebraically closed field, which is much simpler than the theory of curves of genus 1.

Let C be an algebraic curve of genus 0 over a field R. By the Riemann-Roch theorem, its canonical class K_C (which is defined over R) has degree -2, and the class $-K$ gives an embedding $C \hookrightarrow \mathbb{P}^2$ (also defined over R) whose image is a conic. Now, we apply this to the generic curve of a ruled surface X. We see that X is birationally equivalent to a surface given by an equation

$$\alpha x_0^2 + \beta x_1^2 + \gamma x_2^2 = 0 \qquad (\alpha, \beta, \gamma \in k(B)) \tag{1}$$

(provided $\operatorname{char} k \neq 2$); moreover, the birational equivalence is compatible with the projection to the base B. The investigation of conics (1) over an arbitrary field R is a nontrivial task if R is rather complicated (e. g., $R = \mathbb{Q}$ or $R = k(Y)$ where $\dim Y > 1$). However, in the case $R = k(B)$ with $\dim B = 1$, it has a simple solution. The key point is the following lemma.

Lemma. *The conic* (1) *has a point with coordinats in the field* $k(B)$.

The lemma is a special case of Tsen's theorem to the effect that any equation $F(x_0, \ldots, x_n)$ has a nontrivial solution in $k(B)$, where $F \in k(B)[T_0, \ldots, T_n]$ is a form of degree $\leq n$ (CAV, Chap. IV, Sect. 5). The proof of Tsen's theorem is rather elementary. For instance, in case of the conic (1), if $k(B) = k(t)$ is purely transcendental, we can assume that $\alpha, \beta, \gamma \in k[t]$, and write $x_i = \sum_{m=0}^{N} c_{im} t^m$, $i = 0, 1, 2$; then for a sufficiently large N, we get less than $3(N+1)$ relations for the $3(N+1)$ coefficients c_{im}. The general case is reduced to the case $k(B) = k(t)$ by a simple technical trick.

By the lemma, the curve C over $k(B)$ – the generic fiber of the family $X \to B$ of curves of genus 0 – has a rational point P defined over $k(B)$. Again applying the Riemann-Roch theorem, we easily get that $l(P) = 2$, and the linear system $|P|$ defines an isomorphism $C \cong \mathbb{P}^1$, defined over $R = k(B)$. One may view this isomorphism as the projection φ of the conic (1) from its rational point P (Fig. 22).

Therefore $C \cong \mathbb{P}^1$ as a curve over $k(B)$, and we get the following theorem.

Theorem 2. *A ruled surface* $f\colon X \to B$ *is birationally equivalent to the product* $\mathbb{P}^1 \times B$*. Moreover, the birational transformation*

$$X \dashrightarrow \mathbb{P}^1 \times B$$

commutes with the projections to B.

The statement of the theorem as well as its proof yield the corollary.

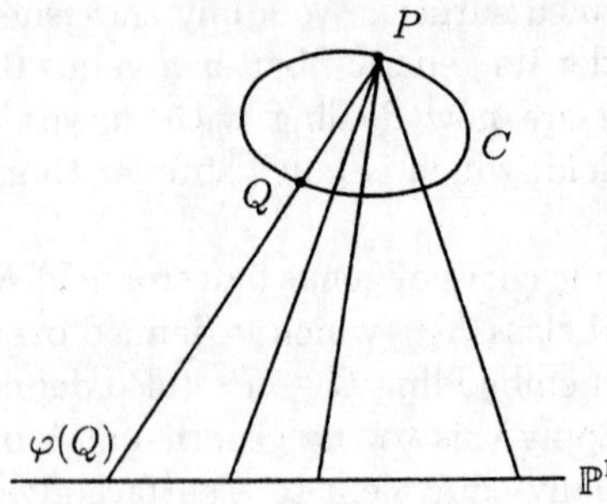

Fig. 22

Corollary. *A ruled surface $f\colon X \to B$ has a section $\sigma\colon B \to X$ with $f\sigma = 1$.*

The rational point P constructe above gives a rational section σ with required properties. Furthermore, a rational map of a smooth curve to a projective surface is in fact a morphism.

One can describe much more explicitly the birational transformation of Theorem 2. Let $S = \sigma(B)$ denote the section, i. e., $S \subset X$ and $(S, F_b) = 1$, where $F_b = f^{-1}(b)$. By general properties of coherent sheaves, the restriction $\mathcal{O}_X(S) \to \mathcal{O}_F(S|_F)$ defines an epimorphism

$$H^0(f^{-1}(U), \mathcal{O}_X(S)) \to H^0(F_b, \mathcal{O}_F(S|_{F_b})) \to 0 \tag{2}$$

for a sufficiently small affine neighborhood U of any point $b \in B$. The divisor $S|_{F_b}$ is a point on F_b and $H^0(F_b, \mathcal{O}_{F_b}(S|_{F_b})) = k \cdot 1 + kx$, where $x \in F_b$ is a "coordinate" along the projective line F_b. It follows from (2) that the function x is a restriction of a certain function φ on X whose restriction to F_b, and consequently to nearby fibers, defines a coordinate on them. Therefore, for a sufficiently small U, we get an isomorphism

$$f^{-1}(U) \xrightarrow{\sim} \mathbb{P}^1 \times U\,.$$

In other words, we get the following theorem.

Theorem 3. *A ruled surface $X \to B$ is a locally trivial bundle (in the Zariski topology) with fiber $\mathbb{P}^1$ (i. e. a $\mathbb{P}^1$-bundle).*

Clearly the group $\operatorname{Aut}\mathbb{P}^1 = PGL(2)$ is the structure group of this bundle.

By working out details of the proof of Theorem 3, we get the following construction. Let $S \subset X$ be s section of a ruled surface $f\colon X \to B$. Then $f_*\mathcal{O}_X(S)$ is a locally free sheaf of rank 2 which coincides with the sheaf of sections of a vector bundle $\mathcal{E} \to B$ of rank 2 over B. We obtain the surface X by replacing each fiber E_b of $\mathcal{E}$ by its projectivization $\mathbb{P}(E_b) \cong \mathbb{P}^1$, i. e. $X \cong \mathbb{P}(\mathcal{E})$. This proves the first half of the following assertion.

Theorem 4. *Any ruled surface can be represented in the form* $\mathbb{P}(\mathcal{E})$, *where* $\mathcal{E}$ *is a vector bundle of rank 2. Furthermore,* $\mathbb{P}(\mathcal{E}_1) \cong \mathbb{P}(\mathcal{E}_2)$ *if and only if* $\mathcal{E}_2 = \mathcal{E}_1 \otimes L$, *where* L *is a line bundle (i. e.* $\operatorname{rk} L = 1$).

Going back to the birational map of Theorem 2 and using the notion of elementary transformation of a ruled surface (Sect. 6.3, Example 5), we obtain the following theorem.

Theorem 5. *A birational transformation* $X \dashrightarrow \mathbb{P}^1 \times B$ *can be obtained as a sequence of elementary transformations.*

Idea of Proof. To fix an isomorphism between a curve and $\mathbb{P}^1$, one has to label the points corresponding to 0, ∞, and 1 on our curve. Similarly, a locally trivial fibration $X' \to B$ is isomorphic to $\mathbb{P}^1 \times B$ if three disjoint sections, S_0, S_∞, and S_1, are given. Every locally trivial fibration $X \to B$ has infinitely many sections (suffice to construct them on an open subset $U \subset B$), in particular, three distinct sections. Now, utilizing elementary transformations, we can obtain three disjoint sections.

For instance, assume that S and S_1 intersect transversely at a point $x \in X$ (Fig. 23a). By blowing up x, we get a curve L

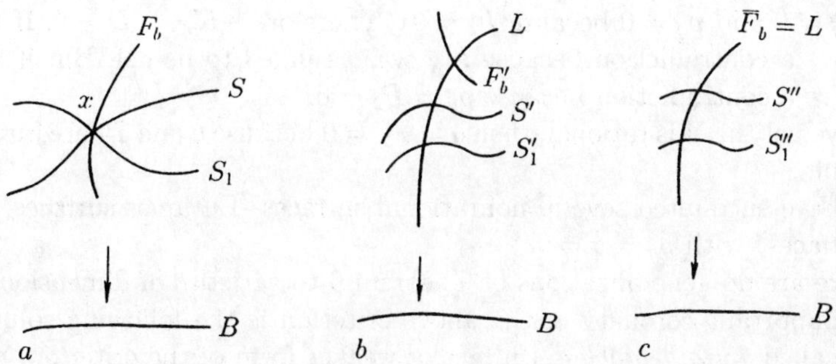

Fig. 23

and pull apart the proper transforms S', S'_1, and F'_b of the sections S, S_1, and the fiber F_b, where $b \in f(x)$ (Fig. 23b and 23c). We then contract the curve F'_b so that L becomes the new fiber, and the images of the sections S and S_1 in the new fiber over b no longer intersect.

If S and S_1 are not transversal at x, then to "pull them apart", we need several elementary transformations.

Corollary. *If* $X \to B$ *is a ruled surface over a curve of genus* g, *then*

$$(K_X^2) = 8(1-g), \quad N_X \cong \mathbb{Z}^2 .$$

Indeed, according to (Sect. 6.1, (2) and (5)), (K_X^2) and the group N_X do not change under one elementary transformation.

13.2. Rational Surfaces. Now we turn to rational surfaces. They are distinguished by the condition $q = 0$ among all minimal surfaces whose canonical class is not nef. Indeed, such a surface, provided it is relatively minimal, is either $\mathbb{P}^2$ or a ruled surface with base $\mathbb{P}^1$; according to Theorem 2, the latter is birationally equivalent to $\mathbb{P}^1 \times \mathbb{P}^1$, i. e., it is rational. In view of Theorem 1, we can even say that rational surfaces are characterized by the conditions $q = 0$ and $P_{12} = 0$. However, a more precise result holds.

Theorem 6 (Castelnuovo-Enriques rationality criterion). *A surface X is rational if and only if $q = P_2 = 0$.*

We can assume that X is a relatively minimal model. By Theorem 1 of Sect. 7.1, it will suffice to show that its canonical class is not nef. Suppose to the contrary that K_X is nef. Then $(K_X^2) \geq 0$. By duality

$$l(2K_X) = \dim H^2(X, \mathcal{O}_X(-K_X)),$$

and from the Riemann-Roch inequality and $P_2 = 0$, we get

$$l(-K_X) \geq (K_X^2) + 1$$

(since $q = 0$ and $p_g = 0$ because $P_2 = 0$). Therefore $-K_X \sim D \geq 0$. If $D \neq 0$, we derive a contradiction because K_X was assumed to be nef. But if $D = 0$, we derive a contradiction because $p_g = P_2 = 0$.

Conversely, if X is rational, then $q = P_2 = 0$ because q and P_2 are birational invariants.

We have encounted several nonrational surfaces (Enriques surfaces, Godeaux surfaces) with $p_g = q = 0$.

There are no generalizations of Theorem 6 to varieties of dimension ≥ 3.

An important corollary of the above criterion is the following solution of the so-called *Lüroth problem* (in fact, a weaker form of the criterion, namely $q = P_{12} = 0$, will suffice).

To begin with, we recall that an algebraic variety Y is said to be *unirational* if there is a surjective rational map $g \colon \mathbb{P}^n \dashrightarrow Y$. In other words, the map $g^*\colon k(Y) \to k(\mathbb{P}^n)$ realizes the field $k(Y)$ of rational functions on Y as a subfield of a purely transcendental extension, i. e. the field of rational functions in n variables $k(x_1, \ldots, x_2) \cong k(\mathbb{P}^n)$. The classical *Lüroth theorem* states that every unirational curve is rational (i. e., every transcendence degree 1 subfield of the field of rational functions in one variable is isomorphic to the field of rational functions in one variable).

The generalization of Lüroth's theorem to the case of dimension 2 follows from the Castelnuovo-Enriques rationality criterion.

Corollary (2-dimensional analog of Lüroth's theorem). *Every unirational surface (over an algebraically closed field of characteristic 0) is rational.*

Indeed, let $g\colon \mathbb{P}^2 \dashrightarrow X$ be a rational surjective map, $\sigma\colon Z \to \mathbb{P}^2$ a resolution of the points of indeterminacy of g, and $h\colon Z \to X$ a surjective morphism such that $g \circ \sigma = h$. Since Z is smooth rational surface, we get $q(Z) = P_2(Z) = 0$. On the other hand, since the map h is separable, the maps

$$h^* : H^0(X, \Omega^1_X) \to H^0(Z, \Omega^1_Z), \qquad h^* : H^0(X, (\Omega^2_X)^{\otimes 2}) \to H^0(Z, (\Omega^2_Z)^{\otimes 2})$$

are inclusions. Hence $q(X) = P_2(X) = 0$, and X is a rational surface by Theorem 6.

The analog of Lüroth's theorem does not hold in higher dimensions ($\dim X \geq 3$). The corresponding examples were obtained in 1971 (Artin-Mumford (1972), Clemens-Griffiths (1972), Iskovskikh-Manin (1971)). For $\dim X = 3$ those examples open a new chapter in algebraic geometry – the theory of varieties similar, in a sense, to rational but not rational. (Among them, the better known are the *Fano varieties* whose class $-K_X$ is ample; in dimension 2, they are called *Del Pezzo surfaces.*) The problem can be also solved by methods of abstract field theory (Saltman (1984)).

Now, we turn to minimal models of rational surfaces. As we have seen, they are isomorphic either to $\mathbb{P}^2$ or ruled surfaces $X \to \mathbb{P}^1$ with base $\mathbb{P}^1$. In the later case, we can apply Theorem 4 of Sect. 13.1 and the fact that rank 2 bundles over $\mathbb{P}^1$ have a very simple structure. We have a general result to the effect that a vector bundle over $\mathbb{P}^1$ admits a unique decomposition in a direct sum of line bundles (Grothendieck (1962)). Since line bundles are determined by elements of $\operatorname{Pic}\mathbb{P}^1 \cong \mathbb{Z}$, i.e., they correspond to sheaves $\mathcal{O}(n)$, Theorem 4 of Sect. 13.1 yields at once the following theorem.

Theorem 7. *Any ruled surface $X \to \mathbb{P}^1$ is isomorphic to $\mathbb{P}(\mathcal{O} + \mathcal{O}(n))$, $n \geq 0$.*

Clearly Theorem 7 means that any ruled surface has two disjoint sections. This way one can obtain another proof of the theorem without appealing to bundles.

One can easily verify that the surfaces $\mathbb{P}(\mathcal{O} + \mathcal{O}(n))$ coincide with the surfaces $\mathbf{F}_n$ constructed in (Sect. 2, Example 4). The surface $\mathbf{F}_n$, $n > 0$, has a unique section S with $(S^2) = -n$. Therefore $\mathbf{F}_1$ is not minimal, namely it is the plane $\mathbb{P}^2$ with a point blown up; the rest of the $\mathbf{F}_n$'s are minimal. Further, $\mathbf{F}_0 \cong \mathbb{P}^1 \times \mathbb{P}^1$ has two distinct structures as a ruled surface. For $n \geq 1$, $\mathbf{F}_n$ has a unique such structure. On can easily verify that

$$\operatorname{Pic}\mathbf{F}_n = \mathbb{Z}f_n + \mathbb{Z}S_n\,, \qquad K_{\mathbf{F}_n} = -(n+2)f_n - 2S_n\,,$$

where f_n is the class of a fiber, and S_n is a section with $(S_n^2) = -n$. The class $af_n + bS_n$ contains an effective divisor if and only if $a \geq 0$ and $b \geq 0$.

The surface $\mathbf{F}_n$ can be represented as a surface of small degree in $\mathbb{P}^N$, namely as a surfaces X with $\deg X = N - 1$.

The linear system $|af_n + bs_n|$ on $\mathbf{F}_n$, where s_n is the class of S_n, has no base points or fixed components for $b \geq 0$ and $a \geq bn$, $a \neq 0$, and is very ample for $b \geq 1$ and $a > bn$. In particular, for $H \sim af_n + s_n$:

$$H^2 = 2a - n\,, \qquad \dim|af_n + s_n| = 2a - n + 1\,.$$

So, for $a > n$, the linear system $|H|$ gives an embedding

$$\varphi_{|H|} : \mathbf{F}_n \to \mathbb{P}^N$$

whose image is a surface of degree $N-1$, where $N = 2a - n + 1$. Moreover, the fibers of π_n: $\mathbf{F}_n \to \mathbb{P}^1$ are mapped to lines that sweep out the surface $\varphi_{|H|}(\mathbf{F}_n)$, which explains the term "ruled surface". In case $a = n$ and $n > 0$, the map

$$\varphi_{|H|} : \mathbf{F}_n \to \mathbb{P}^{n+1}$$

is a birational morphism onto a cone $\varphi_{|H|}$: $\mathbf{F}_n^0 \to \mathbb{P}^{n+1}$ over a rational normal curve of degree n in $\mathbb{P}^n$ that contracts S_n to the vertex of the cone (which is an isolated singular point if $n \geq 2$).

For an arbitrary algebraic variety $X \subset \mathbb{P}^N$, not contained in a hyperplane, we get

$$\deg X \geq \operatorname{codim} X + 1\,. \tag{3}$$

The inequality (3) becomes an equality for $X = \varphi_{|H|}(\mathbf{F}_n)$, $H \sim af_n + s_n$. A classical theorem of Enriques gives a complete classification of all subvarieties $X \subset \mathbb{P}^N$ with $\deg X = \operatorname{codim} X + 1$. In particular, we get the following result for surfaces (Griffiths-Harris (1978)). Let $X \subset \mathbb{P}^N$ be an irreducible surface, not contained in a hyperplane, such that (3) is an equality, i. e. $\deg X = N-1$. Then X is one of the following surfaces:

a) a rational ruled surface, i. e. the image of $\mathbf{F}_n$ under the map

$$\varphi_H : \mathbf{F}_n \to \mathbb{P}^N, \qquad H \sim af_n + s_n\,, \quad a > n\,, \qquad N = 2a - n + 1\,;$$

b) a cone $\mathbf{F}_{N-1}^0$ over a rational normal curve of degree $N-1$ in $\mathbb{P}^{N-1}$; or

c) a Veronese surface $V_4 \subset \mathbb{P}^5$, which is the image of $\mathbb{P}^2$ under the map given by the complete linear system of conics (Griffiths-Harris (1978)).

13.3. Del Pezzo Surfaces. In connection with elliptic surfaces, we have already observed that sometimes it is convinient to drop the minimality assumptions (which is very useful in other instances) in order to preserve other more important properties of surfaces (like the existence of elliptic pencils). Now we will give another example of this kind. It is related to surfaces that are antipodes of surfaces of general type: for them the class $-K$ is ample.

Definition. A nonsingular surface X is called a *Del Pezzo surface* if its anticanonical divisor $-K_X$ is ample.

By the Kodaira vanishing theorem and Serre duality, we deduce from the definition that for all $m \in \mathbb{Z}$:

$$\begin{aligned} h^0(\mathcal{O}_X(mK_X)) &= 0 \quad \text{for } m \geq 1 \\ h^1(\mathcal{O}_X(mK_X)) &= 0 \\ h^2(\mathcal{O}_X(mK_X)) &= h^0(\mathcal{O}_X((1-m)K_X))\,. \end{aligned} \tag{4}$$

Hence, in particular, $h^1(\mathcal{O}_X) = q(X) = 0$, $h^2(\mathcal{O}_X) = p(X) = 0$, and $h^0(\mathcal{O}_X(mK_X)) = P_m(X) = 0$ for $m \geq 1$. Consequently X is a rational surface by Theorem 6. By Noether's formula,

$$1 \leq (K_X^2) = 10 - \operatorname{rk}\operatorname{Pic} X \leq 9,$$

and by the Riemann-Roch formula,

$$h^0(\mathcal{O}_X(-K_X)) = (K_X^2) + 1. \tag{5}$$

The number $d = (K_X^2)$ is called the *degree* of X. The maximal integer $r > 0$ such that $-K_X \sim rH$ for a certain divisor H is called the *index* of X. It follows from the inequality $1 \leq (K_X^2) \leq 9$ that $1 \leq r \leq 3$. We will present several results on Del Pezzo surfaces:

$r = 3 \Leftrightarrow d = 9 \Leftrightarrow X \cong \mathbb{P}^2$; the anticanonical embedding $\varphi_{|-K_X|}\colon X \to \mathbb{P}^9$ realizes X as a surface of degree 9 in $\mathbb{P}^9$;

$r = 2 \Leftrightarrow X \cong \mathbb{P}^1 \times \mathbb{P}^1$; the anticanonical embedding realizes X as a surface of degree 8 in $\mathbb{P}^8$.

Henceforth, we will restrict ourselves to the case $r = 1$; then $1 \leq d \leq 8$.

(i) The anticanonical map $\varphi_{|-K_X|}\colon X \to \mathbb{P}^d$ is an isomorphism onto a surface of degree d in $\mathbb{P}^d$ if $3 \leq d \leq 8$, a double covering branched along a curve of degree 4 if $d = 2$, and a rational map with one base point and irreducible elliptic fibers if $d = 1$.

(ii) For $1 \leq d \leq 8$, the surface X is a blowing-up of $\mathbb{P}^2$ in $9 - d$ points $P_1, \ldots, P_{9-d} \in \mathbb{P}^2$ in general position – the image of X under the anticanonical map $\varphi_{|-K_X|}$ coincides with the image of the rational map $\mathbb{P}^2 \to \mathbb{P}^d$ given by the linear system $|3L - P_1 - \ldots - P_{9-d}|$ of curves of degree 3 through the points $P_1, \ldots, P_{9-d}$, where L is the class of a line in $\mathbb{P}^2$.

Let X_d be a Del Pezzo surface of degree $d = (K_X^2)$, and $\varphi_d\colon X_d \to \mathbb{P}^2$ be a composition of the corresponding $9 - d$ σ-processes. Then, for $3 \leq d \leq 7$, the projection from a sufficiently general point $x \in X_d$ defines a birational equivalence $\psi\colon X_d \dashrightarrow X_{d-1}$, inverse to the σ-process at the point x. Moreover, the diagram

$$\begin{array}{ccc} X_{d-1} & \underset{\psi}{\overset{\sigma}{\rightleftarrows}} & X_d \\ & {\scriptstyle\varphi_{d-1}}\searrow \quad \swarrow{\scriptstyle\varphi_d} & \\ & \mathbb{P}^2 & \end{array}$$

is commutative. Any X_{d-1} can be obtained this way from a suitable X_d.

For $d = 3$, the above diagram degenerates into

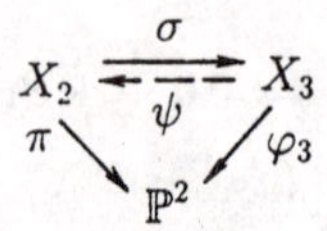

where π is a double covering branched along a curve of degree 4.

(iii) All exceptional curves on a Del Pezzo surface are (-1)-curves (see Sect. 7.1, Lemma 2). Under the embedding φ_{-K} they become lines. There are finitely many such lines. To prove this, we consider an important sublattice $K_X^{\perp} \subset N_X$:

$$K_X^{\perp} = \{x \in N_X \,|\, (x, K_X) = 0\}\,.$$

It is determined by the surface X. By the index theorem, $K_X^{\perp}$ is negative defined. If $x \in N_X$ corresponds to a (-1)-curve on X_d, then

$$(x^2) = (x, K_X) = -1\,, \quad y = x + dK_X \in K_X^{\perp}\,, \quad (y^2) = -d(d+1)\,.$$

Clearly there are finitely many such vectors $y \in K_X^{\perp}$.

It is not hard to prove that the cone $\overline{E}$ is generated by (-1)-vectors for a Del Pezzo surface (compare Sect. 5.2, Examples 5–8). This cone is highly symmetric, and is connected with the root theory. In fact, one can show that for $d \geq 6$, the lattice $K_X^{\perp}$ is generated by the vectors $y \in K_X$ with $(y^2) = -2$. These vectors form a root system R_d, which can be described by the values of $d = (K_X^2)$, $1 \leq d \leq 6$, as follows:

d	1	2	3	4	5	6
R_d	E_8	E_7	E_6	D_5	A_4	$A_1 \times A_2$

Each root $p \in R_d$ defines a reflection $x \to x + (x,p)p$, which is an automorphism of the lattice N_X and translates the class K_X into itself. All the reflections generate a finite group $W(R_d)$, called the *Weyl group* of the root system R_d. It turns out that $W(R_d)$ coincides with the symmetry group of the cone $\overline{E}$. Furthermore, it coincides with the symmetry group of the cone Ω. It also coincides with the permutation group of (-1)-curves that preserve the incidence relations between them. The group $W(R_d)$ is well known in the root theory. Its order is given in the table:

d	1	2	3	4	5	6
$\lvert W(R_d)\rvert$	$2^{14} \cdot 3^5 \cdot 5^2 \cdot 7$	$2^{10} \cdot 3^4 \cdot 5 \cdot 7$	$2^7 \cdot 3^4 \cdot 5$	$2^7 \cdot 3 \cdot 5$	$2^3 \cdot 3 \cdot 5$	$2^2 \cdot 3$

By utilizing the argument employed in the proof of finiteness of the number of exceptional curves, we can also calculate the number itself, denoted by $\Lambda(d)$:

d	1	2	3	4	5	6	7	8
$\Lambda(d)$	240	56	27	16	10	6	3	1

We will consider the case of a cubic in $\mathbb{P}^3$ ($d = 3$) in detail. A configuration of 27 lines on this surface is one of the most classical geometric objects – many articles and several books are devoted to the configuration. Therefore it is interesting to look at the subject from the modern point of view.

It turns out that the geometric properties of the configuration of lines on a Del Pezzo surface can be best understood utilizing the theory of the corresponding root systems R_d. In particular, in case of cubic surfaces we are dealing with the root system E_6.

Recall that to simple roots $\alpha_1, \ldots, \alpha_l$ of the system R there correspond *fundamental weights* $\widetilde{\omega}_i$ ($i = 1, \ldots, l$) defined by the condition $\widetilde{\omega}_i(\alpha_j) = \delta_{ij}$. The most important among them are *microweights.* One may construct them as follows. First, we enlarge R to obtain an extended system $\widetilde{R}$ by adding a root α_0. In case of E_6, we get the system $\widetilde{E}_6$:

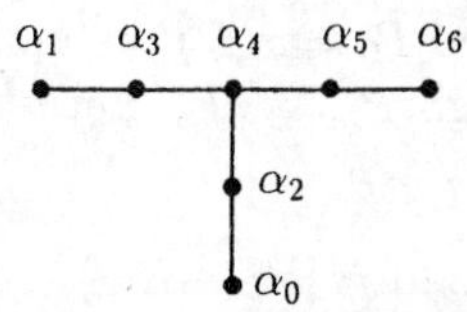

Then a microweight is a fundamental weight $\widetilde{\omega}_i$ whose corresponding simple root α_i is obtained from α_0 by an automorphism of the graph corresponding to R. If we interpret the simple roots α_i as components of a singular fiber of an elliptic pencil, then these are the components that appear in the fiber with multiplicity 1.

In particular, for E_6, the microweights are $\widetilde{\omega}_1$ and $\widetilde{\omega}_6$. The Weyl group $W(R)$ acts on the set of all weights of the system R. In case of E_6, the orbits $W(E_6)\widetilde{\omega}_1$ and $W(E_6)\widetilde{\omega}_6$ of both microweights have 27 elements. Already here the number 27 appear.

To establish the relation with the lines on a cubic, we consider an embedding of the lattice $Q(E_6)$, corresponding to the root system E_6, in a unimodular lattice N. Since the discriminant of $Q(E_6)$ equals 3, it cannot be embedded in a unimodular lattice of the same rank, namely 6. We consider the case $\operatorname{rk} N = 7$. The complement of $Q(E_6)$ in N, denoted by L, must be 1-dimensional with discriminant 3, i. e.

$$L \cong \mathbb{Z}\xi\,, \quad (\xi^2) = 3\,, \quad [N : Q(E_6) \oplus L] = 3\,.$$

One can easily see that there is a unique lattice N with those properties, namely, the lattice S_X of a cubic X. Moreover $\xi = K_X$. The weights $\widetilde{\omega}$ of the system E_6 are determined on the lattice $Q(E_6)$ by vectors $x \in N$: $\widetilde{\omega}(u) = (x, u)$, $u \in Q(E_6)$. In particular, the weights of the orbits $W(E_6)\widetilde{\omega}_1$ and $W(E_6)\widetilde{\omega}_6$ of both microweights are obtained from the vectors $x \in N \cong S_X$ corresponding to the lines l_i, which are interpreted as lines for one orbit and conics for the other orbit (we are unable to intrinsically distinguish between $\widetilde{\omega}_1$ and $\widetilde{\omega}_6$).

There is a similar interpretation for other Del Pezzo surfaces.

In the case $d = 4$, the surface $X \subset \mathbb{P}^4$ is a complete intersection of quadrics. The case $d = 2$ is interesting too. As we have seen before, it is related to the projection $\pi\colon X_3 \dashrightarrow \mathbb{P}^2$ of the cubic onto the plane from a point $x \in X_3$ (one has to choose x outside all the lines). The point x is the point of indeterminacy. Resolving it, we obtain a morphism $\pi'\colon X_2 \to \mathbb{P}^2$ branched along a curve $C_4 \subset \mathbb{P}^2$ of degree 4. Moreover, the exceptional curves on X_2 turn up in pairs from the 27 lines on X_3, as images of a line and a conic passing through a line and the point x. Yet another pair is a line obtained by blowing up x and the image of the intersection of X_3 with the tangent plane at x. Under the morphism π', those pairs of lines merge and are mapped to the famous 28 bitangents of the curve C_4.

In case $d = 1$, the linear system $|-K_X|$ is a pencil of elliptic curves with one base point $P \in X$. Let $\sigma\colon X \to \mathbb{P}^2$ be the blowing-up of eight points in general position, $P_1, \ldots, P_8 \in \mathbb{P}^2$, and $P_0 = \sigma(P)$ the ninth point of intersection of the pencil of cubics in $\mathbb{P}^2$ through $P_1, \ldots, P_8$. Let $\sigma'\colon X' \to X$ be the blowing-up of $P \in X$. Then the linear system

$$|-K_{X'}| = |\sigma'^*(-K_X) - \sigma^{-1}P|$$

has no base points and gives a morphism

$$\varphi_{|-K_{X'}|}\colon X' \to \mathbb{P}^1$$

whose fibers are irreducible reduced (since $-K_X$ is ample) curves of genus 1. We obtain a rational elliptic surface that we have already considered in (Sect. 10.1, Example 1). It contains infinitely many (-1)-curves. The number 8 of points blown up in the case $d = 1$ is extremal – one can show that a surface obtained by blowing up arbitrary 9 points in general position in $\mathbb{P}^2$ contains infinitely many (-1)-curves.

(iv) For a Del Pezzo surface, the group $\operatorname{Pic} X = S_X$ has no torsion, and $\varrho(X_d) = 10 - d$ for $1 \le d \le 8$. For $d \le 7$, it is generated by (-1)-curves. For $d = 3$, if we represent the cubic $X_3 \subset \mathbb{P}^3$ as $\mathbb{P}^2$ blown up in points $P_1, \ldots, P_6$:

$$\sigma\colon X_3 \to \mathbb{P}^2,$$

then all the (-1)-curves (i. e. lines on X_3) can be obtained as follows:

six lines $\sigma^{-1}(P_1), \ldots, \sigma^{-1}(P_6)$;

fifteen lines $\sigma_{\rm pr}^{-1}(L_{ij})$ that are proper transforms of the lines $L_{ij} \subset \mathbb{P}^2$ through the points P_i and P_j, $1 \le i < j \le 6$;

six lines $\sigma_{\rm pr}^{-1}(C_i)$ that are proper transforms of the irreducible conics C_i through the five points $\{P_1, \ldots, P_6\} \backslash P_i$ $(i \le i \le 6)$.

We get similar description for $d = 4$. For $d > 4$, the corresponding description is even simpler (there are no conics), and for $d < 3$, slightly more complicated (for details, see (Kanev (1987)), (Manin (1972))).

13.4. Ruled Surfaces Revisited. The vector bundle technique employed in the derivation of Theorem 7 of Sect. 13.2 can be applied to the investigation of ruled surfaces with a base B of genus > 0. However, in this case, the vector bundle theory is not so easy as for $B \cong \mathbb{P}^1$. Most 2-dimensional vector bundles $\mathcal{E}$ are not sums of line bundles. However, there is always an exact sequence

$$0 \to \mathcal{L} \to \mathcal{E} \to \mathcal{L}' \to 0\,,$$

where $\mathcal{L}$ and $\mathcal{L}'$ are line bundles (Hartshorne (1977)). So, we have to study extensions of line bundles. The group $\operatorname{Ext}(\mathcal{L}, \mathcal{L}')$ of such extensions is a finite dimensional spaces over the field k. Thus 2-dimensional bundles and consequently ruled surfaces are arranged in finite dimensional families.

One may ask when two elements of a family give rise to isomorphic bundles. It is a rather delicate question. In certain cases, the set of "ruled surfaces up to isomorphism" do not admit a structure of an algebraic variety. However, a certain type of bundles as well as ruled surfaces form an algebraic variety. These bundles and surfaces are called *stable*, and they are parametrized by points of certain moduli varieties, which have been studied extensively and posses several beautiful properties (e. g., see (Ramanan (1978))).

§ 14. Complex Analytic Surfaces

In this section, we will describe (more briefly than in the previous ones) how the theory of projective algebraic surfaces can be generalized to compact complex analytic manifolds of dimension 2 (henceforth they are called *complex surfaces*).

14.1. Meromorphic Functions. Let X be a compact complex manifold. A meromorphic function on X is an analog of a rational function. If X is connected, then the meromorphic functions form a field $\mathcal{M}(X)$. We have the following theorem.

Theorem 1. *The field $\mathcal{M}(X)$ is finitely generated, and its transcendence degree over $\mathbb{C}$ is less than or equal to the dimension of X* (Shafarevich (1988)).

We denote this transcendence degree by $a(X)$. So, far a complex surface, $a(X)$ equals 2, 1, or 0.

Theorem 2. *A complex surface X is algebraic if and only if $a(X) = 2$* (Barth-Peters-Van de Ven (1984)).

Let X be a complex surface. The 1-dimensional closed subvarieties of X generate the group $\operatorname{Div} X$ of divisors. To a divisor one can associate a line bundle. However, not every line bundle comes from a divisor. In general, the notion of line bundle is more natural and convenient than that of the divisor. For example, if Ω^1 is the cotangent bundle on X, then $\wedge^2 \Omega^1$ is a line bundle (denoted by ω_X).

The line bundles form a group under tensor multiplication. This group is isomorphic to $H^1(X, \mathcal{O}_X^*)$, where $\mathcal{O}_X^*$ is the sheaf of germs of nonvanishing holomorphic functions on X, and H^1 stands for the cohomology in the complex topology (throughout the section, we are considering only such cohomology). The exact exponential sequence of sheaves

$$0 \to \mathbb{Z} \to \mathcal{O}_X \to \mathcal{O}_X^* \to 0$$

yields the homomorphism $H^1(X, \mathcal{O}_X^*) \to H^2(X, \mathbb{Z})$. Given a line bundle $L \in H^1(X, \mathcal{O}_X^*)$, we denote by $c(L)$ the image of L under this homomorphism.

Theorem 3. *A complex surface is algebraic if and only if it has a line bundle L with $c(L)^2 > 0$* (Barth-Peters-Van de Ven (1984)).

We will present several examples of complex surfaces with $a(X) < 2$.

Example 1. In Example 6 of Sect. 3.3, we have observed that for a "general" choice of a lattice $\Omega \subset \mathbb{C}^2$, every divisor D on a complex torus, in particular, every curve is homologically equivalent to 0. Translating a curve C by a point $a \in X$, we obtain a curve $a + C$ that intersects C. If, in addition, $a + C \neq C$ and $(a + C, C) > 0$, then C cannot be homologically equivalent to 0, a contradiction. But if $a + C = C$ always holds when $(a + C) \cap C \neq \emptyset$, then it is not hard to verify that C is a subgroup in X or a coset of a subgroup. One can easily verify that the torus $\mathbb{C}^2/\Omega$ has no complex analytic subgroups provided the lattice Ω is "general". Therefore X has no curves at all. However, then there are no non-constant meromorphic functions, since the polar divisor $(f)_\infty$ of such a function f consists of curves. Therefore $a(X) = 0$ in this case. On can also construct tori with $a(X) = 1$.

Example 2. Given a torus X with $a(X) < 2$, one can construct a Kummer surface Y as in Example 7 of Sect. 2. Since $\mathcal{M}(Y) \subset \mathcal{M}(X)$ and $[\mathcal{M}(Y) : \mathcal{M}(X)] = 2$, we get $a(Y) = a(X)$. We obtain examples of $K3$ surfaces Y with $a(Y) = 0$ or 1.

Example 3. Let $W = \mathbb{C}^2 \backslash \{0\}$, and G an infinite cyclic group generated by a homotopy $g(z_1, z_2) = (\alpha z_1, \alpha z_2)$, $|\alpha| < 1$. The quotient $X = W/G$ is a complex surface. The map

$$X \to \mathbb{P}^1 , \qquad (z_1, z_2) \mapsto (z_1 : z_2) ,$$

defines a fibration of X whose fibers are isomorphic to

$$(\mathbb{C}\backslash 0)/G' , \quad G' = \{g'\} , \quad g'Z = \alpha z ,$$

i. e. the torus $\mathbb{C}/\Omega'$, where $\Omega' = 2\pi_1 \cdot \mathbb{Z} + \log \alpha \cdot \mathbb{Z}$. It is not hard to verify that X is homeomorphic to $S^3 \times S^1$, hence $b_1 = 1$ and $b_2 = 0$. Thus X is not algebraic (for any algebraic surface, $b_1 = 2q$ is even and $b_2 > 0$, since $(H^2) > 0$ for an ample cycle H so $\langle H \rangle \neq 0$ in $H_2(X, \mathbb{C})$).

Definition. A complex surface X is called a *Hopf surface* if its universal covering is isomorphic to the W. So $X = W/G$.

It was established that a group G acting freely and discretely on W with compact quotient has a cyclic subgroup of finite index. A Hopf surface X is said to be *primary,* if the group $G \cong \pi_1(X)$ is cyclic. If $G = \{T\}$, then one can reduce the automorphism T to a simple normal form by an analytic substitution of coordinates.

14.2. Cohomology. To begin with, we consider an arbitrary compact complex manifold. We get a spectral sequence with the initial term $E_1^{p,q} = H^q(X, \Omega_X^p)$ and the final term $H^{p+q}(X, \mathbb{C})$.

For complex surfaces, this spectral sequence degenerates at E_1. Hence

$$b_1 = \dim H^1(X, \mathcal{O}_X) + \dim H^0(X, \Omega_X^1) , \tag{1}$$

$$b_2 = \dim H^2(X, \mathcal{O}_X) + \dim H^1(X, \Omega_X^1) + \dim H^0(X, \Omega_X^2) . \tag{2}$$

An important question is how much of the Hodge theory holds for an arbitrary complex surface X. Given an arbitrary compact complex manifold, we say that the *Hodge theory holds* for it if a decomposition of m-dimensional differential forms in sums of terms of type (p, q), $p + q = m$, containing p differentials dz_i and q differentials $d\bar{z}_i$, can be extended to cohomology classes. Precisely, every form ω^m with $d\omega^m = 0$ must be homologically equivalent to the form $\sum_{p+q=m} \eta^{p,q}$ with $d\eta^{p,q} = 0$; and if $\sum_{p+q=m} \eta^{p,q} = d\xi^{m-1}$ with $d\eta^{p,q} = 0$, then for all p, $\eta^{p,q} = d\xi_{p,q}^{m-1}$, where $\xi_{p,q}^{m-1}$ is a suitable form.

We then have a decomposition of the cohomology groups

$$H^m = \bigoplus_{p+q=m} H^{p,q} , \tag{3}$$

where the $H^{p,q}$'s are the classes containing closed forms of type (p, q). Clearly $H^{p,q} = \overline{H^{q,p}}$ so, for example, (3) gives $b_1 = 2 \dim H^1$, and in particular b_1 is even. It follows at once from the Hodge theory that $H^{m,0} = H^0(X, \Omega^m)$ is the space of holomorphic forms of degree m and all such forms are closed.

The Hodge theory holds for projective varieties. This follows from the fact that one can introduce a Kähler metric on the projective space and consequently on any projective variety (Griffiths-Harris (1978), Shafarevich (1988)).

However, the availability of the Hodge theory is not necessary connected with the existence of a Kähler metirc, as the example of complex surfaces illustrates.

Theorem 4. *The Hodge theory holds for 2-dimensional cohomology of complex surfaces. Thus*

$$H^2(X,\mathbb{C}) = H^{2,0} \oplus H^{1,1} + H^{0,2}\,, \qquad H^{0,2} = \overline{H}^{2,0}\,. \tag{4}$$

In general, this no longer holds for the 1-dimensional cohomology. However, the following corollaries of the Hodge theory hold:

(i) 1-dimensional holomorphic forms are closed so that $H^0(X,\Omega^1) \subset H^1(X,\mathbb{C})$.

(ii) The images of $H^0(X,\Omega^1)$ and $H^0(X,\overline{\Omega}^1)$ in $H^1(X,\mathbb{C})$ do not intersect (Barth-Peters-Van de Ven (1984), Seminar Palaiseau (1985)).

It follows from (1), (i), and (ii) that

$$2\dim H^0(X,\Omega^1) \le b_1 \le H^0(X,\Omega^1) + q$$

where we set $q = \dim H^1(X,\mathcal{O}_X)$ as usual. Therefore

$$b_1 \le 2q\,. \tag{5}$$

The decomposition (4) easily yields that the product on H^2, given by the formula $(\xi,\eta) = \int_X \xi \wedge \overline{\eta}$, is positive definite on $H^0(X,\Omega^2) \oplus H^0(X,\overline{\Omega}^2)$, hence

$$2p_g \le b_2^+\,, \tag{6}$$

where $p_g = \dim H^0(X,\Omega^2)$, and (b_2^+, b_2^-) is the type of the quadratic form on $H^2(X,\mathbb{R})$ given by the multiplication $H^2 \times H^2 \to H^4$.

The Riemann-Roch theorem holds for complex surfaces, though its proof is different – it depends on the theorem on index of elliptic differential operators. The Riemann-Roch theorem is connected with the *index theorem* that gives an expression for the index $\tau = b^+ - b^-$ in the form $1/3(c(\omega_X)^2 - 2e(X))$. Combining them, we easily get

$$(b^+ - 2p_g) + (2q - b_1) = 1\,. \tag{7}$$

Set $\Delta = 2q - b_1$. It follows from (5), (6), and (7) that $\Delta \ge 0$, hence Δ equals 0 or 1. Therefore we have the following possibilities:

$$\Delta = 0\,, \quad b_1 = 2q\,, \quad b^+ = 2p_g + 1\,,$$
$$\Delta = 1\,, \quad b_1 = 2q - 1\,, \quad b^+ = 2p_g\,.$$

The first case take place for algebraic surfaces and "close to algebraic" ones (precisely, obtained by continuous deformation of their complex structures). The second one characterizes "genuinely non-algebraic" complex surfaces. For example, the Hopf surfaces among them are (Example 3). (For details, see (Barth – Peters – Van de Ven (1984)).)

14.3. Surfaces with $a(X) = 0$ or $a(X) = 1$. Now, we turn to a more systematic description of complex surfaces.

Theorem 5. *A complex surface X with $a(X) = 1$ admits a holomorphic map $f\colon X \to B$ to an algebraic curve B such that for all $b \in B$, except perhaps a finite number, $f^{-1}(b)$ is an elliptic curve. Moreover $\mathcal{M}(X) = f^*\mathcal{M}(B)$* (see (Barth-Peters-Van de Ven (1984))).

The surfaces described in Theorem 5 are called *elliptic.* The major portion of the theory of elliptic algebraic surfaces presented in Sect. 10 can be extended to the complex case, namely, the structure of singular fibers, construction of Jacobian families $J \to B$, and construction of the group $I(J)$. As in the algebraic case, we can get rid of multiple fibers by passing to a finite covering

$$B' \to B, \qquad X' \to X \times_B B'.$$

Henceforth, we will consider only the families without multiple fibers. They form a group denoted by $I_0(J)$. To describe it, we use the analog of the exact exponential sequence (2) in Sect. 3.3.

We will consider two cases: a) J is not isomorphic to $B \times E$, where E is an elliptic curve; and b) $J \cong B \times E$. In the former case, the group $H^2(B, \mathcal{G}^0)$, in the exact cohomology sequence corresponding to the exact sequence (6) of Sect. 10.3, is finite. Further, the group $H^1(B, \mathcal{F})$ is a finite dimensional vector space, and the group $H^1(B, \Gamma)$ is a finitely generated Abelian group.

Theorem 6. *The Jacobian family J is an algebraic surface.*

In Case (a), all the families of the group $I_0(J)$ have $\Delta = 0$; the group $I_0(J)$ is isomorphic to $(\mathbb{C}^N/H) + \mathfrak{U}$, where H is a finitely generated group and $\mathfrak{U}$ a finite Abelian group; and an element $\alpha \in I_0(J)$ gives an algebraic surface if and only if it has a finite order.

In Case (b), the families of the group $I_0(J)$ correspond to surfaces with $\Delta = 0$ or $\Delta = 1$ depending on whether the image of the homomorphism

$$H^1(B, \mathcal{G}^0) \to H^1(B, \Gamma) \cong \mathbb{Z}^2$$

is 0; the elements with $\Delta = 0$ form a group isomorphic to $\mathbb{C}^g/H^1(B, \mathbb{Z})$ ($g = g(B)$) whose elements of finite order correspond to algebraic surfaces (Kodaira (1960–1963), III).

The Hopf surface described in Example 3 gives an example of a surface that belongs to Case (b). In general, a primary Hopf surface X is elliptic if and only if $X = W/G$, where $G = \{T\}$ and the automorphism T has the following form in a suitable coordinate system:

$$T(z_1, z_2) = (\alpha_1 z_1, \alpha_2 z_2), \quad |\alpha_1| < 1, \quad |\alpha_2| < 1, \quad \alpha_1^m = \alpha_2^n, \quad m, n \in \mathbb{Z}.$$

Then the map $(z_1, z_2) \mapsto (z_1^m : z_2^n)$ defines a fibration $X \to \mathbb{P}^1$.

Among elliptic surfaces with $\Delta = 1$, we would like to mention surfaces $X \to B$ with an elliptic base B. They are called *Kodaira surfaces* (these are not the Kodaira surfaces introduced in Sect. 9.1!). They are interesting because their canonical bundle ω_X is trivial. Furthermore, $b_1 = 3$ and $b_2 = 4$. Together with tori and $K3$ surfaces (which have $H^1(X, \mathcal{O}_X) = 0$, $\pi_1(X) = 0$, $e = 24$, and $b_2 = 22$) these are the only complex surfaces with trivial canonical bundle. The universal coverings of the Kodaira surfaces are isomorphic to $\mathbb{C}^2$. They can be represented in the form $\mathbb{C}^2/G$, where G are certain special affine transformation groups of $\mathbb{C}^2$.

Now, we turn to the surfaces with $a(X) = 0$, i. e., the surfaces that do not have non-constant meromorphic functions.

Theorem 7. *If a surface X with $a(X) = 0$ and $\Delta = 0$ has no (-1)-curves, then it is either a complex torus $\mathbb{C}^2/\Omega$ or a complex $K3$ surface* (Kodaira (1960–1963), IV).

So, the remaining surfaces have $a(X) = 0$ and $\Delta = 1$. They are sometimes called *surfaces of Type* VII (according to the number they were assigned in Kodaira's classification). Very little is known about those surfaces.

Theorem 8. *For surfaces of Type* VII, *we get*

$$b_1 = 1\,, \qquad H^2(X, \mathcal{O}) = H^0(X, \Omega^2) = 0\,.$$

One example is provided by non-elliptic Hopf surfaces, for example:

$$X = \mathbb{C}^2 \backslash \{0\}/\{T\}\,, \qquad T\colon (z_1, z_2) \mapsto (\alpha_1 z_1, \alpha_2 z_2)\,,$$

where $\alpha_1^a \neq \alpha_2^b$ for all the integers a and b. There are also other surfaces of Type VII, namely, the *Inoue surfaces.* As for Hopf type surfaces, they have $b_2 = 0$. Moreover, there are surfaces of Type VII with $b_2 > 0$ (Barth-Peters-Van de Ven (1984)). There is no general theory of such surfaces.

The above description of complex surfaces shows the place the algebraic surfaces occupy among them. Two surfaces X and Y are said to be *deformations* one of the other if there is a smooth variety $\mathcal{X}$ and a smooth holomorphic map $f\colon \mathcal{X} \to B$ to a connected variety B such that X and Y are fibers of f, namely: $X = f^{-1}(b_1)$ and $Y = f^{-1}(b_2)$, $b_1, b_2 \in B$. Then the family $\mathcal{X} \to B$ is a locally trivial differential fibration, hence X and Y are diffeomorphic. In particular, two complex surfaces, one with $\Delta = 0$ and the other with $\Delta = 1$, cannot be deformed one into the other. Recall that $\Delta = 0$ for algebraic surfaces.

Theorem 9. *A complex surface X is a deformation of an algebraic surface if and only if $\Delta = 0$ for X. Then one can introduce a Kähler metric on X* (Kodaira (1960–1963), III).

On can illustrate this by an example of complex tori. Every torus has the form $\mathbb{C}^2/\Omega$, and it is uniquely determined by a basis of Ω, i. e. four vectors of $\mathbb{C}^2$ that are linearly independent over $\mathbb{R}$. By taking an appropriate basis in $\mathbb{C}^2$, we can assume that the first two vectors are $(0,1)$ and $(1,0)$. If the third and fourth vectors are (a_1, a_2) and (b_1, b_2), then the condition that they are linearly independent means that

$$\begin{vmatrix} \operatorname{Im} a_1, & \operatorname{Im} a_2 \\ \operatorname{Im} b_1, & \operatorname{Im} b_2 \end{vmatrix} \neq 0$$

and we can assume (by changing the order of vectors) that this determinant > 0. The set of all complex tori is connected, so all 2-dimensional tori form a connected family $\mathcal{M}$ of dimension 4. This family contains also algebraic tori, which fill up countably many 3-dimensional subvarieties, as we have seen in Sect. 11.

It follows from the above description that the algebraic surfaces with non-algebraic deformations are precisely elliptic surfaces, Abelian surfaces, and $K3$ surfaces. In particular, any deformation of a surface of general type is algebraic.

Thus one may (very roughly) summarize the above description of complex surfaces as follows. The majority of surface are algebraic, a smaller part consists of deformations of certain special algebraic surfaces, and even smaller part consists of very special "genuinely complex" surfaces, which are not deformations of algebraic surfaces. The latter admit a topological characterization to the effect that their 1st Betti number b_1 is odd. (For details, see (Kodaira (1960–1963), (1964–1969)).)

14.4. Uniformization. We will conclude this section with a few remarks regarding the uniformization of complex and, in particular, algebraic surfaces. In case of algebraic curves, the uniformization provides a very explicit picture (Shokurov (1988)): A curve S is either isomorphic to $\mathbb{P}^1$, or can be represented in the form $\mathbb{C}/\Omega$ where $\Omega \subset \mathbb{C}$ is a lattice, or in the form D/Γ, where D is the disk $|z| < 1$. This is related to the fact that if $S \not\cong \mathbb{P}^1$, then the fundamental group of S is "very large" and always infinite. Therefore we have to find out, first of all, what kind of groups are the fundamental groups of algebraic surfaces.

The last question is of interest because algebraic surfaces have the same fundamental groups as arbitrary projective varieties of dimension ≥ 2. Precisely, if X is a projective variety of dimension ≥ 2 and H its smooth 2-dimensional section, then the embedding $H \hookrightarrow X$ defines an isomorphism

$$\pi_1(H) \xrightarrow{\sim} \pi_1(X), \tag{8}$$

This follows from the Leftschetz theorem to the effect that for a variety X of dimension ≥ 3 and its hyperplane section H, the embedding $H \hookrightarrow X$ defines the isomorphism (8). The theorem easily follows from basic facts of the Morse theory (Milnor (1963)), or can be derived by an even more elementary argument.

In sharp contrast with algebraic curves, many algebraic surfaces are simply connected. For instance, any surface in $\mathbb{P}^3$ is simply connected. This follows from the above Lefschetz theorem, namely, one can view a surface of degree n as a hyperplane section of the image of $\mathbb{P}^2$ under the Veronese map given by monomials of degree n. One can similarly prove that complete intersections are simply connected. Further, the smooth projective models of the surfaces $z^n = f(x,y)$ are simply connected, provided the curve $f(x,y) = 0$ is smooth or has no too complicated singularities. In fact, one may get an impression that surfaces are simply connected "in general".

The surfaces with finite fundamental groups are similar to simply connected ones. One can easily construct examples of surfaces with an arbitrary given finite fundamental group.

A natural analog of the uniformization of curves of genus $g > 1$ is the representation of surfaces as a quotient U/G, where $U \subset \mathbb{C}^2$ is a bounded homogeneous domain and G a discrete group of its automorphisms. There are two such domains up to isomorphism, namely: the ball $\{|z_1|^2 + |z_2|^2 < 1\}$ and polydisk $\{|z_1| < 1, |z_2| < 1\}$. We have already considered their quotients. They are surfaces of general type. In the "geography of surfaces", they are located on the two lines (see Sect. 9, Fig. 16): $(K^2)/e = 3$ for the first type, and $(K^2)/e = 2$ for the second one, where e is the Euler characteristic. (For details, see (Holzapfel (1980), Van de Geer (1988)).)

However, there are, in addition, nonhomogeneous bounded domains with algebraic surfaces as quotients. For example, there exists a bounded domain U whose automorphism group G is discrete but large so that the quotient U/G is compact and a projective surface (Shabat (1977)).

§ 15. Effects of Finite Characteristic

The geometry over a field of finite characteristic represents a beautiful chapter in algebraic geometry like the geometry of algebraic varieties over complex or real numbers. It should be treated systematically in a separate survey.

In the present section, we will only describe in general terms some typical effects that apppear in the theory of algebraic surfaces over a field of finite characteristic.

15.1. Counterexamples to Bertini's Theorem. Recall that according to that theorem, given a morphism $f\colon X \to Y$ of algebraic varieties over a field of characteristic 0, and X is smooth, there is a nonempty open set $U \subset Y$ such that the morphism $f^{-1}(U) \to U$ is smooth. In particular, all the fibers $f^{-1}(u)$, $u \in U$, are reduced and smooth.

Over a field of characteristic $p > 0$, the Bertini theorem fails for the most typical map in finite characteristic, namely the Frobenius map $x \mapsto x^p$, say, of a line onto itself. In other words, this is a reflection of the fact that in finite characteristic, $f' = 0$ does not imply f is a constant function.

We get a more striking counterexample to the Bertini theorem if we assume that the generic fiber F_ξ of the morphism f is a geometrically irreducible (i. e. irreducible over the algebraic closure $\overline{k(\xi)}$ of the field $k(\xi)$) smooth variety. We first encounter this phenomenon in the theory of surfaces when $\dim X = 2$ and $\dim Y = 1$. The argument, given in our discussion of ruled surfaces in Sect. 13.1, works over fields of arbitrary characteristic and shows that if the genus of the generic fiber of the morphism $f\colon X \to Y$ equals 0, then the analog of Bertini's theorem holds in any characteristic. But if the genus equals 1, then we encounter the first interesting counterexample.

If the characteristic of the ground field equals 2, the counterexample is a surface in $\mathbb{P}^2 \times \mathbb{A}^1$ given an equation

$$\xi_0\xi_2^2 = \xi_1^3 + a(t)\xi_1\xi_0^2 + b(t)x_0^3,$$

where $(\xi_0 : \xi_1 : \xi_2)$ are homogeneous coordinates in $\mathbb{P}^2$, and t a coordinate in $\mathbb{A}^1$. In the coordinates $x = \xi_1/\xi_0$, $y = \xi_2/\xi_0$, it is given by the equation

$$y^2 = x^3 + a(t)x + b(t). \tag{1}$$

A singular point of the surface (1) has coordinate t_0 with $D(t_0) = 0$, where $D(t) = (a')^2 a \times (b')^2$; but for every $t \equiv t_0$, the curve (1) has a singular point $x = a(t_0)^{1/2}$, $y = b(t_0)^{1/2}$, which is a cusp of type $u^2 = v^3$. Therefore, if $D(t) \not\equiv 0$, $Y \subset \mathbb{A}^1$ is defined by the condition $D(t) \neq 0$, X is a portion of the surface (1) given by the condition $D(t) \neq 0$, and $f(x, y, t) = t$, then the Bertini theorem fails.

Similar examples exist in characteristic 3 as well:

$$y^2 = x^3 + a(t). \tag{2}$$

The surfaces (1) and (2) as well as the corresponding fibrations $X \to \mathbb{A}^1$ (and their natural generalizations when a and b are functions on an arbitrary curve C) are called *quasi-elliptic*. If $\pi\colon X \to C$ defines such a pencil, then $mK_X = \pi^*(D)$ where $D \in \operatorname{Div} C$ and $m \geq 0$. One can show that if the generic fiber of a fibration $X \to Y$ (X is a surface and Y is a curve) has genus 1, then the Bertini theorem fails only in characteristic 2 and 3, and the only counterexamples are quasi-elliptic fibrations. However, if the genus of the fiber is arbitrary, such examples exist in any characteristic $p > 0$. For example, consider the surface

$$y^p = x^m + a(t), \qquad m > 1, \quad m \not\equiv 0 (\operatorname{mod} p)$$

(Bombieri-Mumford (1969, 1977, 1976)).

15.2. Quotients by a Nonreduced Group Scheme. We have often employed a helpful method for constructing surfaces as quotients X/G, where

G is a finite group of automorphisms of a surface X (see Sect. 2, Example 3, and so on). In geometry over fields of finite characteristic, one has to consider the cases when G is not necessary reduced finite group scheme too. For instance, over a field of characteristic 2, the method for constructing surfaces described in (Sect. 2, Example 9) does not make sense because $-1 = 1$ in such a field. One has to modify the method.

We consider the group scheme μ_2 defined over the ring $k[\varepsilon]$, $\varepsilon^2 = 1$, by a multiplication map:

$$k[\varepsilon] \to k[\varepsilon] \otimes k[\varepsilon]\,, \qquad \varepsilon \mapsto \varepsilon \otimes \varepsilon\,.$$

Formally, it is similar to a definition of a group of order 2 ($\varepsilon^2 = 1$), except for the property that the ring $k[\varepsilon]$ has nilpotents $k(\varepsilon + 1)$, and the scheme $\operatorname{Spec} k[\varepsilon]$ has a single point but is nonreduced.

Let P_i be quadratic forms in x_0, x_1, x_2, and Q_i quadratic forms in x_3, x_4, x_5 ($i = 1, 2, 3$). Let X be a surface in $\mathbb{P}^5$ given by the equations $P_i + Q_i = 0$ ($i = 1, 2, 3$). Consider the following action of μ_2 on X:

$$(x_0 : x_1 : x_2 : x_3 : x_4 : x_5) \mapsto (\varepsilon x_0 : \varepsilon x_1 : \varepsilon x_2 : x_3 : x_4 : x_5)\,.$$

It is not hard to show that $Y = X/G$ is a smooth surface for a "general" choice of the forms P_i and Q_i, and it is an Enriques surface, namely: $K_Y \neq 0$, $2K_Y = 0$, and $H^1(Y, \mathcal{O}_Y) = 0$ (Bombieri-Mumford (1969, 1977, 1976)).

15.3. Nonreducibility of the Picard Scheme. As we mentioned before, the Picard scheme $\operatorname{Pic} X$ of an algebraic surface X has a connected component $\operatorname{Pic}^0 X$ that is a proper connected, even projective, group scheme over the field k, and the quotient group $S_X = \operatorname{Pic} X / \operatorname{Pic}^0 X$ has a finite number of generators. By the theory of algebraic groups, a proper connected group scheme over a field of characteristic 0 is always reduced, i. e. an Abelian variety (e. g., see (Mumford (1970a))). In general, this no longer holds if $\operatorname{char} k \neq 0$.

As we mentioned in Sect. 3.4, $\dim(\operatorname{Pic}^0 X)_{\mathrm{red}} = \dim \operatorname{Alb} X$. Since the tangent space of the scheme $\operatorname{Pic}^0 X$ (or $\operatorname{Pic} X$) at zero point is isomorphic to $H^1(X, \mathcal{O}_X)$, we get $\dim \operatorname{Alb} X \leq h^{0,1}$ and the scheme $\operatorname{Pic}^0 X$ is reduced if and only if

$$\dim \operatorname{Alb} X = h^{0,1}\,. \tag{3}$$

We will give an example when the equality (3) fails (Igusa (1955)). Assume $\operatorname{char} k = 2$, and C is an elliptic curve over k with a non-zero point c_0 of order 2 (i. e., C is not a supersingular curve). Set $Y = C \times C$. Consider an automorphism s: $(c_1, c_2) \mapsto (c_1 + c_0, -c_2)$, and the quotient $X = Y/G$, $G = \{1, s\}$. Clearly s has no fixed points, so X is smooth and the map π: $Y \to X$ is separable and unramified. It follows that $K_X = 0$ and $p_g(X) = h^{2,0} = 1$. Since $e(Y) = 0$ and $e(X) = e(Y)/2 = 0$, the Noether formula (Sect. 5.1, (2)) gives $\chi(X) = 0$ hence $h^{0,1} = 2$. To calculate $\dim \operatorname{Alb} X$, we recall that $\operatorname{Alb} X$ classifies the morphisms $X \to A$, where A is an Abelian variety. In particular, there is a morphism α: $X \to \operatorname{Alb} X$ such that $\alpha(X)$ generates $\operatorname{Alb} X$.

Now we set $\pi\alpha = \beta\colon C \times C \to \operatorname{Alb} X$. We can assume that $\beta(0) = 0$ by modifying α by a translation of A. By a well-known and elementary properties of Abelian varieties (e.g., (Shafarevich (1988))), such a morphism is a homomorphism, i.e., $\beta = (\gamma_1, \gamma_2)$, where $\gamma_i\colon C_i \to \operatorname{Alb} X$ are homomorphisms for $i = 1, 2$. Clearly $\beta \circ s = \beta$ whence $\gamma_2 = 0$, i.e. $\operatorname{Alb} X = \alpha(X) = \beta(C_1)$, and consequently $\dim \operatorname{Alb} X = 1$. One can construct similar examples over any field of positive characteristic.

15.4. Breakdown of the Symmetry $h^{p,q} = h^{q,p}$. One can find an example among Enriques surfaces. We consider an involution in $\mathbb{P}^5$:

$$s\colon (x_0 : x_1 : x_2 : y_0 : y_1 : y_2) \mapsto (y_0 : y_1 : y_2 : x_0 : x_1 : x_2)\,.$$

The space of invariant quadratic forms has a basis

$$x_i x_j + y_i y_j\,, \quad x_i y_j + y_i x_j\,, \qquad i, j = 0, 1, 2\,.$$

We take three general such forms, F_1, F_2, F_3, and consider the surface Y defined by the equation $F_1 = F_2 = F_3 = 0$. It is a smooth surface, and s is a fixed-point-free automorphism of Y. As we mentioned before, Y is a $K3$ surface and $X = Y/G$, $G = \{1, s\}$, an Enriques surface. Since $H^0(Y, \Omega^1_Y) = 0$ and $\pi\colon Y \to X$ is a separable covering, we get $h^{1,0} = 0$ for X.

On the other hand, under the action of automorphisms $g \in G$, a non-zero form $\omega \in H^0(Y, \Omega^2_Y)$ is multiplied by a character of G. Now we assume that $\operatorname{char} k = 2$. Then G has only a trivial character with values in k, i.e. $g^*(\omega) = \omega$, hence $\omega = \pi^*(\eta)$, $\eta \in H^0(X, \Omega^2_X)$. It follows that $K_X = 0$ (while in the case $\operatorname{char} k \neq 2$, we get $K_X \neq 0$ but $2K_X = 0$). In particular $p_g(X) = 1$. Now, the Noether formula together with the relation $e(X) = e(Y)/2 = 12$ show that $h^{0,1} = 1$ for X.

15.5. Absence of Analogs of the Theorems of Lefschetz and Lüroth. The map $S_X \to H^2(X, \mathbb{Z}_l)$ has a finite kernel (for any prime integer l such that $\operatorname{char} k \neq l$). It follows that the inequality

$$\varrho \leq b_2 \tag{4}$$

holds in arbitrary charcteristic. Already the trivial part of Lefschetz's theorem shows that in characteristic 0, the inequality is strict provided $p_g(X) > 0$ (see Sect. 3.3). If $\operatorname{char} k = p > 0$, (4) can become an equality for X with $p_g(X) > 0$; then the algebraic cycles generate the cohomology group $H^2(X, \mathbb{Q}_l)$, and X is said to be a *supersingular* surface. The latter definitely holds if X is a unirational surface, i.e., there is a rational map $f\colon \mathbb{P}^2 \dashrightarrow X$ with $f(\mathbb{P}^2)$ dense in X. Indeed, we can assume that f is a morphism by resolving the points of indeterminacy. Take $\xi \in H^2(X, \mathbb{Q}_l)$. Then $f_* f^*(\xi) = (\deg f)\xi$, and $\mathbb{P}^2$ is definitely supersingular. Hence $f^*(\xi)$ is a combination of algebraic cycles, so the same holds for ξ as well.

It remains to find an example of a unirational surface X with $p_g(X) > 0$ – a counterexample to the analog of Lüroth's theorem. A beautiful example is provided by a "Fermat surface" with an equation

$$x_0^{p+1} + x_1^{p+1} - x_2^{p+1} - x_3^{p+1} = 0 \tag{5}$$

(see Shioda (1974)). Setting $x_0 = y_0 + y_1$, $x_1 = y_0 + y_1$, $x_2 = y_2 + y_3$, $x_3 = y_2 - y_3$, we can rewrite (5) in the form

$$y_0^p y_1 + y_i^p y_0 - y_2^p y_3 - y_3^p y_2 = 0\,,$$

or, in nonhomogeneous coordinates, in the form

$$x^p y + y^p x = z^p + z\,. \tag{6}$$

Set $y = t^p$. Then (6) gives $(xt - z)^p = z - xt^2$, i.e. $u^p = v$ with $u = xt - z$ and $v = z - xt^{p^2}$. Therefore $k(X) \subset k(x,t,z) = k(u,t)$ so X is unirational. The surface (5) is rational if $p = 2$, and a $K3$ surface if $p = 3$. If $p > 3$, then X is of general type. So, $p_g(X) > 0$ if $p > 2$.

15.6. Failure of the Vanishing Theorem. There are examples in any characteristic $p > 0$, however, we are not going to describe them in detail (see Raynaud's report in (Algebraic Surfaces (1981))). The corresponding smooth surface X is a fibration over a curve C of genus $g > 1$. The fibers of $\pi\colon X \to C$, all singular, have a cusp of type $u^2 + v^p$ (for $p \neq 2$). The normalization of each fiber is rational. Thus the Bertini theorem also fails in this case (compare Sect. 15.1). The surface X is quasi-elliptic for $p = 2, 3$, and of general type for $p \geq 5$.

15.7. Changes in Classification. The classification of algebraic surfaces presented in Sect. 7 and 8 can be extended to fields of finite characteristic with very minor changes in principal results. The corresponding proofs are, however, much more involved. The changes occur only when the canonical dimension κ equals 1 or 0 (Bombieri-Mumford (1969, 1977, 1976)). In the case $\kappa = 1$, we add the quasi-elliptic surfaces to the elliptic ones in characteristic 2 or 3 (compare Sect. 15.1). For $\kappa = 0$, we get the following collections of invariants (see Table 6).

Table 6

b_2	b_1	e	$\chi(\mathcal{O})$	$q = h^{0,1}$	p_g	Type of a surface
22	0	24	2	0	1	$K3$
6	4	0	0	2	1	Abelian
2	2	0	1	0	0	bi-elliptic (classical)
10	0	12	0	0	1	Enriques (classical)
2	2	0	2	1	0	non-classical bi-elliptic
10	0	12	1	1	1	non-classical Enriques

The cases above the line are the same as in characteristic 0. The bi-elliptic surfaces (called *classical* in this case) are quotients of Abelian varieties $E_1 \times E_2$, where E_1 are elliptic curves, by (not necessary reduced) finite groub schemes acting freely on them. All the possible actions are enumerated. The Albanese variety of such a surface has dimension 1 (and is isogenous to E_1 or E_2), and the map $X \longrightarrow \operatorname{Alb} X$ is an elliptic fibration with all its fibers nondegenerate.

The Enriques surfaces (called *classical* in this case) have $K_X \neq 0$ and $2K_X = 0$. They are quotients of $K3$ surfaces by a group of order 2, whereas in characteristic 2 by the group scheme μ_2. Their Picard schemes are reduced in any characteristic, and have torsion $\mathbb{Z}/2\mathbb{Z}$.

The cases below the line occur in characteristic 2 or 3. Non-classical bi-elliptic surfaces are quotients of surfaces of the form $E \times C$, where E is an elliptic curve and C a rational curve with one singular point, namely a cusp (an effect of finite characteristic is that the quotient of the singular surface turns out to be smooth). Again, the Albanese variety is 1-dimensional but the fibration $X \longrightarrow \operatorname{Alb} X$ is quasi-elliptic.

We encounter non-classical Enriques surfaces only in characteristic 2. They have $K_X = 0$, and $\operatorname{Pic}^0 X$ is nonreduced of order 2. If $\operatorname{Pic}^0 X \cong \mu_2$, then X is a quotient of a $K3$ surface by the group $\mathbb{Z}/2\mathbb{Z}$ (see Sect. 2, Example 3). There remains yet another possibility, namely $\operatorname{Pic}^0 X = \alpha_2$ is the kernel (as a scheme) of the morphism $x \mapsto x^2$ of the additive group $\mathbb{G}_a$. Then X is a quotient by a group scheme, isomorphic to α_2, of a certain singular surface whose minimal resolution of singularities is a supersingular $K3$ surface.

All the "non-classical" types of surfaces are connected with one interesting relation between invariants of an arbitrary algebraic surface. Applying Noether's formula (Sect. 5.1, (2)) to the expression $e = 2 - 2b_1 + b_2$, we can rewrite it as follows:

$$10 - 8q + 12p_g = (K^2) + b_2 - 2\Delta\,,$$

where $\Delta = 2q - b_1$. We have encounted it in this form in Sect. 14, where we have seen that for complex surfaces, $\Delta = 0$ or 1 depending on the parity of b_1; so $\Delta = 0$ for algebraic surfaces. The b_1 is even for an algebraic variety over an arbitrary algebraically closed field (Milne (1980)). Moreover, $b_1/2 = \dim \operatorname{Alb} X$ hence Δ is even and $\Delta \geq 0$, with $\Delta = 0$ if and only if $\operatorname{Pic}^0 X$ is reduced. The cases referred to as "non-classical" are precisely the ones with $\Delta > 0$ (with $\kappa = 0$). It was established that $\Delta \leq 2p_g$ for an arbitrary surface. So, in the cases gathered in Table 6, the Δ takes the only possible positive value 2. In any characteristic there are surfaces of general type with $\Delta > 0$ (Bombieri-Mumford (1969, 1977, 1976), I–III).

Bibliography

The surveys (Danilov (1988) and Danilov (1989)) provide a background in algebraic geometry necessary for the present survey. General topics are treated in the books (Hartshorne (1977), Griffiths-Harris (1978), Shafarevich (1988)). Special topics in algebraic geometry are treated in the book (Ahlfors-Bers (1961)) (moduli of Riemann surfaces), the survey (Shokurov (1988)) (algebraic curves and Riemann surfaces), the books (Milne (1980)) (etale cohomology), (Mumford (1970) and Siegel (1949)) (Abelian varieties), (Mumford (1977) and Mumford-Fogarty (1982)) (general topics in moduli theory), and (Grothendieck (1962)) (general topics in scheme theory).

A summary of classical theory of algebraic surfaces is contained in the surveys (Castelnuovo-Enriques (1914), Enriques (1949), and Zariski (1971)). There are several more recent books on the general theory of algebraic surfaces (Algebraic Surfaces (1965), Algebraic Surfaces (1981), Barth-Peters-Van de Ven (1984)) (the latter over the complex numbers only but it also includes the theory of non-algebraic surfaces), (Beauville (1978)), and the survey (Bombieri-Husemöller (1975)) (several results in finite characteristic).

There are several surveys on special topics in the theory of surfaces, namely: divisors and Picard varieties (Mumford (1966)) (the classification of surfaces is contained in all the general surveys mentioned above); surfaces of general type (Bombieri (1973), Gieseker (1977), Reider (1988), Van de Ven (1978)); the recent survey on the "geography of surfaces" (Chen (1987)); elliptic surfaces (Barth-Peters-Van de Ven (1984), Ogg (1962), Shafarevich (1961)); Enriques surfaces (Horikawa (1978), Verra (1983)); $K3$ surfaces (Kulikov (1977), Mukai (1988), Nikulin (1981, 1984), Piatetski-Shapiro – Shafarevich (1971), Seminar Palaiseau (1985)); rational surfaces (Aoki-Shioda (1983), Shabat (1977)); ruled surfaces (Ramanan (1978)); complex analytic surfaces (Barth-Peters-Van de Ven (1984), Kodaira (1960–1963), Kodaira (1964–1969)); effects of finite characteristic are discussed in (Bombieri-Husemöller (1975), Bombieri-Mumford (1969, 1977, 1976), Igusa (1955), Milne (1980)), Raynaud's article in (Ramanujam (1978)), and (Shioda (1974)).

The articles (Artin-Mumford (1972), Clemens-Griffiths (1972), Iskovskikh-Manin (1971), Kawamata-Matsuda-Matsuki (1987), Mori (1987), Saltman (1984)) may help to understand the difficulties one encounters when passing from surfaces to varieties of dimensions 3 and higher.

References[1]

Ahlfors, L., Bers, L. [1961]: Riemann Surfaces and Quasiconformal Maps. Inostranaya Literatura: Moscow, 1961 (Russian)

Algebraic Surfaces [1965]: Tr. Mat. Inst. Steklov. *75* (1965). English transl.: Proc. Steklov Inst. Math. *175* (1967), Zbl. 154, 210

Algebraic Surfaces [1981]: C.I.M. 1977. Liguori: Napoli, 1981

Aoki, N., Shioda, T. [1983]: Generators of the Neron-Severi group of a Fermat surface. In: Arithmetic and Geometry, I. Birkhäuser: Boston, 1983, Prog. Math. *35*, 1–12, Zbl. 586.14028

Artin, M. [1962]: Some numerical criteria for contractibility of curves on algebraic surfaces. Am. J. Math. *84* (1962) 485–496, Zbl. 105, 144

Artin, M., Mumford, D. [1972]: Some elementary examples of unirational varieties which are not rational. Proc. Lond. Math. Soc., III. Ser. *25* (1972) 75–95, Zbl. 244.14017

Artin, M., Winters, G. [1971]: Degenerate fibers and stable reduction of curves. Topology *11* (1971) 373–383, Zbl. 221.14018

Barlow, R. N. [1982]: Some new surfaces with $p_g = 0$. Thesis. Warwick, 1982. [See Duke Math. J. *51* (1984) 889–904], Zbl. 576.14038

Barth, W., Peters, C., Van de Ven, A. [1984]: Compact complex surfaces. Springer-Verlag: Berlin Heidelberg New York, 1984, Zbl. 718.14023

Beauville, A. [1978]: Surfaces algébriques complexes. Asterisque *54* (1978) 1–172, Zbl. 394.14014

Beauville, A. [1983]: Variétés Kählériennes dont la premiere class de Chern est nulle. J. Differ. Geometry *18* (1983) 755–782, Zbl. 537.53056

Bombieri, E. [1973]: Canonical models for surfaces of general type. Inst. Haut. Étud. Sci., Publ. Math. *42* (1973) 171–219, Zbl. 259.14005

Bombieri, E., Husemöller, D. [1975]: Classification and embeddings of surfaces. In: Algebraic Geometry (Arcata 1974). Am. Math. Soc., Proc. Symp. Pure Math. *29* (1975) 329–420, Zbl. 326.14009

Bombieri, E., Mumford, D. [1969, 1977, 1976]: Enriques' classification of surfaces in characteristic p. I. In: Global analysis. Princeton Univ. Press: Princeton, 1969, 325–339. II. In: Complex analysis and algebraic geometry. Cambridge Univ. Press: Cambridge, 1977, 23–42. III. Invent. Math. *35* (1976) 197–232. Zbl. 188 532; Zbl. 348.14021; Zbl. 336.14010

Bourbaki, N. [1968]: Eléments de mathématiques. Groupes et algèbres de Lie. Chapitres 4, 5, 6. Hermann: Paris, 1968, Zbl. 186, 330

Castelnuovo, G., Enriques, F. [1914]: Die algebraischen Flächen vom Gesichtspunkt der birationalen Transformationen aus. In: Enz. der mathematischen Wissenschaften, III, Heft 6, 1914, 674–768, Jbuch 45, 883

Chen, Z. [1987]: On the geography of surfaces. Math. Ann. *277* (1987) 141–164, Zbl. 609.14023

Clemens, C., Griffiths, Ph. [1972]: The intermediate Jacobian of the cubic threefold. Ann. Math., II. Ser. *95* (1972) 281–356, Zbl. 231.14004

[1] For the convenience of the reader, references to review in Zentralblatt für Mathematik (Zbl.), compiled using the MATH database, and Jahrbuch über die Fortschritte der Mathematik (Jbuch) have, as far as possible, been included in this bibliography.

Danilov, V. I. [1988]: Algebraic Varieties and Schemes. Itogi Nauki Tekh. VINITI, Ser. Sovrem. Probl. Mat., Fundam. Napravleniya *23* (1988) 172–302. English transl.: Encycl. Math. Sci. *23*. Springer-Verlag: Berlin Heidelberg New York, 1994, 167–297, Zbl. 787.14001

Danilov, V. I. [1989]: Cohomologies of Algebraic Varieties. Itogi Nauki Tekh. VINITI, Ser. Sovrem. Probl. Mat., Fundam. Napravleniya *35* (1989) 5–130. English transl.: Encycl. Math. Sci. *35*. Springer-Verlag: Berlin Heidelberg New York, 1995 (in this volume)

de Rham, G. [1955]: Variétés differentiables. Hermann: Paris, 1955 (2nd ed. 1960; Zbl. 89, 81)

Dolgachev, I. V. [1966a]: Rational surfaces with a pencil of elliptic curves. Izv. Akad. Nauk SSSR, Ser. Mat. *30* (1966) 1073–1100 (Russian), Zbl. 187, 187

Dolgachev, I. V. [1966b]: On Severi conjecture on simply connected algebraic surfaces. Dokl. Akad. Nauk SSSR *170* (1966) 249–252. English transl.: Sov. Math., Dokl. *7* (1966) 1169–1172, Zbl. 173, 229

Enriques, F. [1949]: Le superficie algebriche. Zanichelli: Bologna, 1949, Zbl. 36, 371

Gieseker, D. [1977]: Global moduli for surfaces of general type. Invent. Math. *43* (1977) 233–282, Zbl. 389.14006

Gizatulin, M. Kh. [1984]: Defining relations for the Cremona transformations of the plane. Izv. Akad. Nauk SSSR, Ser. Mat. *46* (1984) 909–970. English transl.: Math. USSR, Izv. *21* (1983) 211–268, Zbl. 509.14011

Griffiths, Ph., Harris, J. [1978]: Principles of Algebraic Geometry. A Wiley-Interscience Publication: New York, 1978, Zbl. 408.14001

Grothendieck, A. [1962]: Fondements de la Géométrie Algébrique. Secrétariat Math.: Paris, 1962, Zbl. 239.14002

Hartshorne, R. [1977]: Algebraic Geometry. Springer-Verlag: Berlin Heidelberg New York, 1977 (3rd printing 1983 Springer), Zbl. 367.14001

Hirzebruch, F. [1987]: Collected Papers. Vol. I. Springer-Verlag: Berlin Heidelberg New York, 1987, 345–360, Zbl. 627.01044

Holzapfel, R.-P. [1980]: A class of minimal surfaces in the unknown region of surface geography. Math. Nachr. *98* (1980) 211–232, Zbl. 474.14022

Horikawa, E. [1978]: On the periods of Enriques surfaces. I. Math. Ann. *234* (1978) 73–88; II. Math. Ann. *235* (1978) 217–246, Zbl. 371.14019; Zbl. 412.14015

Igusa, J.-I. [1955]: On some problems in abstract algebraic geometry. Proc. Natl. Acad. Sci. USA *41* (1955) 964–967, Zbl. 67, 391

Iskovskikh, V. A. [1985]: A proof of the theorem on relations in the two-dimensional Cremona group. Usp. Mat. Nauk *40* (1985) No. 5, 255–256. English transl.: Russ. Math. Surv. *40* (1985), No. 5, 231–232, Zbl. 613.14012

Kanev, V. [1987]: Spectral curves, simple Lie algebras and Prym-Tjurin varieties. [Preprint. Inst. Math. Bulgarian Acad. Sci., 1987] Proc. Symp. Pure Math. *49* (1989), Part 1, 627–645, Zbl. 711.14026

Kawamata, Y. A. [1982]: A generalization of Kodaira-Ramanujam's vanishing theorem. Math. Ann. *261* (1982) 43–46, Zbl. 488.14003

Kawamata, Y. A., Matsuda, K., Matsuki, K. [1987]: Introduction to the minimal model problem. In: Proc. Symp. Alg. Geom. (Sendai 1985). Adv. Stud. Pure Math. *10*, North-Holland: Tokyo, 1987, 283–360, Zbl. 672.14006

Kodaira, K. [1960, 1963]: On compact analytic surfaces. I. Ann. Math., II. Ser. *71* (1960) 11–152; II. ibid. *77* (1960) 563–626; III. ibid. *78* (1963) 1–40, Zbl. 98, 130; Zbl. 118, 158; Zbl. 171, 196

Kodaira, K. [1964–1969]: On the structure of compact complex analytic surfaces. I. Am. J. Math. *86* (1964) 751–798; II. Am. J. Math. *88* (1966) 682–721; III. Am. J. Math. *90* (1969) 55–83; IV. Am. J. Math. *90* (1969) 1048–1066. I: Zbl. 137, 175; II–IV: Zbl. 193, 377

Kulikov, V. S. [1977]: Degenerations of $K3$ surfaces and Enriques surfaces. Izv. Akad. Nauk SSSR, Ser. Mat. *41* (1977) 1008–1042. English transl.: Math. USSR, Izv. *11* (1977) 957–989, Zbl. 367.14014

Manin, Yu. I. [1972]: Cubic forms: Algebra, Geometry, Arithmetic. Nauka: Moscow, 1972. English. transl.: North-Holland: Amsterdam, 1974, 2nd ed. 1986, Zbl. 255.14002

Milne, J. S. [1980]: Étale Cohomology. Princeton Math. Ser. *33*, Princeton Univ. Press: Princeton, 1980, Zbl. 433.14012

Milnor, J. [1963]: Morse theory. Princeton Univ. Press: Princeton, 1963, Zbl. 108, 104

Miyaoka, Y. [1977]: On the Chern numbers of surfaces of general type. Invent. Math. *42* (1977) 225–237, Zbl. 374.14007

Mori, S. [1982]: Threefolds whose canonical bundles are not numerically effective. Ann. Math., II. Ser. *115* (1982) 133–176, Zbl. 557.14021

Mori, S. [1987]: Classification of higher-dimensional varieties. In: Algebraic Geometry (Bowdoin 1985). Am. Math. Soc. Proc. Symp. Pure Math. *46*, Part 1, 1987, 269–331, Zbl. 656.14022

Mukai, S. [1988]: Finite groups of automorphisms of $K3$ surfaces and the Mathieu group. Invent. Math. *94* (1988) 183–221, Zbl. 705.14045

Mumford, D. [1966]: Lectures on Curves on an Algebraic Surface. Princeton Univ. Press: Princeton, 1966, Zbl. 187, 427

Mumford, D. [1970a]: Abelian Varieties. (Reprint, TIFR, Stud. Math. *5*, Oxford 1985) Oxford Univ. Press: London, 1970, 1974, Zbl. 223.14022/Zbl. 326.14012

Mumford, D. [1970b]: An algebraic surface with K ample, $K^2 = 9$, $p = q = 0$. Am. J. Math. *101* (1970) 233–244, Zbl. 433.14021

Mumford, D. [1977]: Stability of projective varieties. Enseign. Math., II. Ser. *23* (1977) 39–110, Zbl. 363.14003

Mumford, D., Fogarty, J. [1982]: Geometric Invariant Theory, 2nd ed. Springer-Verlag: Berlin Heidelberg New York, 1982, Zbl. 504.14008 (3rd enl. ed. Springer 1993, Zbl. 797.14004)

Nikulin, V. V. [1981]: On quotient groups of automorphism groups of hyperbolic forms by subgroups generated by 2-reflections. Algebra-geometric applications. Itogi Nauki Tekh. VINITI, Ser. Sovrem. Probl. Mat. *18* (1981) 1–114, Zbl. 484.10021. English transl.: J. Sov. Math. *22* (1983) 1401–1475

Nikulin, V. V. [1984]: The $K3$ surfaces with finite automorphism group and rank 3 Picard group. Tr. Mat. Inst. Steklov *165* (1984) 119–142. English transl.: Proc. Steklov Inst. Math. *165* (1985) 131–155, Zbl. 577.10019

Ogg, A. [1962]: Cohomology of Abelian varieties over function fields. Ann. Math., II. Ser. *76* (1962) 185–212, Zbl. 121, 380

Piatetski-Shapiro, I. I., Shafarevich, I. R. [1971]: A Torelli theorem for algebraic surfaces of type $K3$. Izv. Akad. Nauk SSSR, Ser. Mat. *35* (1971) 530–572. English transl.: Math. USSR, Izv. *5* (1971) 547–588, Zbl. 219.14021

Ramanan, S. [1978]: Vector bundles over algebraic curves. In: Proc. Int. Congr. Math., Helsinki 1978, Vol. II. 543–547 (1980), Zbl. 438.14014

Ramanujam, C. P. [1978]: A tribute. Springer-Verlag: Berlin Heidelberg New York, 1978. Tata Inst. Fund. Res., Studies in Math. *8*, Zbl. 401.00005 (entire collection)

Reider, I. [1988]: Vector bundles of rank 2 and linear systems on algebraic surfaces. Ann. Math., II. Ser. *127* (1988) 309–316, Zbl. 663.14010

Saltman, D. J. [1984]: Noether's problem over an algebraically closed field. Invent. Math. *77* (1984) 71–84, Zbl. 546.14014

Seminar Palaiseau [1985]: Géométrie des surfaces $K3$: modules et periodes. Asterisque *126* (1985) 1–192, Zbl. 547.00019 (entire collection)

Serre, J-P. [1970]: Cours d'arithmétique. Presses Univ. France: Paris, 1970. Zbl. 225.12002

Shabat, G. B. [1977]: On complex structure of domains covering algebraic surfaces. Funkt. Anal. Appl. *11* (1977) 67–75. English transl.: Funct. Anal. Appl. *11* (1977) 135–142, Zbl. 355.32031

Shafarevich, I. R. [1961]: Principal homogeneous spaces defined over function fields. Tr. Mat. Inst. Steklov. *64* (1961) 316–346, Zbl. 129, 128. English transl.: Am. Math. Soc. Transl., II. Ser. *37* (1964) 85–114

Shafarevich, I. R. [1988]: Basic Algebraic Geometry, Vol. I, II; 2nd suppl. ed. Nauka: Moscow, 1988. English transl.: Springer-Verlag: Berlin Heidelberg New York, 1994, Zbl. 675.14001 (Zbl. 797.14001, Zbl. 797.14002)

Shioda, T. [1974]: An example of a unirational surface in characteristic p. Math. Ann. *211* (1974) 233–236, Zbl. 283.14009

Shokurov, V. V. [1988]: Riemann surfaces and algebraic curves. Itogi Nauki Tekh. VINITI, Ser. Sovrem. Probl. Mat., Fundam. Napravleniya *23* (1988) 5–171. English transl.: Encycl. Math. Sci. *23*. Springer-Verlag: Berlin Heidelberg New York, 1994, 1–166, Zbl. 787.14022

Siegel, C. L. [1949]: Analytic functions of several complex variables. Lecture Notes. Inst. Adv. Studies.: Princeton, 1949, Zbl. 36, 50

Van de Ven, A. [1978]: Some recent results on surfaces of general type. In: Seminar Bourbaki, Exp. 500, 1976–1977. Lect. Notes Math. 677. Springer-Verlag: Berlin Heidelberg New York, 1978, 155–166, Zbl. 432.14023

Van de Ven, A. [1987]: On the differentiable structure of certain algebraic surfaces. In: Seminar Bourbaki, Exp. 667, 1985–1986. Astérique 145–146 (1987) 299–312, Zbl. 624.57018

Van der Geer, G. [1988]: Hilbert modular surfaces. Springer-Verlag: Berlin Heidelberg New York, 1988, Zbl. 634.14022

Verra, E. [1983]: On the Enriques' surfaces as a fourfold cover of $\mathbb{P}^2$. Math. Ann. *266* (1983) 241–250, Zbl. 519.14027

Viehweg, E. [1982]: Vanishing theorems. J. Reine Angew. Math. *335* (1982) 1–8, Zbl. 485.32019

Vladimirov, V. S., Sergeev, A. G. [1985]: Complex analysis in the future tube. Itogi Nauki Tekh. VINITI, Ser. Sovrem. Probl. Mat., Fundam. Napravleniya *8* (1985) 191–266. English transl.: Encycl. Math. Sci. *8*, 179–253. Springer-Verlag: Berlin Heidelberg New York, 1994, 179–254, Zbl. 787.32001

Yau, S.-T. [1977]: On Calabi's conjecture and some new results in algebraic geometry. Proc. Natl. Acad. Sci. USA *74* (1977) 1798–1799, Zbl. 355.32028

Zariski, O. [1958]: Introduction to the problem of minimal models in the theory of algebraic surfaces. Publ. Math. Soc. Japan *4* (1958) 1–89, Zbl. 93, 339

Zariski, O. [1962]: The theorem of Riemann-Roch for high multiples of an effective divisor on an algebraic surface, Appendix by D. Mumford. Ann. Math., II. Ser. *76* (1962) 560–615, Zbl. 124, 370

Zariski, O. [1971]: Algebraic surfaces. 2nd ed. Springer-Verlag: Berlin Heidelberg New York, 1971, (1st ed. 1935, Zbl. 10, 371), Zbl. 219.14020

Name Index

Subject Index

Printing: Mercedesdruck, Berlin
Binding: Buchbinderei Lüderitz & Bauer, Berlin

《国外数学名著系列》(影印版)

（按初版出版时间排序）

1. 拓扑学Ⅰ:总论　S. P. Novikov(Ed.)　2006.1
2. 代数学基础　Igor R. Shafarevich　2006.1
3. 现代数论导引(第二版)　Yu. I. Manin　A. A. Panchishkin　2006.1
4. 现代概率论基础(第二版)　Olav Kallenberg　2006.1
5. 数值数学　Alfio Quarteroni　Riccardo Sacco　Fausto Saleri　2006.1
6. 数值最优化　Jorge Nocedal　Stephen J. Wright　2006.1
7. 动力系统　Jürgen Jost　2006.1
8. 复杂性理论　Ingo Wegener　2006.1
9. 计算流体力学原理　Pieter Wesseling　2006.1
10. 计算统计学基础　James E. Gentle　2006.1
11. 非线性时间序列　Jianqing Fan　Qiwei Yao　2006.1
12. 函数型数据分析(第二版)　J. O. Ramsay　B. W. Silverman　2006.1
13. 矩阵迭代分析(第二版)　Richard S. Varga　2006.1
14. 偏微分方程的并行算法　Petter Bjørstad　Mitchell Luskin (Eds.)　2006.1
15. 非线性问题的牛顿法　Peter Deuflhard　2006.1
16. 区域分解算法:算法与理论　A. Toselli　O. Widlund　2006.1
17. 常微分方程的解法Ⅰ:非刚性问题(第二版)　E. Hairer　S. P. Nørsett　G. Wanner　2006.1
18. 常微分方程的解法Ⅱ:刚性与微分代数问题(第二版)　E. Hairer　G. Wanner　2006.1
19. 偏微分方程与数值方法　Stig Larsson　Vidar Thomée　2006.1
20. 椭圆型微分方程的理论与数值处理　W. Hackbusch　2006.1
21. 几何拓扑:局部性、周期性和伽罗瓦对称性　Dennis P. Sullivan　2006.1
22. 图论编程:分类树算法　Victor N. Kasyanov　Vladimir A. Evstigneev　2006.1
23. 经济、生态与环境科学中的数学模型　Natali Hritonenko　Yuri Yatsenko　2006.1
24. 代数数论　Jürgen Neukirch　2007.1
25. 代数复杂性理论　Peter Bürgisser　Michael Clausen　M. Amin Shokrollahi　2007.1
26. 一致双曲性之外的动力学:一种整体的几何学的与概率论的观点　Christian Bonatti　Lorenzo J. Díaz　Marcelo Viana　2007.1
27. 算子代数理论Ⅰ　Masamichi Takesaki　2007.1
28. 离散几何中的研究问题　Peter Brass　William Moser　János Pach　2007.1
29. 数论中未解决的问题(第三版)　Richard K. Guy　2007.1

30. 黎曼几何(第二版)　　Peter Petersen　　2007.1
31. 递归可枚举集和图灵度:可计算函数与可计算生成集研究　　Robert I. Soare　　2007.1
32. 模型论引论　　David Marker　　2007.1
33. 线性微分方程的伽罗瓦理论　　Marius van der Put　Michael F. Singer　　2007.1
34. 代数几何 II:代数簇的上同调,代数曲面　　I. R. Shafarevich(Ed.)　　2007.1
35. 伯克利数学问题集(第三版)　　Paulo Ney de Souza　Jorge-Nuno Silva　　2007.1
36. 陶伯理论:百年进展　　Jacob Korevaar　　2007.1